14th International Conference on Axiomatic Design (ICAD 2021)

IOP Conference Series: Materials Science and Engineering
Volume 1174

Online
23-25 June 2021

ISBN: 978-1-7138-4634-5
ISSN: 1757-8981

Printed from e-media with permission by:

Curran Associates, Inc.
57 Morehouse Lane
Red Hook, NY 12571

Some format issues inherent in the e-media version may also appear in this print version.

Institute of Physics
Dirac House, Temple Back
Bristol BS1 6BE UK

Phone: 44 1 17 929 7481
Fax: 44 1 17 920 0979

techtracking@iop.org

Additional copies of this publication are available from:

Curran Associates, Inc.
57 Morehouse Lane
Red Hook, NY 12571 USA
Phone: 845-758-0400
Fax: 845-758-2633
Email: curran@proceedings.com
Web: www.proceedings.com

TABLE OF CONTENTS

IOP Conf. Series: Materials Science and Engineering 1174 (2021) 011001 doi:10.1088/1757-899X/1174/1/011001

Preface

Axiomatic Design (AD) is a general-purpose design theory developed by Nam P. Suh and widely diffused in 1990 by his book "The Principles of Design". Axiomatic Design provides sound science-based criteria to support decision-making within all the steps of the design process. More than just a new design theory, Axiomatic Design implements a unique reasoning framework that allows early rejecting "bad" design solutions, thus helping creative designers rapidly focusing feasible design solutions. Ever since its disclosure, AD is being adopted as a teaching and a working tool.

Since 2000, researchers and engineers start to meet in an international conference to present the latest theoretical developments and examples of application in various areas of engineering. This forum is named International Conference on Axiomatic Design (ICAD) and had its first edition at MIT, Massachusetts, USA in 2000. Since then, it has been held annually or biennially. 13 ICADs have already took place in several countries, namely Australia, China, Iceland, Italy, Korea, Portugal, Romania, and USA.

The 14th International Conference on Axiomatic Design was hosted by NOVA School of Science and Technology (NOVA SST), and the Military University Institute (IUM) in June 23-25, 2021. Due to the current worldwide context related to the COVID-19 pandemic, it was the first ICAD online edition.

ICAD 2021 had the following committees:

Honorary Chairs	
Nam P. Suh	MIT – USA
A.M. Gonçalves-Coelho	FCT/UNL – PT
Conference Chair	
Miguel Cavique	Naval Academy – PT
Conference Co-chairs	
António Mourão	FCT/UNL – PT
João Fradinho	FCT/UNL – PT
Local Organising Committee	
Ang Liu	UNSW - AU
A. Gabriel-Santos	FCT/UNL – PT
Daniel Aelenei	FCT/UNL – PT
Erik Puik	H. Utrecht - NL
Erwin Rauch	Univeristy of Bolzano, IT
Goran Putnik	Univ. Minho - PT
João Ramalho Marreiros	Military University Institute
Miguel Moreira	Naval Academy - PT
Oana Dodun	TU Iasi - RO
Rui Chedas Sampaio	Naval Academy – PT

IOP Conf. Series: Materials Science and Engineering 1174 (2021) 011001 doi:10.1088/1757-899X/1174/1/011001

Scientific Committee

Alessandro Giorgetti	G. Marconi University IT
Amro Farid	Masdar Institute/MIT AE
Ang Liu	UNSW AU
António Gabriel-Santos	FCT NOVA PT
A. M. Gonçalves-Coelho	FCT NOVA PT
António Mourão	FCT NOVA PT
Christopher A. Brown	WPI USA
Dominik Matt	University of Bolzano IT
Erwin Rauch	University of Bolzano IT
Efrén Benavides	UPM ES
Erik Puik	HU University of Applied Sciences Utrecht NL
Gabriele Arcidiacono	G. Marconi University IT
Goran Putnik	Univ. Minho PT
Hilario Oh	MIT USA
Inas Khayal	Dartmouth Institute USA
João Fradinho	FCT NOVA PT
John Thomas	Cognitive Tools Ltd. LLC USA
José Figueiredo	IST UL PT
Joseph Timothy Foley	Reykjavik University IS
Laurenţiu Slătineanu	TUIASI RO
Luc Mathieu	ENS Cachan FR
Mário Simões Marques	Naval Academy PT
Masayuki Nakao	University of Tokyo, JP
Miguel Cavique	Naval Academy PT
Moshe Shpitalni	Israel Institute of Technology – IL
Nam P. Suh	MIT USA
Oana Dodun	TUIASI RO
Paolo Citti	G. Marconi University IT
Petru Duşa	TUIASI RO
Sami Kara	UNSW AU
Taesik Lee	KAIST KO
Vladimir Modrak	TU Kosice SK

ICAD 2021 had 2 tutorial sessions, 7 conference keynotes and 8 paper sessions organized in fields such as Design for Sustainability, Design for Data, Design for Industry and Design for X. The Conference ended with a keynote presentation by the Honorary Chair Professor Nam P. Suh. 32 scientific papers were presented that led to a positive sharing of views among the participants. These works cover a wide range of matters that includes information design, design for sustainable manufacturing, sustainable ship design, building construction, artificial intelligence, hardware improvement, man-machine interface design, manufacturing technology, additive manufacturing, industrial robot design, water scarcity, design for neural networks, design for sustainability and machine design, among others.

The Organizing Committee hopes that the reader could find this volume useful, and gratefully acknowledges the conference sponsors for their valuable support. Furthermore, we are thankful to the paper authors and co-authors, to the Scientific Committee, to the keynote and tutorial presenters, and to the session chairs, whose contributions have made possible the accomplishment of ICAD 2021.

Campus de Caparica, June 8, 2021

Miguel Cavique
António Mourão
João Fradinho

IOP Conf. Series: Materials Science and Engineering 1174 (2021) 011002 doi:10.1088/1757-899X/1174/1/011002

Peer review declaration

All conference organisers/editors are required to declare details about their peer review. Therefore, please provide the following information:

- **Type of peer review: Single-blind / Double-blind / Triple-blind / Open / Other (please describe)**
  Single-blind reviewing.
  29 paper was reviewed twice by the same two reviewers. The acceptance was granted to the papers with the positive reply of both reviewers. The used criteria included technical, presentation, quality, and English language.
  6 papers were reviewed once (4 accepted "as is" on first review, 1 excluded, and 1 withdraw)
  Authors should make sure the submitted papers use the template for this conference.

- **Conference submission management system:**
  EasyChair

- **Number of submissions received:**
  35

- **Number of submissions sent for review:**
  35

- **Number of submissions accepted:**
  28

- **Acceptance Rate (Number of Submissions Accepted / Number of Submissions Received X 100):**
  80 %

- **Average number of reviews per paper:**
  3,6

- **Total number of reviewers involved:**
  17

- **Any additional info on review process:**
  -

- **Contact person for queries:**
  João Fradinho - jmf@fct.unl.pt

Please submit this form along with the rest of your files on the submission date written in your publishing agreement.

IOP Conf. Series: Materials Science and Engineering 1174 (2021) 011002 doi:10.1088/1757-899X/1174/1/011002

The information you provide will be published as part of your proceedings.

IOP Conf. Series: Materials Science and Engineering 1174 (2021) 012001 doi:10.1088/1757-899X/1174/1/012001

Complex information systems configuration using ISACA Design Guide into an Axiomatic Design approach

Eugen Purice[1], Petru Dusa[1,*], Laurentiu Cretu[2], Adrian Cacu[1], Oana Dodun[1], and Laurentiu Slatineanu[1]

[1] "Gheorghe Asachi" Technical University of Iasi, Iasi, Blvd. D. Mangeron, 39 A,700050, Romania
[2] Gentec Data srl, Street Gh. Titeica, 121 C, Bucharest, 020295, Romania

[*]pdusa@tcm.tuiasi.ro

Abstract. This paper describes the process of designing a customized governance solution for an enterprise information system using COBIT framework guidelines and an axiomatic design approach. COBIT (Control Objectives for Information and related Technology) is a generally accepted framework created by the ISACA (Information Systems Audit and Control Association) for governing and managing enterprise information and technology (IT). COBIT framework can be applied to any organization in any industry and was designed to help deliver value while managing better the risks associated with the IT processes. On the other side, the Axiomatic Design (AD) theory involves a continuous interplay between the design objectives (the needs / what we want to achieve) and the means capable of reaching those objectives (how we want to achieve) to determine the best configuration capable of satisfying the design intend. The AD theory requires a description of the design's objectives in terms of specific requirements, called Functional Requirements (FR). The development of a complete solution to a given problem starts by mapping the FRs to Design Parameters (DPs) in the solution domain. ISACA Design Guide proposes ten (10) Design Factors and forty (40) Governance Objectives; each objective (a set of Functional Requirements) can be achieved through several combinations of Design Parameters. From the Axiomatic Design theory perspective, this determines a coupled matrix. To decouple the matrix, the profile of each design factor is drawn, and the sub-factors will be taken only once within the factor where the sub-factor has the highest weight. A case study is presented.

1. Introduction

Information and related information technologies (IT) are ubiquitous and play critical roles in most organizations. IT has become a business facilitator, playing a major role in the efficient delivery of products and services, bringing more value to stakeholders.

Governance of IT enterprises (GEIT), adding value while mitigating risks, has become necessary for meeting strategic objectives and is becoming a fundamental concern of corporate governance.

ISACA (Information Systems Audit and Control Association) has developed COBIT (Control Objectives for Information Technology) as a framework to "help companies implement healthy governance factors," providing a way for organizations to align their business strategy with IT objectives. [1,2]

IOP Conf. Series: Materials Science and Engineering 1174 (2021) 012001 doi:10.1088/1757-899X/1174/1/012001

Over time, the COBIT framework has evolved (several versions are known), and IT specialists have combined this framework with other methods to address specific IT problems.

Research on the use of the COBIT framework is identified from the perspective of axial methodologies combined with other research methods, such as Design Science Research Methodology (DSRM) [3].

There are studies on the integration of the COBIT framework with project management methodologies [4, 5, 6]

The use of the COBIT framework in connection with Sarbanes-Oxley legislation on the security of IT systems in companies is also known [7]

There are approaches to implement sets of good IT management, control, and security practices by unifying the incorporation of ISO 17799, ISO / IEC 38500, Information Technology Infrastructure Library (ITIL), and Capability Maturity Model Integration (CMMI) [8].

The relative importance of different COBIT IT processes in the context of the reliability of corporate financial information has been investigated, which has led to more efficient audits [9].

Based on a literature review on data quality and information integrity, a framework has been created that is considered broader than that provided by COBIT on information integrity [10].

Based on the results obtained in the audit of different systems using the COBIT framework, there are intentions to develop a general theory on internal control applicable to information technology [11]

We have identified studies on the auditing of information systems and regulatory behavior in the field of electricity suppliers [12].

Our research is based on an investigation of a complex IT landscape deployed by a European utilities supplier (electricity and gas) at a time when the company had to undergo an unbundling process [13]. The approach methodology deployed for our investigation used a combination of ISACA and Axiomatic Design [14,15,16] instruments to determine the final state of an IT landscape that would guarantee the most efficient segregation of the IT resources between the unbundled entities.

2. Context of unbundling in the European electricity and gas sector

In the regulation of network industries - such as utilities, e.g., electricity or gas - unbundling refers to the process of separating competitive activities (e.g., energy production and supply) from those in which competition is not feasible or permitted (e.g., transmission and distribution). In the EU, the transport and distribution activities for electricity and gas are delivered by regulated monopolies. To protect the interests of consumers, the EU policies do not permit companies (economic actors) to be engaged simultaneously in both competitive and monopolistic activities.

A company that controls the network and is also involved in the competitive segments of the supply chain has an obvious interest in limiting or denying access to other companies active upstream or downstream. Therefore, EU policies aim to guarantee fair, transparent, and non-discriminatory access to the market for all market players, hence the need for unbundling.

Different degrees of unbundling are possible with varying levels of effectiveness; Figure 1 below is a representation of the various possible degree of activity unbundling.

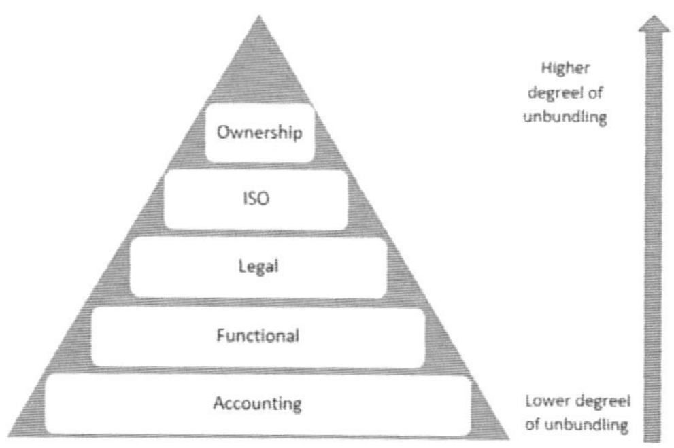

Figure 1. Summary of the different degrees of unbundling [13]

A brief description of each type of unbundling is presented below:

- Account unbundling - the firm is required to separate the bookkeeping of its various activities, separating the costs and revenues that derive from each of them.
- Functional unbundling - the firm is obliged to reorganize its internal structure and split responsibility for the network and competitive activities into different units with independent decision-making structures.
- Legal unbundling - separate legal entities are established, each permitted to operate in only one part of the supply chain.
- Independent System Operator (ISO), not owned by the vertically integrated firm that is tasked with the operation and planning of the infrastructure; however, the ownership of network assets can still remain in the hands of the integrated firm.
- Ownership unbundling - a firm owning and operating a network cannot be active in the competitive segment of the supply chain or have interest in any of the companies involved in those activities.

3. Research background

Our research is based on several concepts and tools that are referred to throughout this paper; for the sake of clarity, these concepts are introduced briefly below.

3.1. COBIT Process Assessment Model (PAM)

The COBIT Process Assessment Model (PAM) provides a number of governance and management objectives should be achieved to contribute to enterprise goals [1].

The governance and management objectives in COBIT are grouped into five domains. The domains have names with verbs that express the key purpose and areas of activity of the objectives contained in them:

- Governance objectives are grouped in one domain: Evaluate, Direct and Monitor (EDM). These governance processes deal with the stakeholder governance objectives - value delivery, risk optimization, and resource optimization - and include practices and activities aimed at evaluating strategic options, providing direction to IT, and monitoring the outcome.
- Management objectives- grouped in four (4) domains that are in line with the responsibility areas of plan, build, run and monitor (PBRM), and they provide end-to-end coverage of IT. Each domain contains a number of processes.

The domains for management objectives are:

- Align, Plan and Organize (APO) - Provides direction to solution delivery (BAI) and service delivery and support (DSS). This domain covers strategy and tactics and concerns identifying the best way IT can contribute to the achievement of the business objectives. The realization of the strategic vision needs to be planned, communicated, and managed from different perspectives. A proper organization, as well as technological infrastructure, should be put in place.
- Build, Acquire and Implement (BAI)—Provides the solutions and passes them on to be turned into services. To realize the IT strategy, IT solutions need to be identified, developed, or acquired, as well as implemented and integrated into the business process. Changes in and maintenance of existing systems are also covered by this domain to ensure that the solutions continue to meet business objectives.
- Deliver, Service, and Support (DSS)—Receives the solutions and makes them usable for end-users. This domain is concerned with the actual delivery and support of required services, which include service delivery, management of security and continuity, service support for users, and management of data and operational facilities.
- Monitor, Evaluate and Assess (MEA)—Monitors all processes to ensure that the direction provided is followed. All IT processes need to be regularly assessed over time for their

quality and compliance with control requirements. This domain addresses performance management, monitoring of internal control, regulatory compliance, and governance.

An image depicting how the five (5) domains are organized in the framework is presented in figure 2 below.

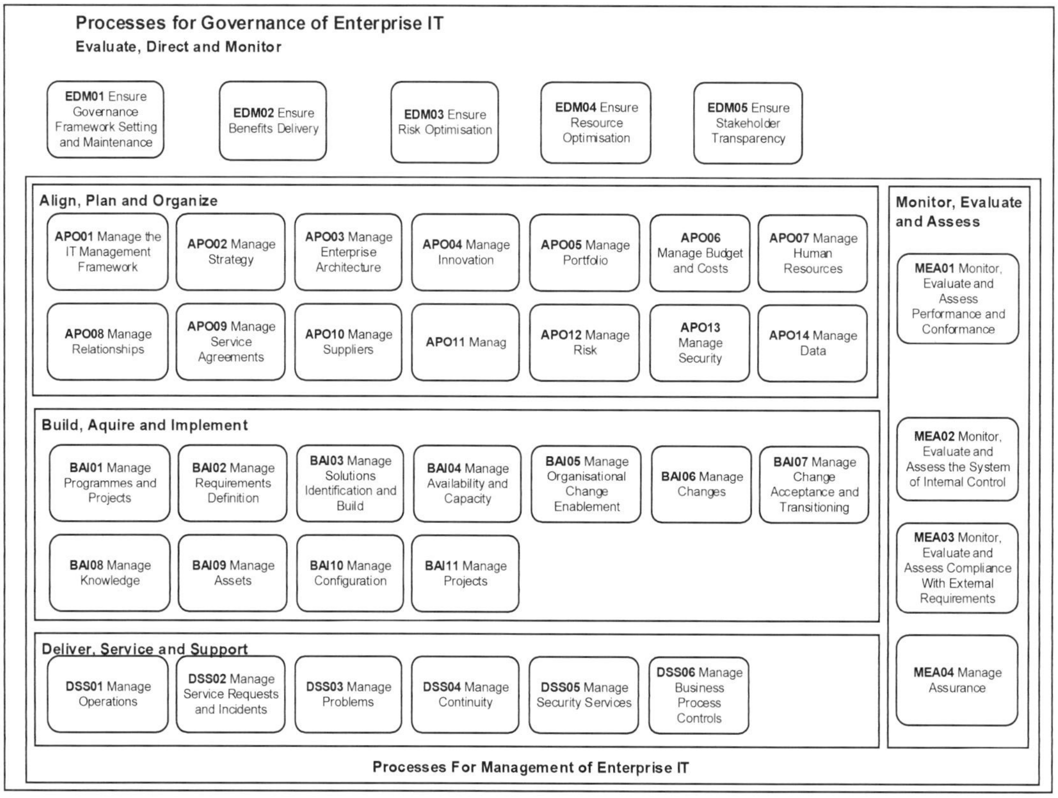

Figure 2. COBIT PAM Overview [1]

Across the five domains (one domain assigned to governance objectives and four domains assigned to management objectives), there are 40 defined IT processes, organized as depicted in figure 2 above.

The process assessment model is a two-dimensional model of process capability; the capability dimension provides a measure of a process's capability to meet an enterprise's current or projected business goals for the process. The process capability is expressed in terms of capability levels. The rating scale involves six capability levels:

- Level 0 Incomplete process - The process is not implemented or fails to achieve its process purpose.
- Level 1 Performed process - The implemented process achieves its process purpose.
- Level 2 Managed process - The previously described performed process is now implemented in a managed fashion (planned, monitored, and adjusted), and its work products are appropriately established, controlled, and maintained.
- Level 3 Established process - The previously described managed process is now implemented using a defined process that is capable of achieving its process outcomes.
- Level 4 Predictable process - The previously described established process now operates within defined limits to achieve its process outcomes.
- Level 5 Optimizing process - The previously described predictable process is continuously improved to meet relevant current and projected business goals

3.2. COBIT 2019 Design Guide and Toolkit. Design Methodology Foundation

To satisfy governance and management objectives, each enterprise needs to establish and maintain a governance system built from a number of components.

- Components are factors that, individually and collectively, contribute to the good operations of the enterprise's governance system over Information and Technology (I&T).
- Components interact with each other, resulting in a holistic governance system for I&T.
- Components can be of different types. The most familiar are processes. However, components of a governance system also include organizational structures; policies and procedures; information items; culture and behavior; skills and competencies; and services, infrastructure, and applications

Design factors are factors that can influence the design of an enterprise's governance system and position it for success in the use of I&T.

The design factors are presented in figure 3 below.

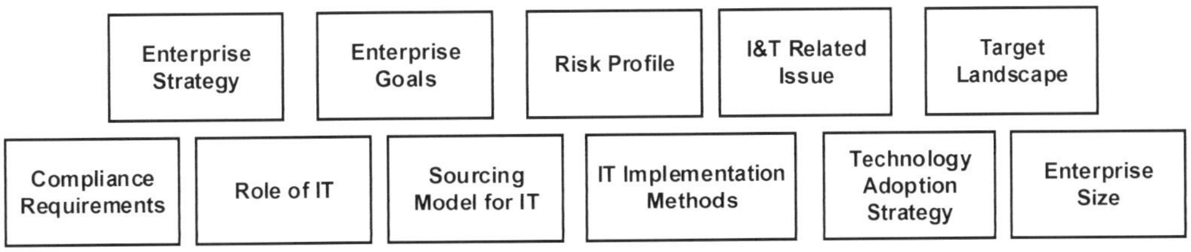

Figure 3. COBIT Design Factors [2]

The different stages and steps in the design process, as illustrated in figure 4, will result in recommendations for prioritizing governance and management objectives or related governance system components, for target capability levels, or for adopting specific variants of a governance system component.

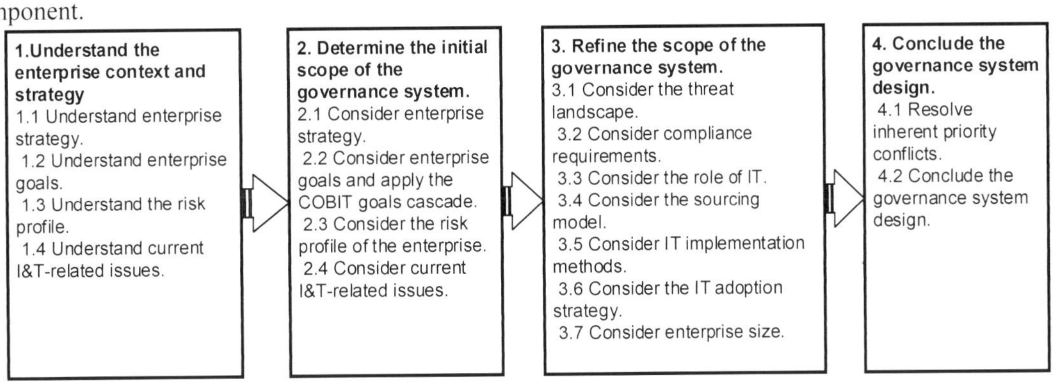

Figure 4. Governance System Design Workflow [2]

The application of the governance system design workflow explained in figure 4 is supported by a COBIT Design Guide toolkit.

3.3. Axiomatic Design Principles

Axiomatic design [14,15, 16] is a design theory and methodology based on the Independence Axiom and the Information Axiom. The methodology contains corollaries and theorems which guide/help designers mapping functional space to physical space (see Figure 5).

IOP Conf. Series: Materials Science and Engineering 1174 (2021) 012001 doi:10.1088/1757-899X/1174/1/012001

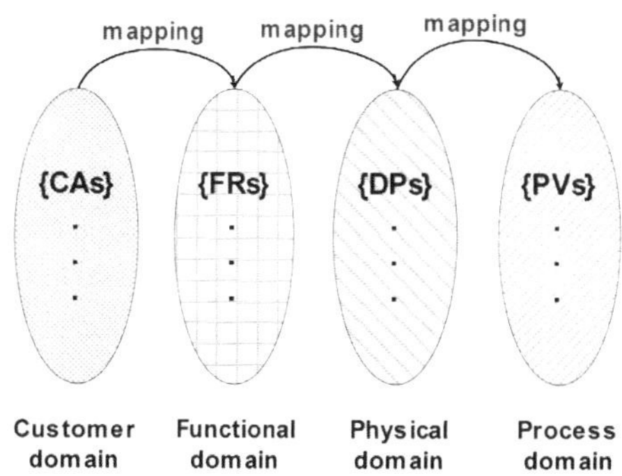

Figure 5. Domains mapping [14,15]

According to Axiomatic Design theory [14,15], design involves a continuous interplay between what is wanted to achieve and how it is wanted to achieve it. It must determine the design's objectives by defining it in terms of specific requirements called functional requirements (FRs). To satisfy functional requirements, a physical embodiment characterized in terms of design parameters (DPs) must be created

The basic assumption of the Axiomatic Design (AD) approach to design is that there exist two axioms that determine good design practice:

Axiom 1: The Independence Axiom (independence of FRs)

Axiom 2: The information Axiom (minimize the information content of the design)

Independence axiom may be mathematically expressed by the design equation (1):

$$\{FR\} = \{A\}\, x\{DP\} \tag{1}$$

Where:

{FR} is the functional requirement vector;

{DP} is the design parameters vector;

[A] is the design matrix.

In a particular case, equation (1) can be written in the form shown in equation (2)

$$\begin{bmatrix} FR_1 \\ FR_2 \\ FR_3 \end{bmatrix} = \begin{bmatrix} A_{11} & A_{12} & A_{13} \\ A_{21} & A_{22} & A_{23} \\ A_{31} & A_{32} & A_{33} \end{bmatrix} X \begin{bmatrix} DP_1 \\ DP_2 \\ DP_3 \end{bmatrix} \tag{2}$$

If all of the elements of the design matrix are nonzero, Axiom 1 cannot be respected, and the design is classified as coupled design.

In the case when only the diagonal elements from the design matrix are different from zero (equation 3), the design satisfies Axiom 1, and the design is classified as an uncoupled design.

$$\begin{bmatrix} FR_1 \\ FR_2 \\ FR_3 \end{bmatrix} = \begin{bmatrix} A_{11} & 0 & 0 \\ 0 & A_{22} & 0 \\ 0 & 0 & A_{33} \end{bmatrix} X \begin{bmatrix} DP_1 \\ DP_2 \\ DP_3 \end{bmatrix} \tag{3}$$

A coupled design can be decupled. A case is presented by equation (4), where the design matrix is a triangular matrix. The independence of FRs can be ensured by adjusting the design parameters in a particular order (DP1 first, DP2 second, and third DP3).

$$\begin{bmatrix} FR_1 \\ FR_2 \\ FR_3 \end{bmatrix} = \begin{bmatrix} A_{11} & 0 & 0 \\ A_{21} & A_{22} & 0 \\ A_{31} & A_{32} & A_{33} \end{bmatrix} X \begin{bmatrix} DP_1 \\ DP_2 \\ DP_3 \end{bmatrix} \quad (4)$$

This system is called decoupled design.

The theory makes use of many theorems and corollaries, and in-depth knowledge can be obtained from de books written by the founder of the theory [14,15].

4. Methodology

The research is based on a case study done on an IT landscape of a large utility supplier and distributor of gas and electricity, which had to undergo unbundling to comply with the applicable policies.

The objective of the study was to identify a configuration for sharing the existing IT landscape post unbundling.

The perspective from which the study was made is that of a functional unbundling (see description in Chapter 2).

4.1. Data Collection

To collect the data, a set of investigation activities were undertaken to determine the As-Is configuration of the existing IT landscape, both the software and the hardware perspective. However, for the purpose of this paper, only software-related data is considered. As part of the baseline investigation, all existing applications deployed on the IT landscape were investigated and recorded in a table with the following structure (fields): Application, Description, Vendor, Version, Environment, Operating System, Application License, Interfaces, No of users, Cores, RAM [GB], Storage [GB], Company entities using the application.

The survey revealed that 73 applications were installed; an excerpt from the application inventory table, including the Application and Description fields, is presented in Table 1

Table 1 Applications Inventory (excerpt)

No.	Application	Description
1	NWM Prod APP	Prod Application mobile workforce management
2	Assets management	Software for assets management
3	Facility Management application	Software for facility management
4	JANTAR	Buildings control access application
...		
71	UCCX	Call center & AR
72	CUCM	Call manager
73	TSM	Backup and Recovery for Power systems

4.2. Data Analysis

The COBIT Process Assessment Model (PAM), described in Section 3.1, was used for data analysis.

The IT landscape (the software component) has been evaluated (audited) rigorously, in line with the COBIT Process Assessment Model methodology, following all forty (40) evaluation processes, grouped into six (6) levels of evaluation, and using specific tools to aid the evaluation process.

The results are shown in Table 2, Process Capability Level columns (an excerpt from the evaluation table).

Table 2: Process Capability and Concluded Scope Table (Excerpt)

	Process Name/Design Factors	Process Capability Level						Concluded Scope: Governance/ Management Objectives Priority Weight
		0	1	2	3	4	5	
	Evaluate, Direct, and Monitor (EDM)							
EDM01	Ensured Governance Framework Setting and Maintenance			x				40
EDM02	Ensured Benefits Delivery				x			30
EDM03	Ensured Risk Optimization			x				65
EDM04	Ensured Resource Optimization				x			0
EDM05	Ensured Stakeholder Engagement			x				20
	Align, Plan and Organize (APO)							
APO01	Managed I&T Management Framework				x			55
APO02	Managed Strategy			x				10
APO03	Managed Enterprise Architecture				x			55
APO04	Managed Innovation			x				60
APO05	Managed Portfolio			x				10
...								
	Deliver, Service, and Support (DSS)							
DSS01	Managed Operations			x				-5
DSS02	Managed Service Requests and Incidents				x			40
DSS03	Managed Problems				x			45
DSS04	Managed Continuity				x			65
DSS05	Managed Security Services					x		60
DSS06	Managed Business Process Controls				x			35
	Monitor, Evaluate and Assess (MEA)							
MEA01	Managed Performance and Conformance Monitoring				x			60
MEA02	Managed System of Internal Control					x		10
MEA03	Managed Compliance with External Requirements				x			25
MEA04	Managed Assurance				x			35

Having the landscape evaluation completed (Process Capability Level established), the problem of improvement is posed by establishing the scope of the governance system.

COBIT 2019 Design Guide [2] explores design factors that can influence governance and includes a workflow for planning a customized governance system for the enterprise (see figure 3 and figure 4).

The application of the methodology described in paragraph 3.2 leads to obtaining Governance and Management Importance scores (for all Design Factors).

The relative importance (of governance and management objectives) is a score that indicates the influence of a certain design factor on the importance of a certain COBIT governance or management objective as compared to the baseline (standard) situation. The number is calculated as a percentage difference between the baseline and the current situation, as determined by the values given to the design factor at hand [2].

The results obtained are presented in Table 2, column titled "Calculated Scope: Governance / Management Objective Priority Weight."

The first conclusion is that priority (importance) scores of the governance objectives are very diverse.

A further review of the data in Table 2, column Calculated Scope: Governance / Management Objective Priority Weight, reveals that Deliver, Service, and Support (DSS) field has the most significant importance. Considering this field, we have assessed further the way in which the processes from this category (field) are satisfied by the portfolio of applications (Table 1) from the perspective of the functional unbundling process.

5. Axiomatic Design Model Development

The Application Inventory (Table 1) revealed an IT landscape consisting of a very heterogeneous and complex set of interdependent applications used non-uniformly at the various operating locations. From the perspective of functional unbundling (as described in Chapter 2), this heterogeneity is a challenge.

In this context, we consider using Axiom 1, The Independence Axiom, from the Axiomatic Design theory to reorganize and redeploy the applications on various existing platforms with the aim to achieve the functional independence objectives. The Independence of Functional Requirements would permit the function segregation of the IT landscape and enable smooth operations of various other organizational groups of the company after unbundling

5.1. Identification of functional requirements

Deliver, Service and Support (DSS) processes can be identified as Functional Requirement, as follows:

DSS01 Manage operations;
DSS02 Manage service requests and incidents;
DSS03 Manage problems;
DSS04 Manage continuity;
DSS05 Manage security services;
DSS06 Manage business process controls;

5.2. Identification of design parameters

The applications identified during the system landscaping investigation can be considered Design Parameters, therefore:

Application 1
Application 2
Application 3
.
.
.
Application 72
Application 73

Note: Because the research was done for a private company, certain information is bound to remain confidential. Accordingly, in this paper, we use generic names for the applications, Application 1, 2,... However, each of the generic names has a correspondence in the Applications Inventory, Table 1; however, this simplification does not invalidate the content or the approach.

By investigating the way how the applications are assigned to different processes, i.e., the connection between Functional Requirements and Design Parameters, the Design Matrix has been generated – Table 3 below. Each design parameter is a module (application) that is related to a functional requirement (or more) on the corresponding point of hierarchy, so there are more DPs than FRs. It can be noted that the matrix design is coupled.

Table 3 Matrix for assigning applications to processes

	Processes / Design Factors	Applications									
	Deliver, Service, and Support (DSS)	1	2	3	4	5	6	7	8	...	73
DSS01	Managed Operations	x				x					
DSS02	Managed Service Requests and Incidents	x	x	x							
DSS03	Managed Problems			x			x		x		
DSS04	Managed Continuity		x					x			x
DSS05	Managed Security Services				x						
DSS06	Managed Business Process Controls						x		x		x

IOP Conf. Series: Materials Science and Engineering 1174 (2021) 012001 doi:10.1088/1757-899X/1174/1/012001

5.3. Identification of design equation

According to Theorem 3 [16], "when there are more DPs than FRs, the design is a redundant design, which can be reduced to an uncoupled design or a decoupled design, or a coupled design."

Theorem 4 [16] states that in an ideal design, the number of DPs is equal to the number of FRs, and the FRs are always maintained independent of each other.

The problem at hand - reconfiguration of the IT landscape to enable an unbundled operation - could be transformed into a model solvable using Axiomatic Design formalism (axioms, corollaries, and theorems).

On the basis of Theorem 4, the first attempt was to decrease the number of design parameters to match the number of Functional Requirements. The result was a deployment of applications portfolio (73 applications) on six (6) platforms; the methodology of assigning the platforms (Design Parameters) to processes (Functional Requirements) is presented in Table 4 below.

Table 4 Matrix for assigning platforms to processes.

	Processes / Design Factors Deliver, Service, and Support (DSS)	Platforms A	 B	 C	 D	 E	 F
DSS01	Managed Operations	x					
DSS02	Managed Service Requests and Incidents		x				
DSS03	Managed Problems			x			
DSS04	Managed Continuity				x		
DSS05	Managed Security Services					x	
DSS06	Managed Business Process Controls						x

Note: The meaning of the platform here is that of a group of applications and technologies that have been used to meet the functional needs of a process. In the sense of Axiomatic Design, the notion of Platform has the meaning of a Parameter Design.

Considering the configuration presented in Table 4 above, the design equation can be written:

$$\begin{bmatrix} DSS01 \\ DSS02 \\ DSS03 \\ DSS04 \\ DSS05 \\ DSS06 \end{bmatrix} = \begin{bmatrix} A_{11} & 0 & 0 & 0 & 0 & 0 \\ 0 & A_{22} & 0 & 0 & 0 & 0 \\ 0 & 0 & A_{33} & 0 & 0 & 0 \\ 0 & 0 & 0 & A_{44} & 0 & 0 \\ 0 & 0 & 0 & 0 & A_{55} & 0 \\ 0 & 0 & 0 & 0 & 0 & A_{66} \end{bmatrix} x \begin{bmatrix} PlatformA \\ PlatformB \\ PlatformC \\ PlatformD \\ PlatformE \\ PlatformF \end{bmatrix} \quad (5)$$

Under these conditions, the independence of Functional Requirements (Axiom 1) is respected, and consequently, it can be concluded that a configuration solution for the IT landscape to support company's operation in an unbundled (Chapter 2) state can be achieved through reconfiguration of existing assets (software and hardware).

6. Conclusions

ISACA (Information Systems Audit and Control Association) has developed COBIT (Control Objectives for Information Technology) as a framework to "help companies implement healthy governance factors," providing a way for organizations to align their business strategy with IT objectives.

Often, however, problems in IT landscapes require extensive approaches using complementary tools, and the literature presents such situations.

This paper, too, demonstrates that the formalism (principles and tools) developed under COBIT Design Framework and Axiomatic Design theory could be deployed complementarily to arrive at a practical solution for a real-life project. After performing the system audit and redesigning it considering the Governance of IT enterprises (GEIT) perspective, Axiomatic Design principles have been applied to derive an acceptable configuration of the IT landscape (software component) of a utility firm, helping it to comply with the applicable market's regulations.

Acknowledgments

Research work was conducted in a consultancy contract facilitated by the Gentec Data Romania. The authors thank the administration for the opportunity.

References

[1] ISACA 2013 *COBIT Process Assessment Model (PAM): Using COBIT 5* (Rolling Meadows: ISACA)
[2] ISACA 2019 *COBIT Design Guide and Toolkit: Designing an Information & Technology Governance Solution* (Rolling Meadows: ISACA)
[3] Amorim A C, Mira da Silva M, Pereira R and Gonçalves M 2020, Using agile methodologies for adopting COBIT, *Information Systems, https://doi.org/10.1016/j.is.2020.101496.*
[4] Bernroider EWN, Ivanov M 2011 IT project management control and the Control Objectives for IT and related Technology (CobiT) framework, *International Journal of Project Management* **29** p 325–336
[5] Gantman S, Fedorowicz J 2016 Communication and control in outsourced IS development projects: Mapping to COBIT domains *International Journal of Accounting Information Systems* **21** p 63–83
[6] Bin-Abbas H, Bakry SH 2014 Assessment of IT governance in organizations: A simple integrated approach *Computers in Human Behavior* **32** p 261–267
[7] Hardy G 2006 Using IT governance and COBIT to deliver value with IT and respond to legal, regulatory and compliance challenges, *Information Security Technical Report II* (Amsterdam: Elsevier) p 55 – 61
[8] Joshia A, Bollena L, Hassinka H, De Haesb S and Grembergenb WV 2018 Explaining IT governance disclosure through the constructs of IT governance maturity and IT strategic role *Information & Management* **55** p 368–380
[9] Kerr D, Murthy US 2013The importance of the CobiT framework IT processes for effective internal control over financial reporting in organizations: An international survey *Information & Management* **50** p 590–597
[10] Boritz JE 2005 IS practitioners' views on core concepts of information integrity *International Journal of Accounting Information Systems* **6** p 260– 279
[11] Tuttle B, Vandervelde SD 2007 An empirical examination of CobiT as an internal control framework for information technology *International Journal of Accounting Information Systems* **8** p 240–263
[12] Wang J, Zhihua Cheng Z 2018 FP-Growth based Regular Behaviors Auditing in Electric Management Information System *Proc. Computer Science* **139** p 275–279
[13] Rosseto N 2020 Unbundling in the European electricity and gas sectors, https://fsr.eui.eu/unbundling-in-the-european-electricity-and-gas-sectors/, accessed: 29.04.2019
[14] Suh NP 1990 *The Principles of Design* (New York : Oxford University Press)
[15] Suh NP 2005 *Complexity.Theory and Applications* (New York : Oxford University Press)
[16] Rodrigues F, Fradinho J, Cavique M, Santos AG and Mourão A 2019 An axiomatic approach to the design and operation of a wood pellet production line, 2019, *Proc. Of The 13th International Conference on Axiomatic Design Sydney, Australia, October 18-20*, MATEC Web Conf, Volume **301**

Modularity Measurement as a Crucial Design Element

V Modrak[1] **and Z Soltysova**[1]

[1]Faculty of Manufacturing Technologies, Technical University of Kosice, Bayerova 1, 080 01 Presov, Slovakia

vladimir.modrak@tuke.sk

Abstract. This paper aims to explore the problem of measurement of assembly process modularity, and to offer a new approach to quantify the relative modularity of different assembly process models. Specifically, it seeks to resolve the problem of the relation between relative modularity measures and optimal modularity measures. This relation brings us more close to find an effective measure for determining whether the process network is optimally modular. In our approach, process modularity expresses the extent to which processes can be decomposed into modules to be executed in parallel, and/or in series. Moreover, it also considers important feature of modular systems - that the great majority of interactions occur within modules and only a few interactions occur between modules. Consequently, the effect of a change in a given module is confined to that module only. This advantage of modular design clearly corresponds with Axiomatic Design theory, especially with its first axiom.

1. Introduction and related work

In general, modularly organized networks are characterized, apart from other properties, by functional segregation and integration [1]. Manufacturing assembly processes that are of interest in this paper belongs to this network type. System modularity problems, equally as system complexity issues are parts of general systems theory, since one can apply them to different kinds of systems, including technical, social, and biological [2, 3]. According to Ulrich [4], a modular architecture is based on a one-to-one mapping functional elements to the physical components of the product, and specifies de-coupled connections between components. Such architecture follows the first axiom in Axiomatic Design defined by Suh [5] specifying that each system function or functional requirement has to be satisfied by an independent design parameter. Subsequently, design parameters are projected onto construction documents, and transformed through the manufacturing process variables into process output. An important feature of the system modularity is that complexity of technical systems can be effectively managed through their modular design [6, 7]. Tate [8] categorized modularity from Axiomatic Design theory perspective into three types: resource, operational, and interfacial one. According to him, the resource modularity can be defined as 'ease of manufacturing'. This definition at least shows a certain connection between process modularity and complexity in system design. According to Mehrsai et al. [9], extension of modularity into processes and resources allows to generate alternative structures of organizations or supply chain networks by splitting their performances into modules and adjust them as required.

Product and process modularity becomes increasingly important over the last decades, especially due to diffusion of mass customization that is considered as an important competition strategy of a lot of businesses [10]. In this nexus, modularity measurement plays a vital role in the product and process design, and need to be continually developed. Existing literature offers several approaches to measure product modularity. A comprehensive overview of them has been offered by Ulrich [11]. From more recent works it can be mentioned work by Hölttä-Otto and De Weck [12] who proposed a relative product modularity metric called the Singular Value Modularity Index. On the other hand, there is a lack of process modularity measures or metrics to quantify a degree of manufacturing system modularity [13]. Selected relevant approaches on process modularity measurement, which in some way relate to our research can be found in works of [14-19].

2. Classification framework of assembly process networks

First, have a look at the simplest structural model of manufacturing assembly process (MAP). As shown in Figure 1, the structure consists of the three obligatory elements, which are inputs elements - at least two, process operation(s), and output element(s). Such a structure also represents simplest assembly module.

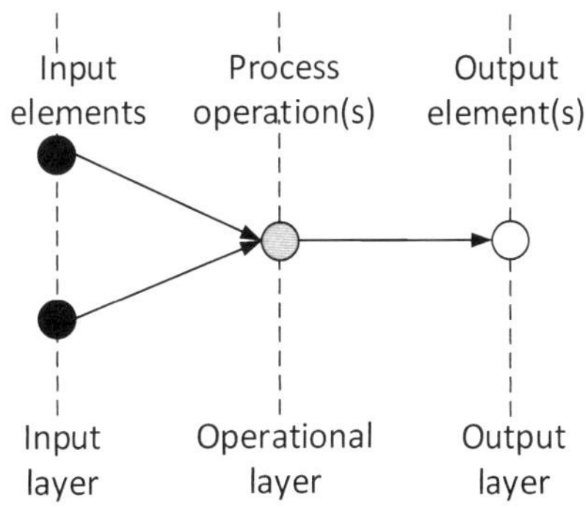

Figure 1. The simplest structural model of manufacturing assembly network.

As topology of assembly process structures varies case by case and depends upon specific factors, it seems to be useful to create working classification of process structures by the number of process layers, and the number of output elements in order to bring out the relevant modularity measures more clearly. Accordingly, we propose to divide assembly process structures into the following classes and subclasses:

1. Single Layer Multi-Product (SLMP) assembly network. This class of MAP structures includes all single-step assembly networks with different numbers of input components, parallel single assembly operations, and output components. Selected alternative MAP structures, when the number of input components equals four, and the number of output components equals six are shown in Figure 2 (see the structures of the type 1). Modularization followed here is based only on vertical fragmentation of the network modules into submodules.

2. Multi-Layer Single-Product (MLSP) assembly network. This class of MAP structures is further divided into two sub-classes, which are:

a) MLSP, when order of input components entering into the assembly operations is not important. All possible alternative MAP structures, when the number of input components equals four, and the number of output component equals one are shown in Figure 2 – see structures of the type 2a. These networks are modeled as single-rooted tree graphs.

b) MLSP, when order of input components entering into the assembly operations is determined. All possible alternative MAP structures, when the number of input components equals four, and the number of output component equals one are shown in Figure 2 – see structures of the type 2b. As it can be seen, the number of all alternative structures is in such case is lower than in case of type 2a. It is specifically because, the structure No. 2.4a does not satisfied the rule of determined order of assembly operations. These networks follow the pattern of single-rooted tree graphs with only single branch of nodes.

3. Multi-Layer Multi-Product (MLMP) assembly network. This class of MAP structures is important in terms of mass customization. Let´s say graph No. 3.1. (see Figure 2) represents originally designed

IOP Conf. Series: Materials Science and Engineering 1174 (2021) 012002 doi:10.1088/1757-899X/1174/1/012002

MAP structure. Moreover, initial elements c, d and e entering the second assembly operation in this structure are optional components. Then, it is topical to create alternative structures (see structures No. 3.2. and No. 3.3.), which better satisfy the requirements of assembly process in terms of mass customization. It is because, assembly processes in terms of mass customization have to be as modular as possible. These networks present multi-rooted tree graphs.

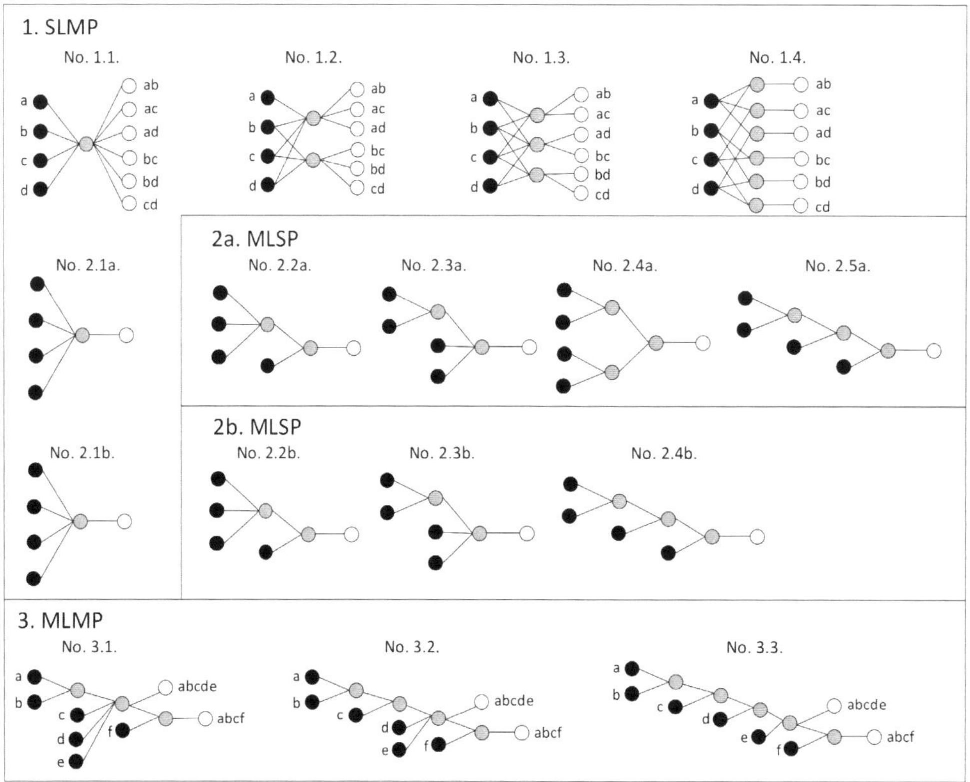

Figure 2. The assembly process networks classification.

3. Proposition of the relative and optimal process modularity measures

3.1. Modularity typology and working definitions

Our interest lies in investigating assembly manufacturing processes, which are generalized as technical systems. According to Li et al. [20] a technical system modularity needs to be divided at least into three basic domains: product, service and supply chain processes, and these three domains fuse into an organic whole, interacting and restraining each other. The same authors argue, that in spite of well-developed product classification, there is absence of service modularity classification and, supply chain modularity classification. In order to identify the object, we want to explore, we classify supply chains into two groups, i.e., general supply chains (interconnecting a group of organizations), and manufacturing assembly supply chains (assembly processes). As there are important differences between the two groups, our research is focused only on the assembly processes modularity.

In generally, modularization of assembly process means subdividing one integral operation into several separate operations/modules, and assembly process modularity indicates the degree of decomposability of a given process into sub-processes without changing the number of inputs and outputs. When process modularity is defined as the degree of decomposability of a given process, then, it reflects so called relative modularity. Efatmaneshnik, and Ryan [21] stated, that modularization remains more of an art than a science, to the extent that the notion of optimal modularity is not applied.

For this reason, the general concept of optimal modularity developed by Newman and Girvan [22] is employed in our approach as complementary indicator along with the proposed relative measure of process modularity. However, adoption of the optimality modularity measure needed in our research to verify its applicability for MAP networks. Its validity for given purpose has been tested through selected alternative MAP structures, while obtained results are presented in section 4 of this paper.

3.2. Description of proposed relative modularity measure

A construct of the proposed process modularity measure is based on a realistic assumption that structural network complexity directly relates to network modularity [6-8]. Then, we further assume that adequate structural complexity measure, which inherently captures the modularity features could be transformed to measure the degree of process modularity. In this order, so called the *vertex degree distribution of a graph* (Ivd) will be used for this purpose, since it intrinsically covers network modularity signs. This indicator was developed by Bonchev and Buck [23], and it measures structural complexity of a general networks. It is formalized as a graph G consisting of a set of V vertices {V}≡{v1, v2, … , vV}, and expressed by formula:

$$Ivd = \sum_{i=1}^{V} deg(v)_i log_2 deg(v)_i, \tag{1}$$

where $deg(v)_i$ is the degree of vertex of $(v)_i$ in G.

To outline mutual relation between network complexity and network modularity, the first type of assembly process networks (SLMP), namely process structures No. 1.1., 1.2., 1.3. and 1.4 will be used. Subsequently, complexity levels of these structures are enumerated by employing formula (1), and obtained values are available in the first line of Table 1.

Table 1. Process complexity values of the selected assembly process networks.

	No. 1.1.	**No. 1.2.**	**No. 1.3.**	**No. 1.4.**
Ivd	33,22	41,16	52,25	47,6
AMC	33,22	20,58	17,42	7,93

By visual comparison of the SLMP models from Figure 2, it can be seen that the structure No. 1.4. is the most modular, and the network No. 1.1 is non-modular. Then, one would expect that complexity of the structure No. 1.4. will be the smallest one and the complexity of the structure no. 1.1. will be the highest one. However, Ivd indicator, which is based on information entropy, expresses absolute complexity measure by prioritization of the highest magnitudes of vertices against overall sum of vertex degrees.

Therefore, Ivd does not satisfy this expectation. For the sake to find out the most appropriate coincidence between the networks complexity and the networks modularity we propose to modify formula (1) to count relative network complexity as the average complexity of its modules. Then, the modified indicator further named as *Average Module Complexity* (AMC) can be expressed by the equation:

$$AMC = \frac{\sum_{i=1}^{V} deg(v)_i log_2 deg(v)_i}{V} \tag{2}$$

An example of enumeration of Ivd and AMC values for the model No. 1.2 is depicted in Figure 3.

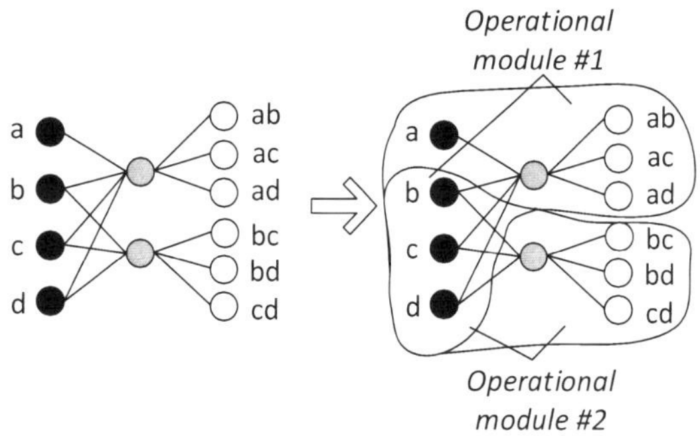

$$Ivd = 1log_2 1 + 2log_2 2 + 2log_2 2 + 2log_2 2 + 7log_2 7 + 6log_2 6 + 1log_2 1 + 1log_2 1 + 1log_2 1 + 1log_2 1 + 1log_2 1 + 1log_2 1 = 41,16$$

$$AMC = \frac{41,16}{2} = 20,58$$

Figure 3. Enumeration of Ivd and AMC values.

By applying AMC indicator for all the SLMP networks from Figure 2, we obtain network complexity values, which are shown in Table 1 – see the second line of the values. Now, we can see that the most modular network is with the lowest complexity and vice versa. Equally, this relation is actual also for networks No. 1.2. and 1.3. Later, it will be showed that this relation is also valid for other types of assembly process structures. Based on these assumptions it is reasonable to quantify relative network modularity through the inverse value of AMC. Then, *relative network modularity* (RNM) can be expressed by formula:

$$RNM = \frac{V}{\sum_{i=1}^{V} deg(v)_i log_2 deg(v)_i}. \tag{3}$$

In the next section, it will be shown how this formula works for other types of networks, i.e., MLSP and MLMP.

3.3. Adaption of optimal modularity measure for assembly process networks

According to [24], network modularity is the degree to which it can be separated into nearly independent sub-networks. Newman and Girvan [22] developed a quantitative measure of optimal modularity (Q). They developed an algorithm finding the division of the nodes into modules that maximizes a measure Q. The formula to quantify optimal level of modularity Q is expressed as follows:

$$Q = \sum_{s=1}^{K} \frac{ls}{L} - \left(\frac{ds}{2L}\right)^2 \tag{4}$$

where:
K - is the number of modules,
L - is the number of edges in the network,
ls - is the number of edges between nodes in module s,
ds - is the sum of the degrees of the nodes in module s.

Adoption of this modularity measure for assembly process networks is explained through the following model of structure No.1.2. (see Figure 4).

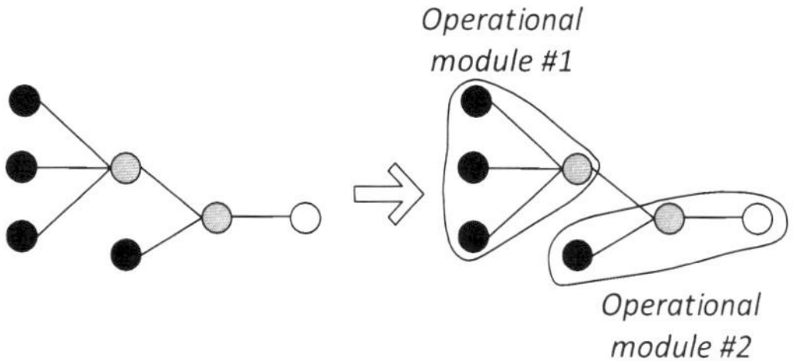

$$Q_1 = \frac{3}{6} - \left(\frac{7}{12}\right)^2 = 0,16$$

$$Q_2 = \frac{2}{6} - \left(\frac{5}{12}\right)^2 = 0,16$$

Figure 4. Enumeration of Q values.

In this way, one can employ formula (4) for our reference models of assembly process networks and compare them with RNM values (see Table 2).

Table 2. The Obtained RNM and Q values.

Type		RNM	Q
1.	**1.1.**	0,03	0
	1.2.	0,049	0,24
	1.3.	0,057	0,25
	1.4.	0,126	0,54
2a	**2.1a**	0,086	0
	2.2a	0,157	0,32
	2.3a	0,157	0,32
	2.4a	0,21	0,38
	2.5a	0,21	0,38
2b	**2.1b**	0,086	0
	2.2b	0,157	0,32
	2.3b	0,157	0,32
	2.4b	0,21	0,38
3.	**3.1.**	0,12	0,425
	3.2.	0,155	0,459
	3.3.	0,185	0,462

As it can be seen, the models No. 1.1., 2.1a. and 2.1b. report zero values of Q. In fact, it is because that these structures are non-modular ones. The remaining Q values of the structures numbered as 1.2., 1.3. and 1.4. have the same tendency as RNM values. It is due to the fact, that optimal network modularity according to indicator Q is dominantly affected by the number of modules in the network. Therefore, SLMP process models are from the optimality viewpoint quite specific, since the most modular structures are at the same time considered as optimal.

Even though, the RNM values and Q values of MLSP and MLMP selected structures (see Table 2), have the same tendency, it is not the rule, but just the specific cases. This situation can be explained by the small number of alternative MAP structures, what is limited in case of MLSP process structures by the number of input components ($i = 4$). In case of MLMP process networks, other MAP alternatives are not taken into account. For this reason, in the next section, the analysis of alternatives process structures with input components $i = 5$ and $i = 6$ will be provided. The aim of this analysis is to prove or disprove our above mentioned assumption that structural network complexity directly relates to network modularity in this way that higher network modularity reduces its complexity.

4. Testing of the modularity indicators

In this section, MAP structures of type 2 - MLSP with five and six input components for both subclasses, will be tested using the selected modularity indicators with the aim to find the optimal modularity values.

4.1. Testing MAP structure with five input components

Firstly, we start with the first subclass of MLSP type of the MAP structures - when order of input components entering into the assembly operations is not important. Then, it is possible to deterministically generate twelve possible alternative MAP structures (see Figure 5a).

IOP Conf. Series: Materials Science and Engineering 1174 (2021) 012002 doi:10.1088/1757-899X/1174/1/012002

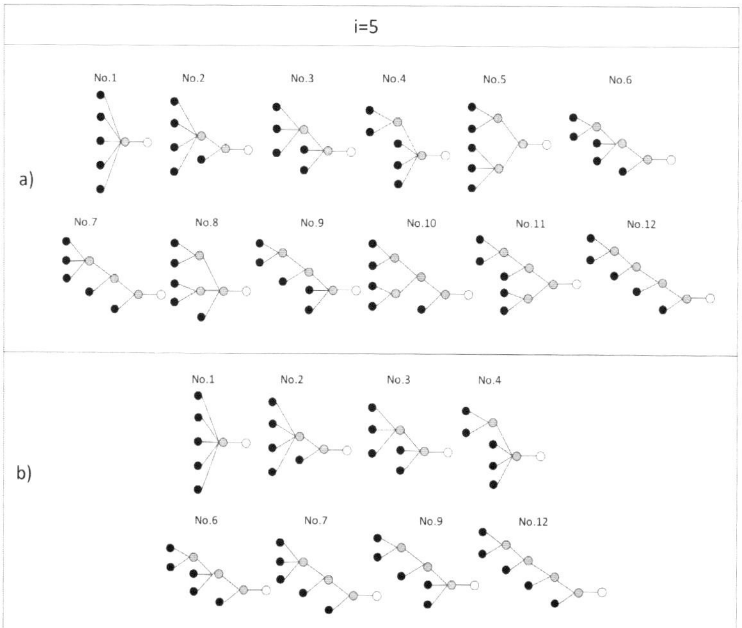

Figure 5. All possible MLSP process structures with five input components a) when order of input components entering into the assembly operations is not important, b) when order of input components entering into the assembly operations is important.

After generating of the all possible MAP structures for i = 5, RNM values using equation (3) and Q values using equation (4) are calculated, ordered by the increasing indicator RNM, and shown using graphical representation in Figure 6a.

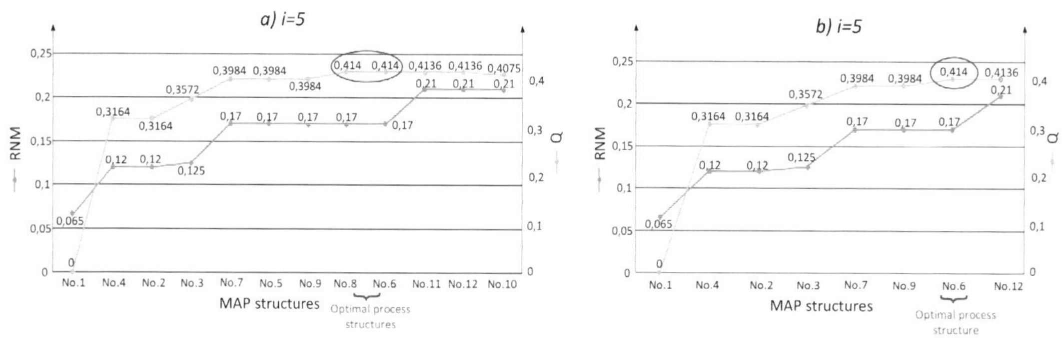

Figure 6. MLSP process structure with five input components a) when order of input components entering into the assembly operations is not important, b) when order of input components entering into the assembly operations is important.

As can be seen from Figure 6a, two optimal modular MAP structures, namely, No. 6 and No.8 are found.

Now, we continue with the second subclass of MLSP of the MAP structures - when order of input components entering into the assembly operations is important. Then, it is possible to exactly generate eight possible alternative MAP structures (see Figure 5b).

After generating of all possible MAP structures for i = 5, RNM values using equation (3) and Q values using equation (4) are calculated, and ordered by the increasing indicator RNM as shown in

Figure 6b. As can be seen in this figure, the MAP structure No. 6 is accordingly identified as optimal modular structure.

4.2. Testing MAP structure with six input components

Let us start in this subsection with the first subclass of MLSP type of the MAP structures - when order of input components entering into the assembly operations is not important. Then, it is possible to generate 33 possible alternative MAP structures (see Figure 7a).

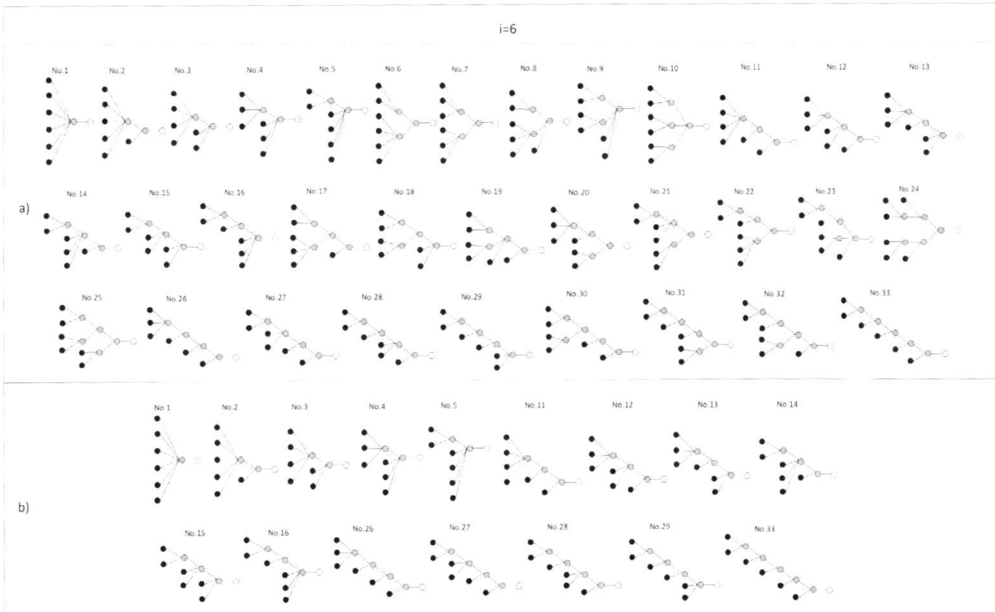

Figure 7. All possible MLSP process structures with six input components a) when order of input components entering into the assembly operations is not important, b) when order of input components entering into the assembly operations is important.

Subsequently, RNM values and Q values are calculated for all possible MAP structures. Then, the MAP structures are ordered by the increasing indicator RNM, as shown in Figure 8a.

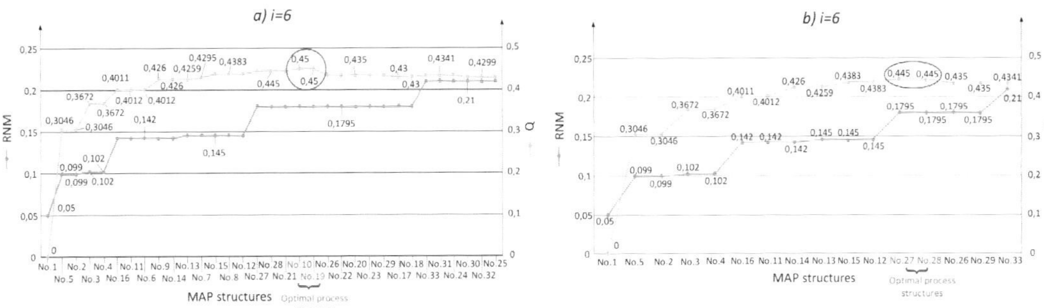

Figure 8. MLSP process structure with six input components a) when order of input components entering into the assembly operations is not important, b) when order of input components entering into the assembly operations is important.

Accordingly, the optimal modular MAP structures are identified, which are namely structures No. 10 and No. 19.

In the case of the second subclass of MLSP type of the MAP structures - when order of input components entering into the assembly operations is important, it is possible to generate 16 possible alternative MAP structures (see Figure 7b).

Then, RNM values and Q values are calculated and presented in Figure 8b, where the structures are ordered by the increasing values of RNM indicator. The MAP structures No. 27 and No. 28 are considered as optimal ones.

5. Conclusions

The presented results from computational experiments in section 4 indicate promising potential of the two modularity measures, namely, optimal level of modularity Q and relative modularity network RNM to manage modularity issues in early stage of assembly process designing. This two complementary indicators allowing better understanding of relation between the relative modularity concept and the optimal modularity method.

There is no doubt that the problem of detecting structural modularity is one of the most challenging issues in the study of network systems. Even though, the method of optimal modularity developed by Newman and Girvan [22] is outstanding algorithm for the optimization of the network modularity that is applicable to a many real-world networks, its application to the new fields requires certain adaptation and at least theoretical validation. Hopefully, this paper contributes to the knowledge in this domain, and after further investigation will offer a useful tool for manufacturing process designers.

Acknowledgement:
This research was funded by the KEGA project No. 025TUKE-4/2020 granted by the Ministry of Education of the Slovak Republic and by the European Union's Horizon 2020 research and innovation program under the Marie Skłodowska-Curie grant agreement No. 734713.

References

[1] Hilger K, Ekman M, Fiebach C J, Basten U 2017 Intelligence is associated with the modular structure of intrinsic brain networks *Scientific reports* **7(1)** 1-12
[2] Schilling M A 2000 Toward a general modular systems theory and its application to interfirm product modularity *Academy of management review* **25(2)** 312-334
[3] Baldwin C Y and Clark K B 2003 Managing in an age of modularity *Managing in the modular age: Architectures, networks, and organizations* **149** 84-93
[4] Ulrich K 1995 The role of product architecture in the manufacturing firm *Research policy* **24(3)** 419-440
[5] Suh N P 1990 The Principles of Design *New York, Oxford University Press* 418
[6] Ethiraj S K and Levinthal D 2004 Modularity and innovation in complex systems *Management science* **50(2)** 159-173
[7] Parker D B 2010 Modularity and complexity: An examination of the effects of product structure on the intricacy of production systems *Michigan State University* 119
[8] Tate D 1999 A Roadmap for Decomposition: Activities, Theories, and Tools for System Design Ph.D. Thesis *Massachusetts Institute of Technology* 210
[9] Mehrsai A, Karimi H R, Thoben K D 2013 Integration of supply networks for customization with modularity in cloud and make-to-upgrade strategy *An Open Access Journal Systems Science & Control Engineering* **1(1)** pp 28-42
[10] Modrak V (Ed.) 2017 Mass customized manufacturing: theoretical concepts and practical approaches *CRC Press* 313
[11] Ulrich K 1994 Fundamentals of Product Modularity Management of Design *Dordrecht: Springer* 219–231
[12] Holtta K, Suh E S, de Weck O 2005 Tradeoff between modularity and performance for engineered systems and products ICED 05: 15th International Conference on Engineering Design: Engineering Design and the Global Economy *Engineers Australia* 449-450

[13] Calcagno M 2002 Dynamics of Modularity: A Critical Approach *Euram Conference* **9(11)** 1-11
[14] Vickery S K, Koufteros X, Dröge C, Calantone R 2016 Product modularity, process modularity, and new product introduction performance: does complexity matter? *Production and Operations management* **24(4)** 751-770
[15] Blackenfelt M 2001 Managing complexity by product modularisation *Doctoral dissertation, Maskinkonstruktion* 100
[16] Maier J F, Eckert C M, Clarkson P J 2017 Model granularity in engineering design–concepts and framework *Design Science* **3** 1-29
[17] Frenken K and Mendritzki S 2012 Optimal modularity: a demonstration of the evolutionary advantage of modular architectures *Journal of Evol Econ* **22** 935–956
[18] Modrak V and Soltysova Z 2018 Process modularity of mass customized manufacturing systems: principles, measures and assessment *Procedia CIRP* **67** 36-40
[19] Kusiak A 2002 Integrated product and process design: a modularity perspective *Journal of Engineering Design* **13(3)** 223-231
[20] Li H, Yang M, Evans S 2019 Classifying different types of modularity for technical system *International Journal of Technology Management* **81(1-2)** 1-23
[21] Efatmaneshnik M and Michael J R 2016 On optimal modularity for system construction Complexity *Complexity* **21** 176-189
[22] Newman M E J and Girvan M 2004 Finding and evaluating community structure in networks *Physical Review E* **69(2)** 1-16
[23] Bonchev D and Buck G A 2005 Quantitative measures of network complexity *Complexity in chemistry, biology, and ecology* 191-235
[24] Alon U 2019 An introduction to systems biology: design principles of biological circuits *CRC press* 343

IOP Conf. Series: Materials Science and Engineering 1174 (2021) 012003 doi:10.1088/1757-899X/1174/1/012003

Data-driven sustainable ship design using Axiomatic Design and Bayesian Network Model

G Fardelas[1,2] **and S G Kim**[1,3]

[1] Department of Mechanical Engineering, Massachusetts Institute of Technology, Cambridge, MA 02139, USA
[2] System Design and Management Program, Massachusetts Institute of Technology, Cambridge, MA 02139, US
[3] Corresponding author's email: sangkim@mit.edu

Abstract. Environmental sustainability, as well as social and economic well-being, must be considered in every stage of a product lifecycle, from conceptual design to its retirement. Even though this sustainability-centric approach represents a critical driver for innovation, it also increases the design complexity. Nowadays, the maritime transport accounts for a large share of transport demand, and the importance of sustainable ship design is increasingly growing, not only for ethical and legislative but also for competitive reasons. The design of a sustainable ship considering all those aspects is a complex process in this regard. One way to manage the complexity is to identify and avoid the functional couplings at the early stage of the design process. This paper presents the conceptual design of a merchant ship's conventional propulsion system with a view to the Axiomatic Design framework and known sustainable engineering principles. We also explore the Bayesian machine learning interface to propose a data-driven method for calculating the probability of achieving specific sustainability-related functional requirements. Data-driven Bayesian reasoning can also be used to select the best design parameter among the proposed alternatives as well as to identify hidden design couplings that have not identified by the designers in the conceptual design stage.

1. Introduction

The design of ships has been a process that dates back thousands of years. In the past, it was more of an art than science, based on heuristic experience and trial-and-error design methods. Given the importance of solving the ship design problem, trial-and-error methods have been replaced by a system-based approach, which considers a ship as a system integrating various subsystems and their components (e.g., ship propulsion, navigation systems, and arrangements). The objective of this work is to present the conceptual design of a merchant ship's conventional propulsion system with a view to known sustainable engineering principles in the Axiomatic Design (AD) framework.

Ships are large systems not only because they are physically large in size, but also, they have various FRs to be met throughout their life cycle. Ships are also time-variant systems as they have many FRs at the highest level, but only a subset of these FRs must be satisfied at a given time, while different subsets need to be satisfied at different points in time. As a result, a ship as a system must reconfigure itself to satisfy different subsets of FRs throughout its life cycle.

The design of ship has been an iterative decision-making process that often reflects the conflicting interests of the various stakeholders. Additionally, the surging trend of digitalization in the maritime sector and the urgent requirement for a more sustainable world push for designs that can reduce the complexity of ship design in the context of ethical responsibility, legislative compliance, and fair market competition. One way to manage the complexity and control the design of a system is to identify and manage the potential functional couplings at the early stage of the design process [1]. The evolution of Systems Engineering has encouraged the development of various design methods applicable to ship design, such as the "set-based design" and the "analytic hierarchy process (AHP)". In practice, these methods drive to an ad hoc approach to the conceptual design or give priority only to some FRs without checking their dependency, resulting in a complex design that can negatively affect system's quality attributes such as reliability, maintainability, and flexibility. Nevertheless, sustainable design requires for managing or even restricting design's complexity. The sustainability imperative requires insight into a Human–Environment system [2], which would increase the design's complexity as it demands in-depth knowledge of new interrelationships between science, society, and environment. The holistic approach of AD is ideal to accommodate the challenges related to sustainable ship design.

The paper shows a scalable example for the total ship design following sustainable engineering principles in two aspects: 1) Axiomatic Design as a methodology to control the complexity of sustainable ship design and 2) Bayesian machine learning technique as a supportive tool for improving system's architecture and assessing system's sustainable impact.

2. Intersection of sustainability initiative and ship design

Ships are built to cover society's needs through the provision of specific (commercial or noncommercial) services. Simultaneously, demand for maritime transport grows in parallel with the world population, industrial activity, and trade growth. Maritime transport will continue to account for the largest share of demand, with a contribution of 75% in 2050 [3]. Changes in the structure of energy markets and fuel industry, the volatility of energy prices, piracy and security concerns, automation, big data technology, and shipping profitability along with the 2030 Agenda for Sustainable Development (ASD), and the Paris Agreement on climate change require a balanced approach between contradicting economic, social and environmental objectives.

According to the book, "Our Common Future" from the UN committee, sustainable development is defined as the development that meets the needs of the present without compromising future generations' ability to meet their own needs [4]. Within this scope, a sustainable design outcome should consider all environmental impacts throughout its life cycle (from concept design to recycling), including social and economic well-being, without compromising cost, appearance, and quality attributes. This approach also applies to ship design. This paper focuses on designing at concept level a ship's conventional propulsion system that is compatible with the fuels currently prevalent in the maritime sector, providing at the same time an environment-friendly and energy-efficient solution without sacrificing the profitability of the shipping industry.

3. Application of Axiomatic Design on sustainable ship design

3.1. Conceptual design of a merchant ship's conventional propulsion system: Design hierarchies

Once the ship's total resistance has been estimated, the next steps of designing a vessel's propulsion system are the choice of the propulsor and its matching with the hull and the prime mover (propulsion engine). Assuming a fixed-pitch propeller, the decomposition of the design problem results in the FR and DP hierarchies as shown in Table 1. The system architecture is described by the design equations (1) to (8).

The design of a ship's propulsion system in the context of the Axiomatic Design begins with analyzing the operational requirements that are common to each type of propulsion plant. At the highest level of design, we need a system that can move the ship in the water, overcoming its total resistance. In addition to producing thrust, the propulsion system consists of individual parts whose function is to seal the system with the marine environment and hold the system in place under extreme weather conditions. The shaft system must transmit the power generated by a sustainable main engine to the propeller, whose rotational motion is converted into thrust. The propulsion system's components must also be supported and interconnected and must comply with the existing regulations of the International Maritime Organization (IMO).

Table 1. FRs and DPs of the sustainable propulsion system.

ID	FR	DP
1	Move through water	Propulsion system
1.1	Produce thrust	Drivetrain with engine
1.1.1	Convert power into thrust	Propeller
1.1.2	Generate sustainable rotary power	Sustainable main engine
1.1.2.1	Machinery system that meets NOx regulation	Dual fuel engine
1.1.2.2	Machinery system that meets SOx regulation	Low sulfur combustion system
1.1.2.3	Machinery system that meets EEDI regulation	Fuel injection system
1.1.3	Transmit power	Shaft system
1.1.3.1	Transmit power from main engine to thrust shaft	Engine-thrust shaft coupling
1.1.3.2	Transmit power from main engine-thrust shaft coupling to propeller	Shaft arrangement
1.1.3.2.1	Receive the rotational motion from the crankshaft	Thrust shaft
1.1.3.2.2	Receive the rotational motion from the thrust shaft	Intermediate shaft
1.1.3.2.3	Receive the rotational motion from the intermediate shaft	Tail shaft
1.2	Seal components	Stern tube sealing arrangement
1.3	Fix components	Mounting system
1.3.1	Fix main engine	Main engine mounts
1.3.2	Support shaft's load	Bearings system
1.3.2.1	Support thrust shaft's load	Thrust blocks
1.3.2.2	Support intermediate shaft's load	Intermediate shaft mounted bearings
1.3.2.3	Support tail shaft's load	Stern tube
1.3.3	Connect adjacent shafts	Shaft couplings

$$\left\{FR_1\right\} = \left(X\right)\left\{DP_1\right\} \tag{1}$$

$$\begin{Bmatrix} FR_{1.1} \\ FR_{1.2} \\ FR_{1.3} \end{Bmatrix} = \begin{pmatrix} X & O & O \\ O & X & O \\ O & O & X \end{pmatrix} \begin{Bmatrix} DP_{1.1} \\ DP_{1.2} \\ DP_{1.3} \end{Bmatrix} \tag{2}$$

$$\begin{Bmatrix} FR_{1.1.1} \\ FR_{1.1.2} \\ FR_{1.1.3} \end{Bmatrix} = \begin{pmatrix} X & O & O \\ X & X & O \\ X & X & X \end{pmatrix} \begin{Bmatrix} DP_{1.1.1} \\ DP_{1.1.2} \\ DP_{1.1.3} \end{Bmatrix} \tag{3}$$

$$\begin{Bmatrix} FR_{1.3.1} \\ FR_{1.3.2} \\ FR_{1.3.3} \end{Bmatrix} = \begin{pmatrix} X & O & O \\ O & X & O \\ O & O & X \end{pmatrix} \begin{Bmatrix} DP_{1.3.1} \\ DP_{1.3.2} \\ DP_{1.3.3} \end{Bmatrix} \tag{4}$$

$$\begin{Bmatrix} FR_{1.1.2.1} \\ FR_{1.1.2.2} \\ FR_{1.1.2.3} \end{Bmatrix} = \begin{pmatrix} X & O & O \\ X & X & O \\ X & O & X \end{pmatrix} \begin{Bmatrix} DP_{1.1.2.1} \\ DP_{1.1.2.2} \\ DP_{1.1.2.3} \end{Bmatrix} \tag{5}$$

$$\begin{Bmatrix} FR_{1.1.3.1} \\ FR_{1.1.3.2} \end{Bmatrix} = \begin{pmatrix} X & O \\ O & X \end{pmatrix} \begin{Bmatrix} DP_{1.1.3.1} \\ DP_{1.1.3.2} \end{Bmatrix} \tag{6}$$

$$\begin{Bmatrix} FR_{1.1.3.2.1} \\ FR_{1.1.3.2.2} \\ FR_{1.1.3.2.3} \end{Bmatrix} = \begin{pmatrix} X & O & O \\ O & X & O \\ O & O & X \end{pmatrix} \begin{Bmatrix} DP_{1.1.3.2.1} \\ DP_{1.1.3.2.2} \\ DP_{1.1.3.2.3} \end{Bmatrix} \quad (7)$$

$$\begin{Bmatrix} FR_{1.3.2.1} \\ FR_{1.3.2.2} \\ FR_{1.3.2.3} \end{Bmatrix} = \begin{pmatrix} X & O & O \\ O & X & O \\ O & O & X \end{pmatrix} \begin{Bmatrix} DP_{1.1.2.1} \\ DP_{1.1.2.2} \\ DP_{1.1.2.3} \end{Bmatrix} \quad (8)$$

3.2. The module junction structure diagram and the data flow chart

Another means of representing the architecture of the system are the module junction structure diagram and the data flow chart [5]. The module junction diagram (Figure 1a) is composed of modules and junctions, which represent FR leaves and their vertical integration respectively. The module junction diagram can be converted to the data flow chart (Figure 1b), which can also express the system architecture showing the data stream flow among modules. The module junction diagram and data flow chart, can be used for understanding of the overall ship design process in terms of the data and information flow and, therefore, can promote sustainable ship design and operation in multiple ways considering its whole life cycle. First, the diagrams can be used in conjunction with failure diagnosis algorithms and can help engineers trace causes of failures, saving sources and time. Second, in project management, the diagrams can provide the roadmap for task coordination and project execution, highlighting how a change in design decisions propagates and affects the system. Third, if the modules are correctly arranged and satisfy the Independence Axiom, a modular ship design can foster professional diversity, as the diagrams can provide the roadmap for stakeholders of different backgrounds located all over the world. Finally, in case the modules, as "providers" of design information, can be linked to sustainability metrics, the diagrams can be the main component of a dynamic sustainability assessment tool for the vessel's whole life cycle.

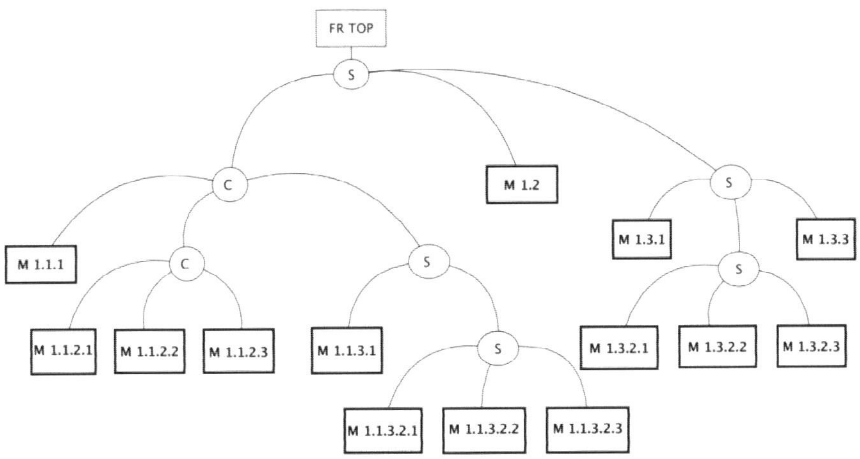

(a)

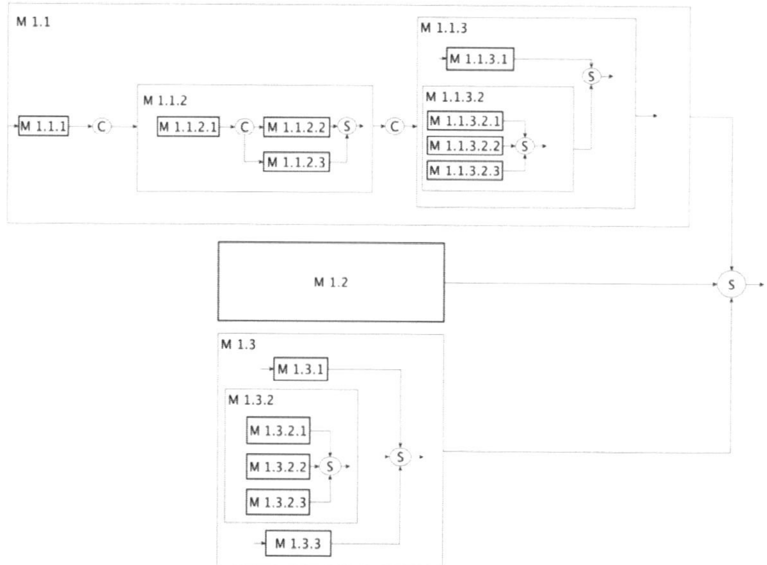

(b)

Figure 1. The module junction diagram (a) and the data flow chart (b) for the design described by the equations (1) to (8).

4. The interface of Bayesian Networks and Axiomatic Design

4.1. Bayesian Networks overview

Bayesian network (BN) is a probabilistic graphical model that reflects the states of some part of a world being modelled, and describes how probabilities relate to those states. It uses Bayesian inference in order to model conditional dependence, and therefore causation. A Bayesian network is a directed acyclic graph that consists of directed nodes (parent and child nodes) structured hierarchically. The directed arrows connecting the nodes have the probability dependencies between them represented in condition probabilities tables [6].

First, Bayesian networks are useful as they help us model systems of reasonable complexity with a significant computational effort saving, because there is no need to store all possible configurations of states, but only all possible combinations of states between sets of family (parent and child) nodes. After training, the network can help the user identify new or wrong relationships, which can be used for process improvement [7]. Second, a BN is a tool for hybrid intelligence as they combine human learning, in terms of heuristic knowledge provided by domain experts, and machine learning in terms of knowledge learned from data [8]. Third, BNs can support both causal and evidential reasoning, as Bayesian inference allows users to make a prediction of the effects given some causes (causal reasoning) and determine the causes given an observation of the effects (evidential reasoning) [6].

Since AD method and BNs share a similar hierarchical parent-child structure and use the concept of identifying couplings between modules and nodes respectively, we believe that BNs can support AD methodology concerning the design of complex systems that promote sustainability.

4.2. Develop a Bayesian Network for improving system's architecture and assessing its sustainable impact

The information garnered from the decomposition and mapping process between the functional and physical domain can help designers understand design couplings and tradeoffs. However, this information can be evaluated and verified using a data-driven approach to become more useful. Figure 2 shows a compiled BN created using Netica software 6.07 (copyright 1992-2019 by Norsys Software

Corp.) [9]. The network was built using domain knowledge and insight obtained through the decomposition of FRs and DPs (Table 1). Then, it was trained with 750 real cases and its performance was tested with a different dataset of 250 real cases. Each of those 1000 cases corresponds to a ship with a separate propulsion system and operational profile and includes data, which were recorded using sensors [10], [11] or retrieved from ship's technical documentation [11], [12]. The probabilistic relationships among data are represented in the belief bars of the nodes (Figure 2).

The BN does not reflect precisely the hierarchical relationships between FRs and DPs, but being built with knowledge from the decomposition and mapping process, aims to capture the couplings between variables related to FRs, DPs, and system's function, so that it can be used as a supportive tool for designers to ensure or explore the system's sustainable compliance and impact. The network's goal is to help designers quantify some of the elements of the conceptual design matrices and calculate the probability of success for the requirements FR 1.1.2.1, FR 1.1.2.2, and FR 1.1.2.3. These FRs are described by the nodes "NOx Compliance Tier III", "SOx Compliance", and "EEDI Compliance Phase 1" respectively, in the BN of Figure 2.

4.2.1. Bayesian Inference and Independence Axiom. The probability relationships of the nodes of a BN can be validated through the technique of sensitivity analysis. Sensitivity analysis in BNs is a tool to quantify how much a particular node is affected by other nodes. Several advantages have been identified using the sensitivity analysis of a BN, which is built with knowledge obtained from the AD approach. First, the sensitivity analysis can identify design couplings wrongly judged in design matrices or detect hidden design couplings that the designers have not recognized. Second, some sensitivity analysis results, representing the degree of connectivity between nodes related to FRs or DPs, can work as a quantitative measure of the design matrix elements. Table 2 shows the sensitivity analysis results for the "NOx Compliance Tier III" node (relationship with FR 1.1.2.1). Third, the sensitivity results provide a metric of coupling of specific FRs. Table 2 illustrates, for example, the degree of connectivity (equal to 3.9 %) between "NOx Compliance Tier III" (relationship with FR 1.1.2.1) and "SOx Compliance" (relationship with FR 1.1.2.2).

4.2.2. Bayesian inference and Information Axiom. BNs are used by definition to predict the effects given some causes (causal reasoning) and determine the causes given an observation of the effects (evidential reasoning). In the first case, if a node corresponds to a specific FR, BNs can be used to calculate its probability of success and information content, Ii. In the case of evidential reasoning, the Bayesian inference can help designers determine the most possible causes that lead to satisfaction or not of an FR. Figure 3 shows a causal reasoning example, where the network calculates the probability of achieving FR 1.1.2.1 (NOx Compliance Tier III), which equals 0.783, given that the main engine operates with Liquid Natural Gas (LNG) and its rated speed is LOW (80rpm-700rpm for the available data set). The choice of using LNG as fuel presumes the installation of a dual fuel engine (DP 1.1.2.1), while the option of using DIESEL as fuel excludes the dual fuel applications and presumes a diesel single fuel engine. Therefore, the network, using as input the choice of a design parameter (DP 1.1.2.1), returns the probability of achieving its respective functional requirement (FR 1.1.2.1). In the case of evidential reasoning, Figure 4 shows that if we know that the propulsion system is not compliant with the IMO NOx limits (FR 1.1.2.1), then the most probable causes should be the "HIGH" main engine's specific fuel consumption (relationship with DP 1.1.2.3) or /and the "DIESEL" as used fuel (relationship with DP 1.1.2.1). In addition, the module junction diagram and data flow chart, as illustrated in Figure 1, can be used synergistically with the BN casual and evidential reasoning process. Considering the module diagram, we could pinpoint a specific module of interest to apply a BN in quantifying the probability of success in achieving sustainability-related requirements, and at the same time, the data flow chart can be used to guide designers and engineers to where sustainable ship design effort must be focused.

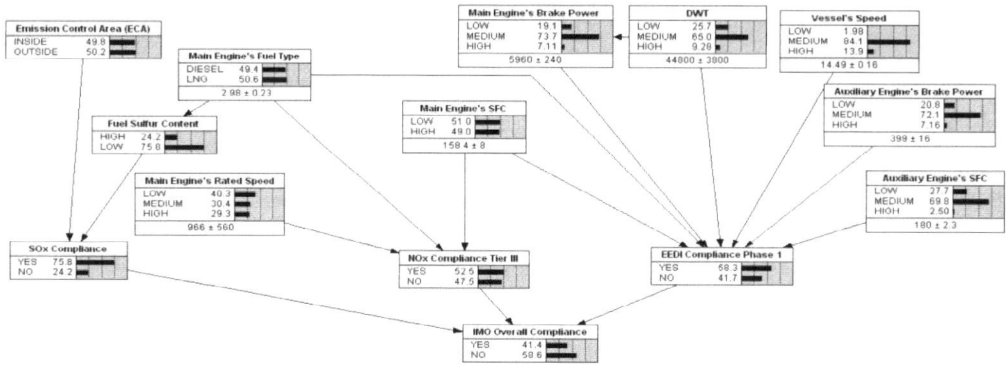

Figure 2. Screenshot of compiled BN for sustainability-related requirements.

Table 2. Sensitivity analysis outcome for the node "NOx Compliance Tier III".

Sensitivity of "NOx Compliance Tier III" due to finding at node:	Percentage
Main Engine's Fuel Type	17
Main Engine's SFC	14
Fuel Sulfur Content	4.33
SOx Compliance	3.9
Main Engine's Rated Speed	0.288
EEDI Compliance Phase 1	0.0257
Auxiliary Engine's SFC	0
Auxiliary Engine's Brake Power	0
Vessel's Speed	0
DWT	0
Main Engine's Brake Power	0
Emission Control Area (ECA)	0

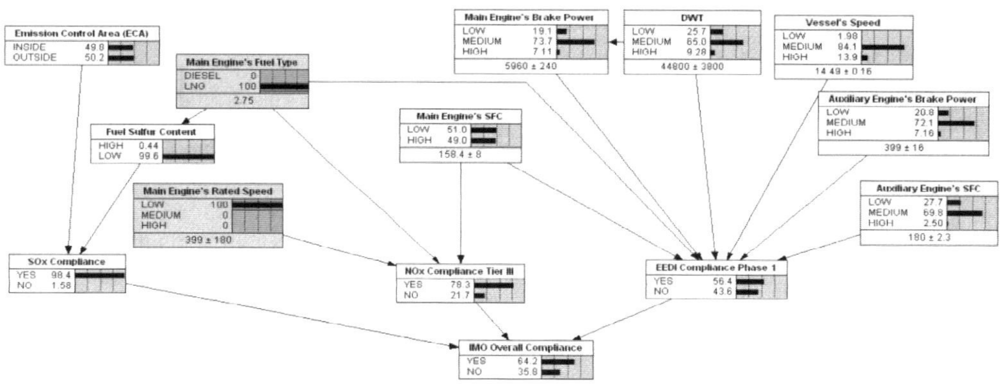

Figure 3. Screenshot of BN when the "Main Engine's Fuel Type" is LNG and "Main Engine's Rated speed" is LOW (causal reasoning).

IOP Conf. Series: Materials Science and Engineering 1174 (2021) 012003 doi:10.1088/1757-899X/1174/1/012003

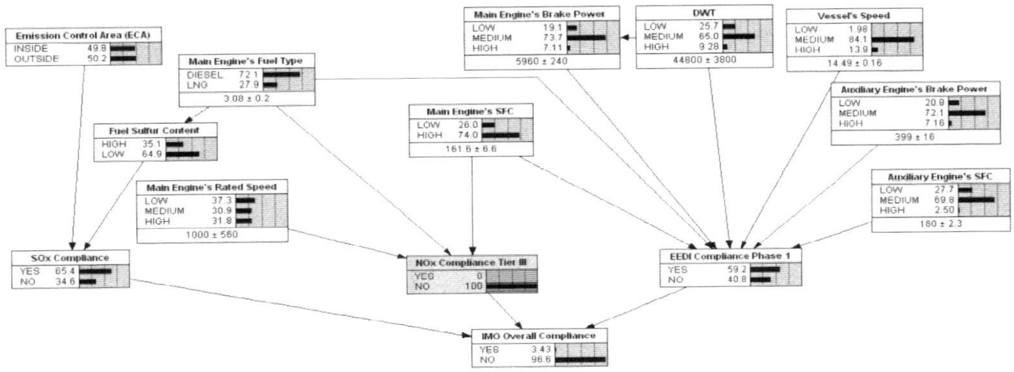

Figure 4. Screenshot of BN when the system does not meet the NOx regulation (evidential reasoning).

5. Conclusion

Sustainable design can increase a system's complexity, which is an inherent challenge in system architecture and is driven by the interconnections and the functional relationships of the system's entities. This paper shows that AD and BNs can be synergistically combined to promote sustainable ship design in two ways. Firstly, the FR-DP diagrams can be used to demonstrate how a change in one of the system's entities affects the others concerning the vessel's sustainable footprint along the life cycle of them. Secondly, the process of forming and mapping the hierarchical structures can be used as the roadmap of building a BN, which after training, can provide data-based feedback for the system architecture and calculate the probability of achieving sustainability-related functional requirements.

References

[1] Suh N P 2001 *Axiomatic Design: Advances and Applications* (New York: Oxford University Press)

[2] Jonker G and Harmsen J 2012 *Engineering for Sustainability: A Practical Guide for Sustainable Design* (London: Elsevier Science) p 6

[3] Psaraftis H N 2019 *Sustainable Shipping: A Cross-Disciplinary View* (Switzerland: Springer International Publishing)

[4] United Nations World Commission on Environment and Development 1987 *Report of the World Comission on Environment and Development: Our Common Future* (London: Oxford University Press)

[5] Kim S J, Suh N P and Kim S G 1991 Design of software systems based on axiomatic design *Robot. Comput. Integr. Manuf.* **8** 4, pp 243–55

[6] Barber D 2016 *Bayesian Reasoning and Machine Learning* (Cambridge, England: Cambridge University Press)

[7] Ni Z, Phillips L D and Hanna G B 2011 *Evidence Synthesis in Healthcare: Exploring Bayesian Belief Networks Using Netica®* (London: Springer London) pp 293–318

[8] Li Y, Zhang H, Roy U and Lee Y T 2017 *Proc. 2017 IEEE Int. Conf. on Big Data (Boston)* **1** (Piscataway, NJ: IEEE) pp 1736-1745

[9] Norsys Sofware Corporation 2021 *Netica Tutorial* (Vancouver: Norsys Software Corporation)

[10] Stenersen D and Thonstad O 2017 *Report for GHG and NOx Emissions from Gas Fuelled Engines* (Trondheim: SINTEF Ocean AS)

[11] International Maritime Organization 2016 *Report of the Marine Environment Protection Committee: Amendments to the 2014 Guidelines on the Method of Calculation of the Attained Energy Efficiency Design Index (EEDI) for New Ships* (London: IMO)

[12] Woud H K and Stapersma D 2002 *Design of Propulsion and Electric Power Generation Systems* (London: Institute of Marine Engineering, Science and Technology)

IOP Conf. Series: Materials Science and Engineering 1174 (2021) 012004 doi:10.1088/1757-899X/1174/1/012004

Decision support systems in building construction – an Axiomatic Design approach

Carmen Marcher[1,2], Erwin Rauch[1], Andrea Giusti[2] and Dominik T Matt[1,2]

[1] Free University of Bolzano, Faculty of Science and Technology, Piazza Università 5, 39100 Bolzano, Italy
[2] Fraunhofer Research Italia s.c.a.r.l., Via A. Volta 13 A, 39100 Bolzano, Italy

E-mail: carmen.marcher@natec.unibz.it

Abstract. The introduction of Industry 4.0 leads to new decision-making scenarios in the construction industry. The construction sector still hesitates to adopt emerging technologies and many construction companies require support in the field of digitalisation and automation of processes. One way to support this sector is the use of decision support systems that perform a careful assessment of the feasibility for replacing traditional systems with automated ones. However, the design and the development of decision support systems is a complex and time-consuming task. In this work we apply Axiomatic Design to define design guidelines for the development of decision support systems in construction equipment selection. The needs of the construction industry for decision support in these scenarios are derived from existing literature in the field of construction and the key characteristic of decision-theoretic expert systems. These needs are analysed and translated into functional requirements. Then, through a mapping and decomposition process, the design solutions for the development of a decision support system are deduced. Functional requirements and design solutions are transformed into design guidelines that support the systematic development and implementation of decision support systems for the construction phase of a building.

1. Introduction

While the industrial sector is radically changing with the increasing adoption of Industry 4.0 [1], the construction sector still hesitates to adopt emerging technologies. To overcome this gap, it is necessary to support the construction sector in the field of digitalisation and automation [2]. One way to assist this sector in meeting this challenge is the development and employment of systems that support decision makers in the choice of the most appropriate technological solution to be adopted on the construction site [3]. Decision support systems (DSS) are widely employed interactive computer-based systems that assist users in decision making [4], and the need for adopting such systems in construction is constantly increasing [5]. In recent years, the interest towards the development and the adoption of DSS in construction has grown substantially, and most of the DSS are proposed for the construction phase [6]. Although formal methods for decision making are the most employed methods within DSS in construction [5,6], approaches that account for uncertainty are preferred by decision makers as they reflect reality more accurately [7]. In this work, we apply Axiomatic Design (AD) [8] to define a design guideline for developing and implementing DSS for equipment selection in the construction phase of a building. The development of the guideline is based on the fundamental characteristics of decision theoretic expert systems and an analysis of relevant literature in the field of construction equipment selection. The contribution of this paper is to structure a complex development and implementation process by dividing the process into small and manageable development and implementation packages.

This work is structured as follows: In Section 2 we present the related background. In Section 3 we describe the application of the AD-based design approach and the resulting design guidelines. In Section 4 we draw the conclusions.

2. Background

2.1. Bayesian decision theory and decision-theoretic expert systems

DSS based on probabilistic networks are called decision-theoretic expert systems. Bayesian networks are probabilistic networks that allow reasoning under uncertainty and the representation of uncertain knowledge. Bayesian networks that incorporate also actions and utilities are called Bayesian decision networks or influence diagrams [9]. They allow one to represent and describe decision problems both quantitatively and qualitatively [10], and evaluate preferences and the probability that a specific outcome occurs. They follow the Principle of Maximum Expected Utility (MEU) [11] to compute the action that yields the highest expected utility [9,10]. In summary, decision-theoretic expert systems perform rational decision making based on available evidence and preferences within a given area of expertise [9] by means of probabilistic networks [10]. The applicability of Bayesian decision theory for developing and implementing DSS in the field of construction equipment selection has been shown by means of an exemplary case study that emulates the choice between a conventional scaffold and a mast climber platform for façade installation [3].

2.2. Literature survey on DSS for construction equipment selection

The relevant literature was mapped by performing a manual search of articles and conference papers on Elsevier's database Scopus, a peer-review database in the field of engineering sciences. The search is limited to English articles in the publication period from 2016 to November 2020, and to articles that are part of the subject areas Engineering or Business, Management and Accounting. The search keywords are *equipment* AND *construction* AND (*selection* OR *choice*). The resulting 451 items are reduced to 36 items by searching the keyword "decision support" within the results. Lastly, the items are reduced to 17 items by selecting only items that contain the exact keywords "Equipment" OR "Construction Equipment" OR "Equipment Selection". The screening of the abstracts is performed to exclude articles with no to low relevance to our field of study. This reduces the data to seven highly relevant items further analysed to extract requirements that must be considered when designing DSS in the field of construction equipment selection.

Table 1. Results of the literature analysis.

Requirement	[12]	[13]	[14]	[15]	[16]	[17]	[18]
Define all relevant decision criteria	x	x	x			x	x
Include expert knowledge		x	x	x			x
Weight the criteria		x				x	x
Consider project needs		x		x	x		
Consider preferences	x					x	

Table 1 presents the results of the literature analysis by highlighting the major requirements that are summarised in the following. DSS are used for the selection of the optimal solution between a set of available alternatives. The major requirement is the identification of all the criteria or variables that are relevant for the selection problem [12–14,17,18]. Here, also the definition of the weights of the criteria that have impact on the selection is necessary [13,17,18]. In many selection problems it is essential to involve domain experts and use their knowledge for the development of DSS [13–15,18]. The selection of the best alternative must be based on the conditions and needs of a specific project [13,16,18]. Another requirement is the integration of preferences [12] or the definition of preference functions [17] to assess alternatives.

3. AD-based design of DSS in building construction

3.1. Preliminaries on AD

AD is a systematic procedure for design and was developed by Suh in the 1970s. In AD costumer, functional, physical and process domain are used to address a specific problem. Each of the domains contains associated design elements namely the customer needs (CNs), the functional requirements (FRs) and design constraints (Cs), the design parameters (DPs) and finally the process variables (PVs). AD theory is based on two axioms [8]: 1) The Independence Axiom requires that functional independence between functional elements must be maintained, i.e. by avoiding unnecessary interactions between FRs and DPs; 2) The Information Axiom requires that the information content of the design must be minimized, i.e. by selecting the solution with the slightest information content.

3.2. Identification of costumer needs (CNs)

Based on the requirements collected in Section 2, five major customer needs are derived for design of DSS for construction equipment selection: CN1 - DSS shall give advice on the choice between a set of available alternatives; CN2 – DSS shall consider all the relevant criteria of the problem domain and weight them; CN3 – DSS shall exploit expert knowledge; CN4 – DSS shall consider available information and project-specific needs; CN5 – DSS should consider preferences of users in terms of utilities.

3.3. Top-level functional requirements (FRs) and design parameters (DPs)

Based on the identified CNs, the top-level FRs and DPs are defined. FR0 represents the main objective: "Provide support to make rational decisions in construction equipment selection". To satisfy FR0 the design solution at the highest hierarchical level, DP0 is defined: "DSSs for construction equipment selection". The top-level decomposition tree represented in Figure 1 links the top-level FRs to DPs. Design fields (DF) are defined to allow an efficient presentation of the results.

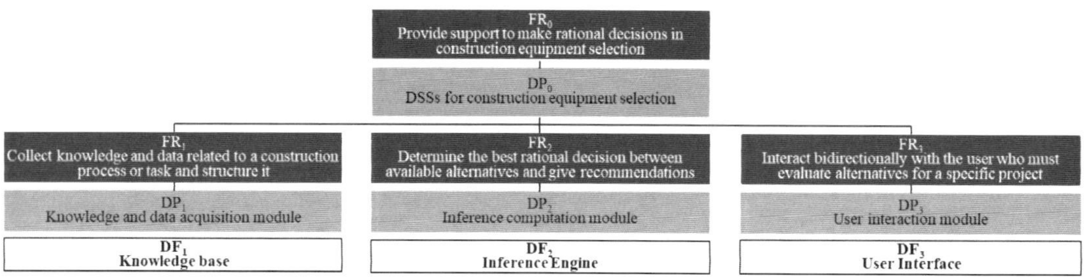

Figure 1. Top-level decomposition tree (FR-DP).

$$\begin{Bmatrix} FR_1 \\ FR_2 \\ FR_3 \end{Bmatrix} = \begin{bmatrix} X & 0 & 0 \\ X & X & 0 \\ X & X & X \end{bmatrix} \begin{Bmatrix} DP_1 \\ DP_2 \\ DP_3 \end{Bmatrix} \tag{1}$$

The design matrix at the top level (1) has a decoupled design, showing the interdependencies between DPs and FRs, and establishes a defined order of implementation of the design solutions. In summary, the top-level pairs indicate the following requirements: 1) to exploit expert knowledge for defining alternatives and decision criteria, and to structure and store available knowledge; 2) to compute the best rational decision between a set of available alternatives and give recommendations; 3) to use available project information, to insert preferences of the users and to get recommendations.

3.4. Decomposition and mapping of FRs and DPs

The top-level FR-DP pairs are further decomposed to transform the abstract requirements into more tangible design solutions that are relevant for the development and the implementation of DSS to be used for construction equipment selection. The results of the decomposition and mapping process are presented in Table 1, 2 and 3.

Table 1. DF_1 Knowledge base

FRs		DPs	
1.1	Elicit the qualitative part of the model	1.1	Decision network modelling tool
1.1.1	*Define the variables*	*1.1.1*	*Variable elicitation tool*
1.1.2	*Define the relations between the variables*	*1.1.2*	*Causal relations definition tool*
1.2	Elicit the quantitative part of the model	1.2	Decision network implementation tool
1.2.1	*Define the probabilities*	*1.2.1*	*Conditional probabilities elicitation tool*
1.2.2	*Define the utility function*	*1.2.2*	*Utility function definition tool*
1.3	Store data	1.2.3	Database

Table 2. DF_2 Inference Engine

FRs		DPs	
2.1	Compute which decision yields the MEU	2.2	MEU computation module
2.2	Evaluate information and give recommendations	2.1	Value of Information computation module

Table 3. DF_3 User Interface

FRs		DPs	
3.1	Visualise decision network	3.1	Decision network visualisation tool
3.2	Insert available project information	3.2	Evidence entry field
3.3	Insert user preferences	3.3	Preferences entry field
3.4	Answer to recommendations	3.4	Question & answer tool
3.5	Get advice on the best rational decision	3.5	Advice display

3.5. Design guidelines and summary

Based on the mapping and decomposition process of the DFs we can now derive the guidelines for designing DSS in the construction phase (see Figure 2). The AD-based approach provides developers with an instrument sequencing the development of the single software modules and components. In our field of study, the problem domain concerns the selection of the construction process or task that could be enhanced by using advanced equipment.

The first main component of the DSS is represented by the knowledge base. Here, existing knowledge of the problem domain is used to i) define the qualitative part of the decision model, and ii) to implement the quantitative part of the decision model. Once the construction process or task for which the most adequate equipment must be selected is defined, the qualitative part of the decision model is implemented by identifying alternatives, variables, and causal relations between them. The alternatives are the available decisions, i.e., the equipment that can be employed for executing a

IOP Conf. Series: Materials Science and Engineering 1174 (2021) 012004 doi:10.1088/1757-899X/1174/1/012004

specific task or process. The variables represent the criteria that affect the decision between the different alternatives. To define alternatives, variables, and causal relations the analysis of literature, data, and the expert interviews should focus on exploring the following main groups of criteria: i) Technical criteria [12–14,16–18]: specifications, characteristics, and performance of the equipment; ii) Economic criteria [12–14,17,18]: renting costs or lifecycle costs of the equipment; iii) Project specific criteria [13–16,18]: available evidence like budget, time and site constraints of a specific project; iv) Safety and risk criteria [13,14,16]: health, safety and risk issues related to the operation of equipment; v) Environmental criteria [14,15]: environmental impacts or damages caused by the equipment. Once the qualitative part of the decision model is defined, the model can be implemented by defining the numbers that are needed for performing the computations. The implementation of the numbers concerns the elicitation of conditional probabilities, that can be defined by mapping verbal statements of domain experts [10], or from data when available, and the definition of the utility function.

The second main component of the DSS is the inference engine, where computations are performed. The expected utility for every alternative is computed also based on weights defined by the user. According to the principle of MEU the system selects the alternative that yields the highest expected utility and provides this information to the user interface. Furthermore, value of information theory is exploited to determine which variables have the highest impact on the decision. This information is also sent to the user interface to give recommendations on which information should be provided or acquired to improve the decision.

The third main component of the DSS is the user interface. This component supports users in the selection of the most appropriate equipment for a specific construction project. Within the user interface the decision network on which the computations are based is visualised. The user has the possibility to insert the project specific information, any available evidence, or specific needs of the user or the construction site. In addition, the user can also state his subjective preferences, by weighting the parameters that are contained in the utility function. The user can enhance the results by following the recommendations of the inference engine by providing additional information on variables. Lastly, the user interface displays which alternative should be chosen.

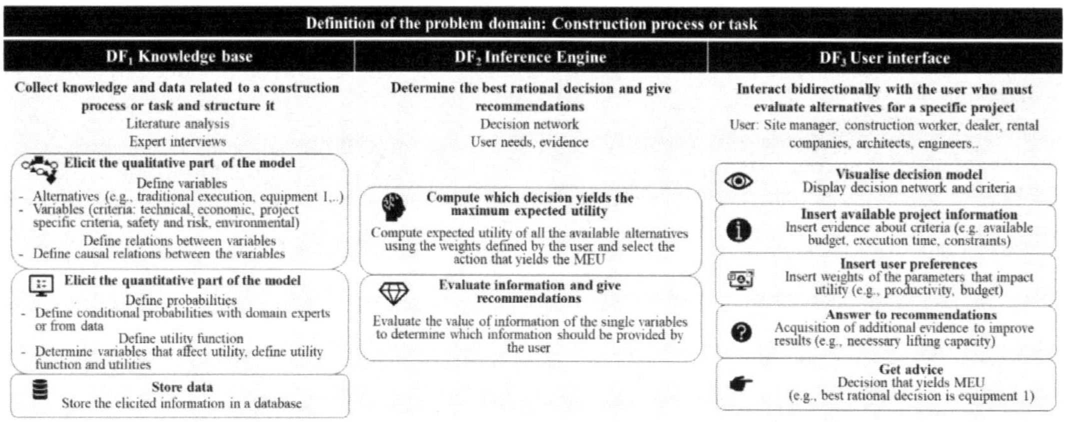

Figure 2. Design guidelines for implementing DSS for construction equipment selection.

4. Conclusion

This paper presents an AD-based approach for designing guidelines for the implementation of DSS for construction equipment selection. CNs are gathered from the characteristics of decision theoretic expert systems and an analysis of literature. Subsequently, FRs and DPs are derived by using AD decomposition and mapping. Finally, FRs and DPs are transformed into the design guidelines represented in Figure 2. The use of the guideline to implement DSS for construction equipment selection can aid the introduction of technological solutions in the construction industry.

References

[1] Rauch E, Vickery A, Garcia M, Rojas R and Matt D T 2018 Axiomatic Design based Design of a Software Prototype for Smart Shopfloor Management ed E C N Puik, J T Foley, D S Cochran and M L Betasolo *MATEC Web Conf.* **223** 01012

[2] Pasetti Monizza G, Bendetti C and Matt D T 2018 Parametric and Generative Design techniques in mass-production environments as effective enablers of Industry 4.0 approaches in the Building Industry *Autom Constr* **92** 270–85

[3] Marcher C, Giusti A, Schimanski C P and Matt D T 2019 Application of Decision Support Systems for Advanced Equipment Selection in Construction *Cooperative Design, Visualization, and Engineering* Lecture Notes in Computer Science vol 11792, ed Y Luo (Cham: Springer International Publishing) pp 229–35

[4] Power D J 2002 *Decision support systems: concepts and resources for managers* (Greenwood Publishing Group)

[5] Jato-Espino D, Castillo-Lopez E, Rodriguez-Hernandez J and Canteras-Jordana J C 2014 A review of application of multi-criteria decision making methods in construction *Autom Constr* **45** 151–62

[6] Marcher C, Giusti A and Matt D T 2020 Decision Support in Building Construction: A Systematic Review of Methods and Application Areas *Buildings* **10** 170

[7] Hodgett R E 2016 Comparison of multi-criteria decision-making methods for equipment selection *Int J Adv Manuf Technol* **85** 1145–57

[8] Suh N P 1998 Axiomatic Design Theory for Systems *Res Eng Des* **10** 189–209

[9] Russell S J, Norvig P and Davis E 2010 *Artificial intelligence: a modern approach* (Upper Saddle River: Prentice Hall)

[10] Kjærulff U B and Madsen A L 2013 *Bayesian Networks and Influence Diagrams: A Guide to Construction and Analysis* vol 22 (New York, NY: Springer New York)

[11] Horvitz E J, Breese J S and Henrion M 1988 Decision theory in expert systems and artificial intelligence *International Journal of Approximate Reasoning* **2** 247–302

[12] Jankovic I, Djenadic S, Ignjatovic D, Jovancic P, Subaranovic T and Ristovic I 2019 Multi-criteria approach for selecting optimal dozer type in open-cast coal mining *Energies* **12**

[13] Alshibani A, Elassir H, Al-Najjar M and Hamida H 2019 AHP based approach for crane selection of building construction in Saudi Arabia: A case study *Proc Annu Conf Can Soc Civ Eng* 2019 Canadian Society for Civil Engineering Annual Conference, CSCE 2019 vol 2019-June (Canadian Society for Civil Engineering)

[14] Waris M, Panigrahi S, Mengal A, Soomro M I, Mirjat N H, Ullah M, Azlan Z S and Khan A 2019 An Application of Analytic Hierarchy Process (AHP) for Sustainable Procurement of Construction Equipment: Multicriteria-Based Decision Framework for Malaysia *Mathematical Problems in Engineering* **2019** 1–20

[15] Zeynalian M and Dehaghi I K 2018 Choice of optimum combination of construction machinery using modified advanced programmatic risk analysis and management model *Sci. Iran.* **25** 1015–24

[16] Peng L and Chua D K H 2017 Decision Support for Mobile Crane Lifting Plan with Building Information Modelling (BIM) *Procedia Eng.* 7th International Conference on Engineering, Project, and Production Management, EPPM 2016 vol 182, ed Halicka K., Nazarko L., and Wasiak A. (Elsevier Ltd) pp 563–70

[17] Temiz I and Calis G 2017 Selection of Construction Equipment by using Multi-criteria Decision Making Methods *Procedia Eng.* Creative Construction Conference, CCC 2017 vol 196 (Elsevier Ltd) pp 286–93

[18] Marzouk M and Abubakr A 2016 Decision support for tower crane selection with building information models and genetic algorithms *Autom Constr* **61** 1–15

IOP Conf. Series: Materials Science and Engineering 1174 (2021) 012005 doi:10.1088/1757-899X/1174/1/012005

Artificial Intelligence Tools for Better Use of Axiomatic Design

Haluk Akay, Sang-Gook Kim

Department of Mechanical Engineering, Massachusetts Institute of Technology
Cambridge MA, United States

sangkim@mit.edu

Abstract: Axiomatic Design (AD) provides a powerful thinking framework for solving complex engineering problems through the concept of design domains and diligent mapping and decomposition between functional and physical domains. Despite this utility, AD is yet to be implemented for widespread use by design practitioners solving real world problems in industry and exists primarily in the realm of academia. This is due, in part, to a high level of design expertise and familiarity with its methodology required to apply the AD approach effectively. It is difficult to correctly identify, extract, and abstract top-level functional requirements (FRs) based on early-stage design research. Furthermore, guiding early-stage design by striving to maintain functional independence, the first Axiom, is difficult at a systems level without explicit methods of quantifying the relationship between high-level FRs and design parameters (DPs). To address these challenges, Artificial Intelligence (AI) methods, specifically in deep learning (DL) assisted Natural Language Processing (NLP), have been applied to represent design knowledge for machines to understand, and, following AD principles, support the practice of human designers. NLP-based question-answering is demonstrated to automate early-stage identification of FRs and to assist design decomposition by recursively mapping and traversing down along the FR-DP hierarchical structure. Functional coupling analysis could then be conducted with vectorized FRs and DPs from NLP-based language embeddings. This paper presents a framework for how AI can be applied to design based on the principles of AD, which will enable a virtual design assistant system based on both human and machine intelligence.

1. Introduction

Since first introduced to the ASME and CIRP community in the late 1970s, Axiomatic Design (AD) has provided principles for complex systems design and has been advanced as a powerful methodology for reducing complexity and formalizing the process by which latent needs are translated into concrete functional requirements (FRs) and systematically addressed by design parameters (DPs) [1-3]. The range of AD applications has been very wide and broad. AD has been successfully applied to product design in various fields including automotive [4], electronics [5], manufacturing equipment [6], and MEMS devices [7]. Another popular area of application is manufacturing processes [8] and manufacturing systems design [9]. More recent applications include large, complex socio-technical systems such as healthcare systems [10], enterprise systems, transportation systems, supply chain management, and information system architecture. Although AD has its root in manufacturing and product design, it has also been applied to software design and software development [11].

Despite the value that AD brings to engineering design and production practice, it has not seen widespread adoption in industry, with training in AD remaining limited to mostly academic settings [12]. While AD is a useful thinking framework for expert designers, extensive familiarity with its methodology is required to implement it in design practice. Correctly identifying FRs and abstracting

IOP Conf. Series: Materials Science and Engineering 1174 (2021) 012005 doi:10.1088/1757-899X/1174/1/012005

them into top-down hierarchical structure while mapping them to the physical domain is difficult for junior designers with little exposure to this methodology. Even for seasoned designers, identifying functional coupling, extracting FR-DP structures, and analyzing design matrices become challenging when a problem scales up and decomposition goes beyond a few levels along the hierarchy [13].

At the time of AD's conception in the 1970s and until quite recently, the paradigm for computational design aids has been rule-based. However, with the current resurgence of AI bolstered by low-cost computing hardware, novel neural network architectures, and an abundance of digitalized data, the move away from rule-based methods towards models built on learned parameters from data (so-called *Deep Learning*) has proven transformational for certain fields. The field of computer vision, which historically relied on image filters carefully designed by experts, was revolutionized when convolutional neural networks (CNN) trained on examples of handwriting to learn these filters and outperform the status quo [14]. Similarly, the field of natural language processing (NLP) has benefitted tremendously from the application of deep neural networks trained on massive text datasets to learn representations of language [15] which can be used to perform a number of automated language tasks.

While design has benefited from computational aids such as CAD/CAM tools and use of machine learning for generative design models [16], the field of design has yet to be transformed by AI in the manner by which computer vision and NLP have been. By representing design knowledge for machines to comprehend for storage, manipulation, and retrieval, in a similar way to how language has been represented in NLP, the field of design can truly begin to benefit from methods in AI. In this paper, we propose and show how Axiomatic Design with AI tools provides the necessary framework for formalizing design knowledge representations which has the potential to transform AD thinking from a framework to an accessible methodology.

2. Background

2.1. Axiomatic Design relates the functional domain to the physical domain by modeling interdependencies in design with a matrix framework where functional requirements and design parameters can now be represented as vectors in design space with AI-based Natural Language Processing (NLP) technology. This section provides an overview on these design knowledge representations as well as background on recent advances in AI and NLP.

2.2. Functional Structure in Axiomatic Design

The first step of applying Axiomatic Design (AD) to a problem is identifying the functional requirements (FRs) which must be addressed by design parameters (DPs). AD starts by identifying the highest-level, most overarching FR (what) of a problem, and mapping it to the highest-level DP (how). The highest FR-DP pair need to be decomposed top-down as shown in Figure 1. This FR-DP tree structure is a useful framework for representing design not only because it preserves the hierarchical relationships between perceived needs and conceptualized solutions, but also because it lays a foundation for mapping designs with computational methods.

However, in practice, identifying the key FR and conducting a correct decomposition has been challenging. Differentiating FRs from DPs can be difficult without training in AD theory, and for a complex problem, a multi-level functional hierarchy may need to be abstracted to extract the highest-level FR. As problems scale in breadth, domain expertise across a wide range of fields may be required for a designer to comprehensively identify all necessary FRs and matching DPs for a design concept. A thorough AD analysis is very valuable but very laborious in terms of the human expertise required to complete it.

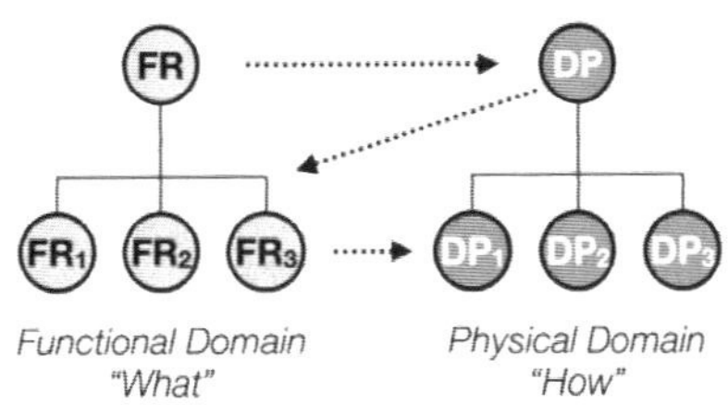

Figure 1: Structured decomposition of functional requirements and design parameters

2.3. Evaluating Functional Coupling

Once the FRs and DPs of a design are identified, their functional coupling needs to be evaluated to assure and maintain the functional independence of the design; this is a critical metric for a good design. One of the fundamental relationships defined in AD is between the functional domain (what) and the physical domain (how). FRs are related to DPs through the design equation, shown in (1), which represents one layer of the hierarchy introduced in Figure 1.

$$\begin{Bmatrix} FR_1 \\ \vdots \\ FR_n \end{Bmatrix} = \begin{bmatrix} A_{11} & \cdots & A_{1n} \\ \vdots & A_{ij} & \vdots \\ A_{n1} & \cdots & A_{nn} \end{bmatrix} \begin{Bmatrix} DP_1 \\ \vdots \\ DP_n \end{Bmatrix} \tag{1}$$

The terms of the design matrix A_{ij} represent the effect each DP has on each FR. Terms along the diagonal where $i = j$ should equal 1 as a given DP_n is designed to address FR_n. For a design exhibiting perfect functional independence, where each DP addresses and affects only one FR, the number of DPs will correspond to the number of FRs, and the matrix A will be diagonal, where $A_{ij} = 0$ when $i \neq j$. Such functionally independent designs are the holy grail of systems engineering. As designs scale in number of FRs and complexity, coupling between the functional and physical domain may result in performance inefficiency, as in the case of pre-industrial revolution steam engines [17], and even catastrophic safety issues when designs are updated without an understanding of functional interdependencies. Conversely, functionally independent systems are much simpler to modify, and often result in better performance even experienced by end users as in the classic AD example of the uncoupled faucet where separate vertical and horizontal levers to control water flow rate and temperature are easier to use than dual knob faucets [18].

Identifying and measuring functional coupling is not trivial in practice, however, especially for systems containing many FRs and DPs. Furthermore, many real-world design cases are not easily characterized into the discrete categories of "uncoupled," "decoupled," or "coupled." To address this issue, metrics of functional independence have been developed [19-20] based on the matrix relationship in (1). These metrics seek to characterize not only if a system is coupled or not, but also *how* coupled a system may be. Based on the values of the design matrix A, the metrics of *Reangularity* (R) and *Semiangularity* (S), expressed in (2), (3) reflect the degree to which DPs affect each other, and affect the set of FRs defining the functional domain. R and S values close to 1 indicate ideal functional independence, and values close to 0 indicate worst-case total functional coupling.

$$R = \prod_{\substack{i=1,n-1 \\ j=1+i,n}} \left(1 - \frac{\left(\sum_{k=1}^{n} A_{ki}A_{kj}\right)^2}{\left(\sum_{k=1}^{n} {A_{ki}}^2\right)\left(\sum_{k=1}^{n} {A_{kj}}^2\right)}\right)^{1/2} \qquad S = \prod_{j=1}^{n} \left(\frac{|A_{jj}|}{\left(\sum_{k=1}^{n} A_{kj}^2\right)^{1/2}}\right) \tag{2}(3)$$

The challenge with implementing such metrics to evaluate real-world designs is that design matrix elements are difficult to quantify, especially in the case of heterogeneous units of measure and qualitative FRs in text form. For example, in the classic AD case of faucet design, if one FR pertains to flow rate, measured in units of volume or mass over time, and another FR pertains to temperature, quantifying how a physical DP affects each FR is not straight forward. Furthermore, while simple systems of few FRs and DPs with quantifiable values can be characterized using designers' intuition, when problems scale up with numerous FRs, accurately computing measurements like R and S can be challenging, which is why such metrics, while theoretically powerful in characterizing systems, have not seen widespread use.

For such challenging cases, AI and NLP models can now be used as a method for representing FRs and DPs in a multi-dimensional vector space where R and S can be quantitatively measured, and thereby functional independence can be quantitatively determined. In our study with NLP modeling, 2^{10} dimensional vector space is used to represent FRs and DPs.

2.4. AI-based Natural Language Processing

In early-stage design, user needs are often communicated with words and sentences, and upstream design efforts such as stakeholder interviews, process proposals, and design specifications are often documented and stored in textual format. In order to analyze such textual data with methods beyond conventional keyword searches relying on rule-based taxonomies, AI methods specifically in natural language processing (NLP) may be applied.

A key functionality in NLP is representing language sequences in a manner which is computationally accessible. Strings of alphabetical characters have no inherent quantitative meaning, but neural networks trained on large document datasets can be trained to embed semantic meaning of language into vectors based on probabilistic modeling tasks [21]. State of the art language models utilize deep neural network architectures trained on billions of words to comprehensively encode language meaning and may be further trained or "fine-tuned" to perform specific tasks. *Bidirectional Encoder Representations from Transformers* (BERT) is one such language model developed by Google AI [15].

BERT is a machine learning-based language model which is able to represent sequences of language (words in sentences) by producing vector representations where words which have similar meaning are placed nearby each other in a multi-dimensional semantic feature space. Vector representations from BERT are contextually dynamic in that the same word's vector will change slightly depending on the context it exists in (neighboring language). BERT is trained in two steps. The first pre-training step is a general training process and is "unsupervised" meaning the model learns from un-annotated text documents, training on tasks such as predicting the identity of randomly masked words in a sentence, and if two sentences should follow one another consecutively. The second fine-tuning step depends on the application of BERT and is usually supervised learning meaning that example inputs and target output pairs are provided. Due to the manner in which design decomposition can be conducted by extracting the answers to "what" and "how" type questions, BERT fine-tuned on the NLP task of question-answering can be applied as a form of *Hybrid Intelligence* [22] to automate many of the steps in AD. We have previously demonstrated how highest-level FRs can be extracted with question-answering [23], how *Design Reading* can be automated with recursive question-answering to extract and structure large numbers of FRs for a design case [24], and that such extracted functional hierarchies demonstrate agreement with the judgements made by human designers [25]. In this work, we demonstrate how applying such AI-based NLP models can aid with challenging tasks in Axiomatic Design to make it more accessible to the engineering design community and for wider-spread use in industry.

3. Method

The key tasks of a designer using Axiomatic Design involve identifying functional requirements (FRs), decomposing FRs from the top-down to produce a hierarchical structure, and determining functional interdependencies to evaluate a designed system. Representations of language from AI-based models can be used to automate the difficult aspects of these tasks.

3.1. Question-Answering for Identifying and Structuring FRs

Detailed designs may be developed by identifying a key FR and zigzagging between the functional and physical domains to identify corresponding DPs and decompose to a hierarchical structure. Depending on the complexity of the problem and the designer's expertise in AD theory, this can be a difficult process, but AI-based NLP can be leveraged to provide aid.

AD clearly defines the functional domain as the "what" and the physical domain as the "how" of design. For example, FRs are "what we want to achieve" and DPs are "how we achieve them" [1]. While such prompting statements are simple, answering them accurately to identify FRs and DPs can be challenging for those less familiar with AD. However, if sufficient documentation about a design space is available, the extraction and structuring of FRs and DPs can be automated using NLP-based question-answering. Language models such as BERT can be fine-tuned on the task of question-answering, which is a standard information retrieval task in the field of NLP. This task considers a context document and a question as inputs, and probabilistically determines the span of text within the context which has the maximum likelihood of being the answer to the input question, and returns the indices of the answer as output. BERT can be trained on this task with a dataset of crowd-sourced context-question-answer examples, such as the Stanford Question Answering Dataset (SQuAD) [26] and applied to AD decomposition. By choosing "what" and "how" type questions that build in specificity with information content previously retrieved, high-level FRs and DPs may be decomposed automatically into detailed lower-level FRs, where the questions prompting their retrieval are informed by a previously extracted hierarchy, as visualized in Figure 2.

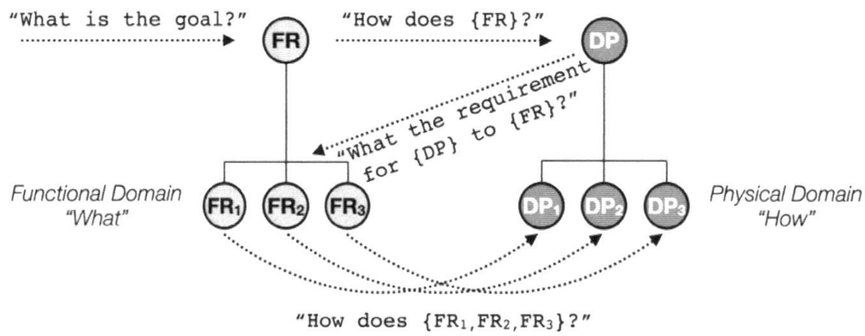

Figure 2: Question-Answering for Axiomatic Design Decomposition

An example of the automated decomposition steps described in Figure 2 can be considered to further elucidate the power of question-answering for application to AD. While actual design specification documents from industry are the ideal target of deploying this automated decomposition method, we will demonstrate an example based on the abstract of an academic paper describing the design of a bio-inspired compliant locking device [27]. Paper abstracts are not only publicly available and distributable, but also can be expected to be rich with high-level FRs and DPs and make a dense testing ground for demonstrating an extraction method.

As a result of this recursive application of question-answering, we can automatically produce an AD decomposition and a set of structured FRs and DPs without any prior taxonomical information about this design. Due to the extractive nature of the information retrieval, the grammatical syntax of some of the questions juxtaposed with FRs are nonstandard, such as "How does stability?" but they elicit DPs from the context, i.e., "split locking mechanism." With this method, design documentation can be processed to determine an Axiomatic Design functional structure, allowing designers to experience the benefits of this analysis without expertise in AD theory that would have been required to manually produce these decompositions.

Table 1: Recursive Question-Answering Example of Axiomatic Design Decomposition

Design context document [27]: A device reminiscent of the mammalian spine has been designed and built with the ability to lock each individual joint in a string of ball joints. The assembly may be controlled in a manner similar to other hyper-redundant robots, with the added advantage of locking in a straight or axial position. Locking is achieved by orienting two mating collars in a singular configuration that forces compression against neighboring collars and prohibits bending or rotation. Locking is desirable for added strength in supporting objects, as well as for stabilization and power efficiency when bending is not necessary. The split locking mechanism represents a biologically inspired structure with added strength and stability for use in robotics. **Legend:** Functional Requirement Design Parameter	Q: What is the goal? **FR**: added strength in supporting objects Q: How does {added strength in supporting objects}? **DP**: locking Q: What is the requirement for {locking} to {added strength in supporting objects}? **FR1**: stability Q: How does {stability}? **DP1**: split locking mechanism **FR2**: prohibits bending or rotation Q: How does {prohibits bending or rotation}? **DP2**: forces compression against neighboring collars **FR3**: straight or axial position Q: How does {straight or axial position}? **DP3**: locking **FR4**: string of ball joints Q: How does {string of ball joints}? **DP4**: spine

3.2. Axiomatic Design Analysis with Language Vector Embeddings

Once the FRs and DPs of a designed system have been defined, such a system may be evaluated in Axiomatic Design with respect to the first axiom regarding maintenance of functional independence. While the definitions of a "coupled" versus "uncoupled" system are quite clear, identifying functional interdependencies and measuring them can be difficult in practice when units of measure do not align and FRs are qualitative in nature. For such challenging cases, NLP models may be used as a method for representing FRs and DPs in a feature space where interdependencies may be quantitatively measured. We have previously demonstrated [28-29] how metrics of functional independence may be accurately estimated for the classic AD example of coupling in faucet design, and we will show an extension of this case with the following study.

The pre-trained language model BERT (fine-tuned previously on the task of question-answering) may also be used to produce vector representations of language. Consequently, sentence spans describing FRs and DPs may be converted into vector form by averaging the vectors of words making up the descriptions. Such representations are designed to encode the semantic meaning of language into vector space such that more similar words occur closer together. As a result, semantic similarity can be quantified in terms of the cosine similarity between word vectors. For example, the semantic similarity between the words "motor" and "engine" is 0.834, while the similarity between words "motor" and "donut" is 0.383, using vector representations from BERT. If we consider the case of faucet design, we can note that the FRs of the design refer to controlling water (1) "temperature" and (2) "flow rate." If we consider descriptions of DPs for the coupled case in Design A, we can note that both DPs reference temperature ("hot" or "cold") and "flow rate". However, the DPs of uncoupled Design B are also semantically uncoupled in that there is no mention of "flow rate" in DP1 and no mention (explicit or similar to) "temperature" in DP2. To quantify these semantic relationships in the functional domain, we can obtain vectors of each FR and DP description and use these quantitative representations to estimate the design matrix A relating the functional to physical domain in AD. In turn, metrics for coupling *Reangularity (R)* and *Semiangularity (S)* may be computed. We can reproduce our previous study in [29] to visualize how differently phrased descriptions, all of which are similar to those explicitly stated in the left of Figure 3, describing faucet designs. The descriptions of the designs cluster around R and S values accordingly, and the semantic representations reflect the metrics of functional independence from Axiomatic Design.

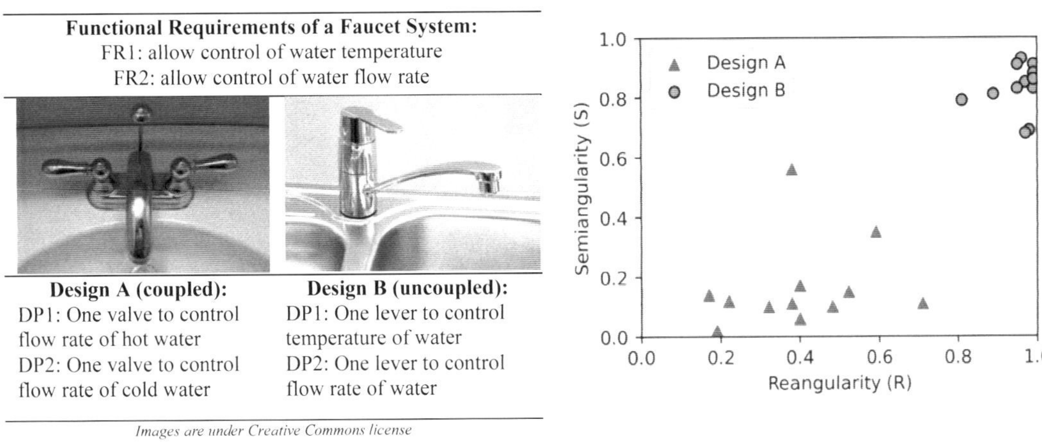

Figure 3: left: images and descriptions of faucet design; right: plot of R and S metrics of coupling for the faucet design based on vectorized descriptions of FRs and DPs from [29]

4. Discussion and Conclusion

Early-stage design can be subjective, requiring expertise to practice. Axiomatic Design provides a powerful framework for resolving customer needs into concrete functional requirements, and mapping these to the physical domain, but actually implementing these steps can be challenging, and training junior designers to practice AD is difficult. While understanding a design space and answering the key questions of AD decomposition is challenging for junior designers, NLP models trained specifically on retrieving information in a question-answering format are able to perform the identification of FRs, mapping to DPs from textual design descriptions, and decomposing them to lower levels. A recursive implementation of Google's language model BERT, fine-tuned for question-answering demonstrated that NLP was applied to automate the fundamental steps of AD. This automated decomposition method can be extended beyond the functional and physical domains to link process variables to early-stage design and create a holistic, digitalized AD-based design knowledge of production steps. The difficulty of manually producing such design decompositions has hindered the implementation of AD in industry, but using the AI-based automated or human-assisted hybrid methods, industry may finally truly benefit from the power of Axiomatic Design thinking.

By producing vector representations of designs based on textual descriptions, semantic similarity can be translated into AD metrics for functional independence. If the detection of functional coupling could be automated using the NLP-based knowledge representation methods shown in this work, functional interdependencies of large-scale systems could be quickly assessed to highlight bad designs and to avoid any potentially catastrophic safety issues.

Axiomatic Design has provided powerful thinking framework for designers but has been also challenging to use and implement at industrial practices. By applying AI-based models to perform the most challenging steps of Axiomatic Design, namely extracting and structuring functional requirements, and evaluating designed systems on the basis of functional independence, we have shown that design knowledge can be represented in a form that machines can understand. We believe that this will enable Axiomatic Design to finally become an accessible method to benefit practitioners of design in a range of industries.

5. Acknowledgements

This work was supported by the National Science Foundation (NSF) Leading Engineering for America's Prosperity, Health, and Infrastructure (LEAP HI) program, award number 1854833.

References

[1] Suh NP, 1990. *The Principles of Design*. New York: Oxford University Press.
[2] Suh NP, Bell AC, Gossard DC, 1978. On an axiomatic approach to manufacturing and manufacturing systems.
[3] Jahanmir S, Saka N, Tucker III C, Kim SG, 2015. *Advances in Multidisciplinary Engineering*, ASME Press, NY.
[4] Bae S, Lee JM, Chu CN. Axiomatic design of automotive suspension systems. CIRP Annals. 2002 Jan 1;51(1):115-8.
[5] Schuh G, Rudolf S, Breunig S. Modular platform design for mechatronic systems using axiomatic design and mechatronic function modules. Procedia CIRP. 2016 Jan 1;50:701-6.
[6] Babic B. Axiomatic design of flexible manufacturing systems. International Journal of Production Research. 1999 Mar 1;37(5):1159-73.
[7] Schlipf M, Bathurst S, Kippenbrock K, Kim SG, Lanza G. A structured approach to integrate MEMS and Precision Engineering methods. CIRP Journal of Manufacturing Science and Technology. 2010 Jan 1;3(3):236-47.
[8] Salonitis K. Design for additive manufacturing based on the axiomatic design method. The International Journal of Advanced Manufacturing Technology. 2016 Oct;87(1):989-96.
[9] Gu P, Rao HA, Tseng MM. Systematic design of manufacturing systems based on axiomatic design approach. CIRP Annals. 2001 Jan 1;50(1):299-304.
[10] Peck J, Nightingale D, Kim SG. Axiomatic approach for efficient healthcare system design and optimization. CIRP annals. 2010 Jan 1;59(1):469-72.

[11] Kim SJ, Suh NP, Kim SG. Design of software systems based on axiomatic design. Robotics and Computer-Integrated Manufacturing. 1991 Jan 1;8(4):243-55.
[12] Nordlund M, Tate D, Suh NP. Growth of axiomatic design through industrial practice. In3rd CIRP Workshop on Design and the Implementation of Intelligent Manufacturing Systems 1996 Jun 19 (pp. 77-84).
[13] Brown CA. Teaching axiomatic design to engineers—Theory, applications, and software. Journal of Manufacturing Systems. 2005 Jan 1;24(3):186-95.
[14] LeCun Y, Bottou L, Bengio Y, Haffner P. Gradient-based learning applied to document recognition. Proceedings of the IEEE. 1998 Nov;86(11):2278-324.
[15] Devlin J, Chang MW, Lee K, Toutanova K. Bert: Pre-training of deep bidirectional transformers for language understanding. arXiv preprint arXiv:1810.04805. 2018 Oct 11.
[16] Chen W, Ahmed F. PaDGAN: Learning to Generate High-Quality Novel Designs. Journal of Mechanical Design. 2021 Mar 1;143(3).
[17] Cavique M, Fradinho J, Gabriel-Santos A, Mourão A, Gonçalves-Coelho A. Recovery of waste heat from engines: an AD view. InMATEC Web of Conferences 2018 (Vol. 223, p. 01007). EDP Sciences.
[18] Foley JT, Puik E, Cochran DS. The Faucet Reloaded: Improving Axiomatic Design by Example. InMATEC Web of Conferences 2017 (Vol. 127, p. 01009). EDP Sciences.
[19] Rinderle JR. *Measures of functional coupling in design* (Doctoral dissertation, Massachusetts Institute of Technology).
[20] Suh NP, Rinderle JR. Qualitative and quantitative use of design and manufacturing axioms. Cirp Annals. 1982 Jan 1;31(1):333-8.
[21] Bengio Y, Ducharme R, Vincent P, Janvin C. A neural probabilistic language model. The journal of machine learning research. 2003 Mar 1;3:1137-55.
[22] Kim SG, Yoon SM, Yang M, Choi J, Akay H, Burnell E. AI for design: Virtual design assistant. CIRP Annals. 2019 Jan 1;68(1):141-4.
[23] Akay H, Kim SG. Extracting Functional Requirements from Design Documentation using Machine Learning. Procedia CIRP. 2021.
[24] Akay H, Kim SG. Reading functional requirements using machine learning-based language processing. CIRP Annals. 2021 Apr 24.
[25] Akay H, Yang M, Kim SG. Automating Design Requirement Extraction from Text with Deep Learning. ASME IDETC Proceedings. 2021.
[26] Rajpurkar P, Zhang J, Lopyrev K, Liang P. Squad: 100,000+ questions for machine comprehension of text. arXiv preprint arXiv:1606.05250. 2016 Jun 16.
[27] Kern NI, Majewski TJ, Triolo RJ, Kobetic R, Quinn RD. A locking compliant device inspired by the anatomy of the spine. Journal of Mechanical Design. 2009 Jan 1;131(1).
[28] Akay H, Kim SG. Measuring functional independence in design with deep-learning language representation models. Procedia CIRP. 2020 Jan 1;91:528-33.
[29] Akay H, Kim SG. Design transcription: Deep learning based design feature representation. CIRP Annals. 2020 Jan 1;69(1):141-4.

IOP Conf. Series: Materials Science and Engineering 1174 (2021) 012006 doi:10.1088/1757-899X/1174/1/012006

Continuous Product Concept Improvement for Integrated Systems: A 3-D Printer Hardware Improvement Case

Yash Vashi, Chu-Yi Wang*, Rutuja Karampure, Stephen Lu

Viterbi School of Engineering, University of Southern California, Los Angeles, CA 90089-1453, USA

* chuyiwan@alumni.usc.edu

Abstract. The advent of globalization and technology has led to proliferation of integrated hardware and software systems such as personal computers, and smartphones while legacy systems such as automobile are increasingly incorporating software technology into their products. The demanding consumer requirements compel companies to deliver improved products on short timelines. Therefore, a partial product improvement process would assist designer in meeting this demand. This paper focuses on uncovering the drawbacks of an integrated system and suggests a methodology to improve its shortcomings. We develop a conceptual framework that highlights the key steps that a designer can consider when performing continuous product development: functional model construction, concept improvement, and improvement strategy. This paper uses a case of an FFF 3D Printer system to demonstrate the method.

1. 1. Introduction

Global markets, rise of internet and more demanding customers has led to increase in complexity of products. To meet these demands, the companies have a large product mix and model mixes. Apple Inc. introduced the iPhone in 2007, with one color and two sizes (4GB and 8GB). In 2020, they have seven different variations of iPhone: five colors and three sizes (64, 128 and 256 GB) or about 105 different devices. This trend is mirrored across various markets - automotive, power tools, biomedical etc.

There is a considerable advantage for firms who have short product development lead times, which can introduce many successive versions of the same product line with increased performance levels. Two important techniques used by firms are product modularization and product platform development, which allow firms to reach high levels of product variety, and at the same time, keep complexity and its related costs at a limited level.

Even with this high demand, there is not structured methodology a designer can utilize to analyze the systems and improve them partially. This paper works within product modularization strategy, as it introduces a new tool which would assist firms to analyze and improve products fractionally. The research outlines three steps to fractional product improvement - Functional Modeling using Design Matrix, Analysis using Design Coupling Sequence, and Improvement Suggestion.

The paper follows the structure as explained further. The background is discussed in detail in Section 2: Research Background. Then the framework and methodology used to understand the couplings and improve the software design is explained in Section 3: Method. Section 4: Case Study: Hardware Design Concept Improvement of 3-D Printer, discusses how the methodology explained in the Method section can be applied to practical use and the improvements that can be made using this methodology. The Section 5: Results and Discussions discusses and compare the different results obtained using this methodology and how this methodology helps in software conceptual design. Finally, the paper is concluded and the future work that can be done in this domain is suggested in Section 6: Conclusions and Future Work.

Nomenclature

DCS	Design Coupling Sequence
DM	Design Matrix
FR	Functional Requirements
DP	Design Parameters
FFF	Fused Filament Fabrication
FDM	Fused Deposition Modeling

2. Research Background

Product development is a network of activities that transform a market opportunity into a product that meets the customer's needs and the strategic goals of a company [1,2]. Firms consider product development as their competitive lever to survive in a fast-evolving market [3,4]. With the advent of faster manufacturing techniques, global supply chain and internet, there is a need for quicker introduction of better and less costly products [5]. With the advent of multiple generation product lines in many Business to Consumer industries like for instance the technology and automotive, there is constant pressure on the design teams to improve their current products.

The aim of this paper is to establish a systematic analysis for finding improvement opportunities in existing designs. Axiomatic Design Theory (ADT) [2,6], introduced and developed by Nam Suh in 1990 at MIT, is the framework within which the proposed approach has been developed. The AD separates design into four domains – Customer, Functional, Physical and Process.[16]

This framework operates in the design space between the functional and physical domains, called Conceptual Design[15]. Zhang et.al. [17] define conceptual design as activities to develop the technical specification of a design problem based on customer requirements and provide solutions to satisfy the customer requirement. At this stage, designers are determining functional design of the product and have a large design space to test new innovations and improvement to them.

In the following section, different Axiomatic Design Theory approaches used to perform continuous improvement are discussed.

The ADT articulates that an ideal design has minimal complexity [9]. A pathway to improve existing products is by transforming existing design matrices using matrix manipulation techniques and minimize the complexity. This method cannot be effectively deployed for most real-life products with highly coupled matrices.

In their paper, Qiang et. al. [10] use ADT and Design Structure Matrix to modularize products, which would allow designers to substitute sub-assemblies to improve the products. They decomposed a product hierarchically into its functional, physical, and process domains using axiomatic design method. After transformation from design matrix to design structure matrix, suitable design structure matrices of design parameters describing function, structure, and manufacturing process are constructed. This method cannot be effectively deployed more than once for a single product or for a modular product.

Zhong-hang et.al. [11] focused on improving the module interface design using TRIZ and AD. It uses AD to analyze the coupling relationship between the module interface and the functional requirements and then uses the conflict solution tool in TRIZ to decouple and propose a suitable connection structure for modular product.

Wang et. al. [13,14] introduced the Design Coupling Sequence (DCS), a four-step process to reduce the design complexity during the concept improvement stage. The DCS provides a hierarchical model of design parameters (DPs) to identify the level of coupling in a design matrix. This hierarchical model has two major applications – improving existing design and modularization of system. It defines two functional sets viz. the independent U-set: which is the collection of functionally dependent concepts, and the coupled C-set: which is the collection of the strongly coupled concepts. Improvements in the conceptual design and partial system conceptual designs can be obtained by using TRIZ or other principles to resolve the complexity by following the execution sequence suggested by the DCS methodology which shows the functional schematics of the design concept.

3. Methodology

The objective of this study is to outline a framework which would enable a designer to improve a system partially. This section outlines the proposed method for partial concept improvement of hardware in an integrated system based on the Design Coupling Sequence (DCS) approach. The procedure is shown in Fig. 1.

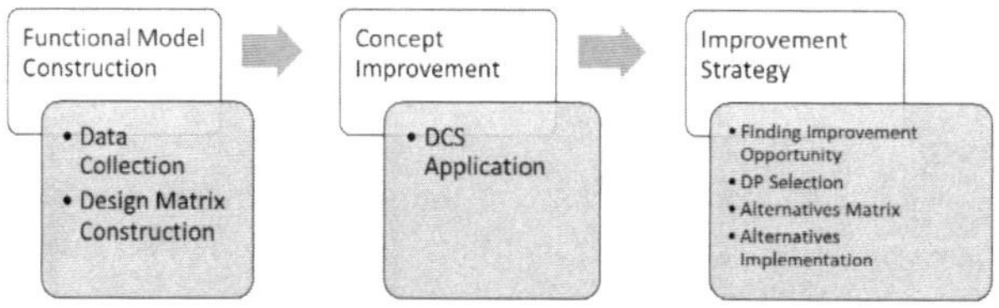

Figure 1. Methodology of Hardware Partial Improvement.

The objective of this study is to come up with a framework which would enable a designer to improve a system partially." should be shifted before picture 1, expanding the argument and explaining from a general standpoint how the proposed method pursues the declared aim.

3.1. *Functional Model Construction*

DCS is a functional model analysis tool, and the first step is to construct the functional model of the system. A functional model is a structured representation all the components, their functions and their relationship with each other within the modeled system. The functional model can be of a new system or an existing one. In DCS, a design matrix, a type of functional model based on Functional Requirements and Design Parameters is used.

3.1.1. Data Collection. The data collection step is aimed to collect the information about the system, its components, their functions and relationship among them. At the end of data collection phase, the designer should have enough information about the system to construct a design matrix and perform DCS analysis on it.

The four steps of data collection are -

- Identification of major system components.
- Defining the functions of components.
- Relating the components via functional relationships
- Isolating Performance Parameters

3.1.2. Design Matrix (DM) Construction. The Functional Requirements and Design Parameters are described in Axiomatic Design mathematically as a vector [2]. The Design Matrix [DM] describes the relationship between FRs and DPs in a mathematical equation (1)

$$\{\boldsymbol{FR}\} = [\boldsymbol{DM}]\{\boldsymbol{DP}\} \tag{1}$$

One can construct the design matrix through such relationship.

3.2. *Concept Improvement*

The multifaceted dynamics of components often may lead to heavily coupled design matrices for complex systems. They can be made simpler or of higher quality using improvement methodologies like

Design Coupling Sequence (DCS). The application of DCS algorithm [13] on a DM can be done in 4 steps.

- Counting the number of 'X'/Couplings in each FR
- Listing the DPs according to increasing number of couplings
- Examining coupled DPs
- Arranging DPs into sets.

The outcome of the DCS algorithm is the Executable Module (EM), which informs the designer about the precedence and functional sets of different DPs. To execute the improvement of DPs one by one or module by module according to the sequence shown in EM, the design can be improved with the minimal design complexity.

3.3. *Improvement Strategies*

In the previous section, the EM of the system was obtained, laying out the precedence of the DPs. In this section, the steps a designer needs to take to improve the system are outlined.

3.3.1. Finding Improvement Opportunity. According to DCS approach [13], one improvement opportunity is to resolve strongly coupled DPs, which are indicated by the C-set in the EM. The other improvement opportunity is to break down U-set in the EM to have more U-sets of a design.

3.3.2. DP Selection. The DP selection basically follows the sequence shown in the EM. In each selection, designer needs to think some alternatives of the DP or a group of unsolvable DPs (wrapped by C-set) and should select the design whose system range falls in the design range completely. By using Probability Density Function (Fig. 2), designer can check between system range and design range. If none of alternatives falls in the design range completely, the designer may choose the one which meets the design range most, i.e., the one with the most common area (Acr) shown in Fig. 2.

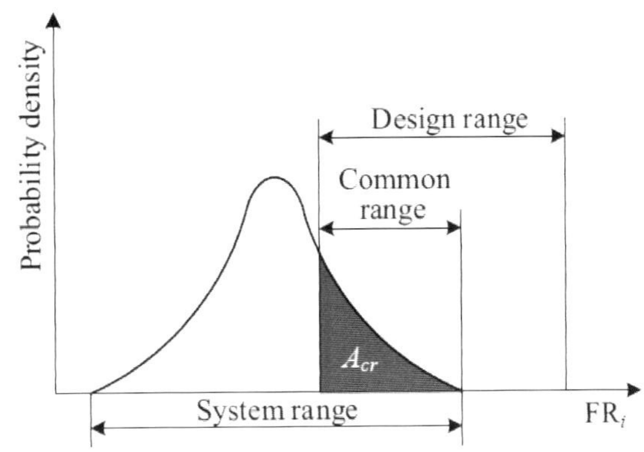

Figure 2. Probability Density Plot between FR and DP depicting System and Design Range, and Common Area.

3.3.3. Alternatives Matrix. An alternatives matrix is a tool to compare and assess different alternatives of DPs/C-sets to review the DP selection results. It includes selected DP/C-set, design range, expectation, system range, and the reason of chosen the alternative. An example of the alternatives matrix is shown in Table 7.

3.3.4. Partial Improvement. Since U-sets in EM can be seen as independent design modules, the concept improvement of a large/complex system can be partially improved by improving smaller system separated by U-set. Besides, the system can also be partially improved by fixing the higher precedence DPs. This strategy is explained using the example in Table 1. The designer wants to improve DP_{21}. The

DP_{21} belongs to the U-set U_1 and is presided by two other DPs, DP_{12} and DP_{22}. The order of improvement for the designer would be DP_{21}, to DP_{12} and then to DP_{22}.

Table 1. Executable Module of a Sample System

U-set	Row	Sequence		
U_1	1	DP_{22}	$\rightarrow DP_{12}$	$\rightarrow \{DP_{11}, DP_{31}\}$
	2		$\searrow DP_{21}$	
U2	3	DP_{32}		

The designer would now select the DPs from the alternative's matrix. When the sequence is followed, the design range for the DP22 would be highest, followed by DP12 and DP21. This means that progressively as a designer makes decisions, the options of alternatives decrease for DPs with lower precedence.

This step would result in one of the two possibilities –

1. Satisfaction of all DPs – If the designer is able to satisfy the selection of alternative DPs using the above-mentioned framework, they can implement partial changes in the system.

2. Unable to satisfy all DPs – If the designer is unable to satisfy the selection of alternative DPs, they would have to change the U-Set to implement the changes. It can be either done by choosing a different set of DPs or by changing the ES of the U-Set.

The designer has to select one Execution Sequence (ES) from all the possible EM.

4. Application of Hardware Improvement in a 3-D printer case

This application shows how to improve an FFF Printer like Prusa i3 MK3S within the proposed method. The selection was made due to expertise of authors in field, and availability of documentation of its components and assembling process

4.1. Functional Model Construction

4.1.1. Data Collection

In the first step, all the system components were identified, listed and categorized according to the function they performed. The inferences for the components were based on system architecture diagrams as shown in Fig. 3. These components are classified as Design Parameters or DPs. In the following table, the highest-level DPs have been outlined.

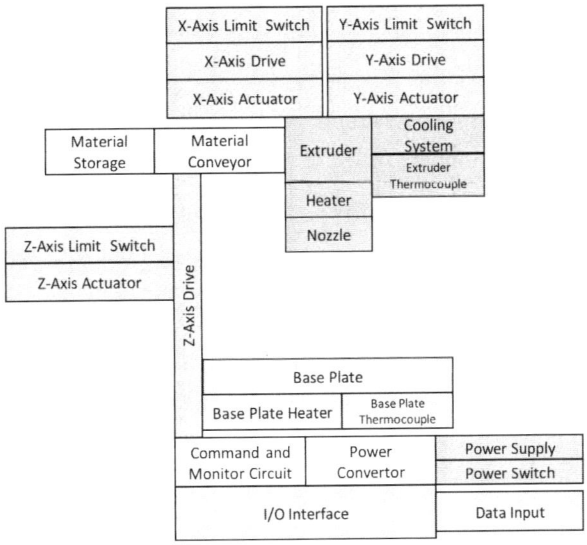

Figure 3. System Architecture of FFF 3D Printer.

Table 2. List of Modules (DPs) of FFF Printer

#	Name	#	Name
1	Power Module	5	Base Plate Module
2	Controller	6	Extruder Module
3	Display Module	7	Control Module
4	Filament Module		

The third step consists the identification of the relationship between FRs and DPs according to their interaction (i.e., functional coupling) including physical, material, information and energy with other components. In the 3-D printer case, the functional couplings were verified by interview with subject matter expert.

In the fourth step, the performance parameters of each component or DP were identified. They would be used in the latter half while identifying improvement strategies.

4.1.2. Design Matrix Construction

The data collection resulted in a rectangular design matrix for the FDM 3D Printer of 25 by 25 (Table 3). There are 139 couplings – 26 diagonal and 113 off-diagonal elements (Fig. 4).

Table 3. List of Functional Requirements and Design Parameters

#	FR	DP
11	Supply Power to System	Power Supply
12	Turn the system on or off	Power Switch
21	Control Mechanical Commands and Verify Sub-System Status	Command and Monitor Circuit
22	Supply Power to Printer Components	Power Convertor
31	Display Printing Information	I/O Interface
32	Enable User to input Information	Data Input
33	Send print data	
41	Store raw material	Material Storage Space
42	Transport material	Material Conveyor
51	Feed Material	Filament Extruder
54	Shape and Extrude Melt	Nozzle Head

52	Melt Material	Heater
55	Cool the Hot Side	Cooling System
53	Measure Temperature of Hot Side	Extruder Thermocouple
611	To move along X-Axis	X Positioning Drive
612	Restrict Motion in X axis	Limit Switch
613	To trigger motion of extruder in X-Axis	Actuator
621	To move along Y-Axis	Y Positioning Drive
622	Restrict Motion in Y-Axis	Limit Switch
623	To trigger motion of extruder in Y Axis	Actuator
631	To move along Z-Axis	Z positioning drive
632	Restrict Motion in Z-Axis	Limit Switch
633	To trigger motion of extruder in Z Axis	Actuator
71	Support printing product	Base Plate
72	Prevent Warping	Base Plate Heater
		Base Plate Thermocouple

DP ↓	Design Matrix of FFF 3D Printer																								
FR →	11	12	21	22	31	33	41	42	51	54	52	55	53	611	612	613	621	622	623	631	632	633	71	72	73
11	X		X	X																					
12	X	X																							
21			X	X	X	X			X		X	X	X	X	X	X	X	X	X	X	X	X		X	X
22	X		X	X	X	X			X		X	X	X		X	X		X	X		X	X		X	X
31			X		X																				
32			X		X																				
33			X		X	X																			
41							X																		
42							X	X	X																
51			X				X	X	X																
54									X	X	X														
52			X	X					X		X														
55			X	X						X	X	X													
53			X							X	X		X												
611									X	X	X			X		X									
612			X											X	X										
613			X						X	X	X					X									
621									X	X	X						X		X						
622			X														X	X							
623			X						X	X	X								X						
631									X	X	X									X		X			
632			X																	X	X				
633			X						X	X	X											X			
71																							X		
72			X	X																			X	X	X

Figure 4. Design Matrix for FFF 3D Printer

Since this case was designed to show the partial improvement (the hardware improvement only) of a FFF printer, the DM was modified by following the proposed method. We analysed the couplings among the modules within the DCS method to determine the precedence and obtained the results as shown in Fig. 5.

DP ↓	Modified Design Matrix of FFF 3D Printer																					
FR →	11	12	31	33	41	42	51	54	52	55	53	611	612	613	621	622	623	631	632	633	71	72	73
11	X																						
12	X	X																					
21			X	X			X		X	X	X	X	X	X	X	X	X	X	X	X		X	X
22	X		X	X			X		X	X	X		X	X		X	X		X	X		X	X
31			X																				
32			X																				
33			X	X																			
41					X																		
42					X	X	X																
51					X	X	X																
54							X	X	X														
52							X		X														
55								X	X	X													
53								X	X		X												
611							X	X	X			X		X									
612												X	X										
613							X	X	X					X									
621							X	X	X						X		X						
622															X	X							
623							X	X	X								X						
631							X	X	X									X		X			
632																		X	X				
633							X	X	X											X			
71																					X		
72																					X	X	X

Figure 5. Modified Design Matrix for FFF 3D Printer

4.2. Concept Improvement

Once the design matrix was obtained, the DCS methodology was applied. The executable modules of the FFF 3-D printer were obtained (Table 4) to show the functional structure of the modified design matrix.

The execution module gives us 4 U-sets and 2 C-sets. The 4 U-Sets allow the designers to compartmentalize design improvement process. Effectively, a designer target specific U-sets to improve without sacrificing the performance of other U-sets. From the U-sets, it is clearly visible that U2, U3, and U4 containing – Base Plate Module, Display Module and Power Module can be improved independently whereas all other modules are encompassed within U1.

The execution module gives us 4 U-sets and 2 C-sets. The 4 U-Sets allow the designers to compartmentalize design improvement process. Effectively, a designer target specific U-sets to improve without sacrificing the performance of other U-sets. From the U-sets, it is clearly visible that U2, U3, and U4 containing – Base Plate Module, Display Module and Power Module can be improved independently whereas all other modules are encompassed within U1.

Table 4. Executable Module of the DM in Fig. 5.

U-Set	Row	Sequence
U1	1	DP41 →{DP51, DP52} →DP42
	2	↘DP611 →DP612
	3	↘DP621 ↘DP622
	4	↘DP631 ↘DP632
	5	↘DP54 →DP613
	6	↘DP623
	7	↘DP633
	8	↘DP53
	9	↘DP55
U2	10	DP31 → DP33
U3	11	DP71 → {DP72, DP73}
U4	12	DP11 → DP12

4.3. Improvement Strategy

4.3.1. Finding Improvement Opportunity

The improvement opportunities for FFF printers were outlined in research papers. The primary drawbacks of FFF printing are low dimensional tolerance of final parts and slow printing times.

Therefore, our improvement objective is to improve the printer which produces prints with high dimensional tolerance at a faster rate.

As shown in Table 4, there are two C-sets ({DP51, DP52} and {DP72, DP73}) in the EM, and they are the improvement opportunities of this design. In addition, U1 should be the main system to be improved as it has the most coupled DPs to be break down.

4.3.2. DP Selection

The next step was to identify the DPs whose performance needs to be improved. A Root Cause Analysis of an FFF printer was analyzed to understand the parameters which affect the build time and dimensional accuracy. The major parameters from RCA which can be improved by altering the system are –Layer Thickness, Air Gap, Extrusion Temperature, and Print Speed.

The DPs which affect the parameters are Filament Module, Extruder Module and Control Module. In Table 5, the parameters and their interaction with DPs are outlined.

Table 5. Design Parameter Selection.

		Performance Factors	
	DP ↓	**Printing Time**	**Dimensional Accuracy**
42	Material Conveyor	Performance affects printing time	No Affect
51	Filament Extruder	Component Weight affects Print Time	Component Weight affects Dimensional Accuracy
54	Nozzle Head	Component Weight affects Print Time	Component Weight affects Dimensional Accuracy
52	Heater	Component Weight affects Print Time	Component Weight affects Dimensional Accuracy
55	Cooling System	Component Weight affects Print Time	No Affect
53	Extruder Thermocouple	Component Weight affects Print Time	No Affect
61	X Positioning Drive	Component Performance affects printing time	Performance affects dimensional accuracy
62	Y Positioning Drive	Component Performance affects printing time	Performance affects dimensional accuracy

Then the design range were determined as shown in Table 6. According to the design ranges of the sequenced DP/C-est, the designs of each alternatives were selected.

Table 6. Design Parameter Target and Design Ranges for DP Selection

Sequence	DP	Target Parameter	Design range
41	Material Storage Space	Ease of accessibility	90-100%
{51, 52}	Filament Extruder / Heater	Feed rate / Efficiency	80-100%
42	Material Conveyor	Tension in the filament	80-100%
611	X Positioning Drive	Precision	-0.05 to + 0.05 mm
612	X Limit Switch	Responsiveness	5-10 micron
621	Y Positioning Drive	Precision	+/- 0.01 - +/- 0.05
622	Y Limit Switch	Responsiveness	5-10 micron
631	Z Positioning Drive	Precision	+/- 0.01 - +/- 0.05
632	Z Limit Switch	Responsiveness	5-10 micron
54	Nozzle Head	Withstand high temperature	200 - 400 deg C
613	X Actuator	Speed	1000-2000 rpm
623	Y Actuator	Speed	1000-2000 rpm
633	Z Actuator	Speed	1000-2000 rpm
53	Extruder Thermocouple	Accuracy and Response time	1deg/ 5 sec
55	Cooling System for Heater	Heat dissipated	5 - 15W/deg C

4.3.3. Alternatives Matrix

An Alternatives Matrix was built with Target Parameters and Design Range. The alternatives matrix with all the possible alternatives, is listed in Table 7. The alternatives selection table outlines the alternatives selected and the reason for selection.

As shown in Table 7, the scope of improvement would be to increase the dimensional accuracy of the printer, and as a result, the DPs affected, in sequence of precedence are – DP611, DP612, DP613, DP621, DP622, DP623, DP51, DP54, DP52.

Table 7. Alternatives Matrix of the FFF 3-D Printer Hardware System

Sequence	**DP41**	**{DP51, DP52}**	**DP42**
Selection	Mounted Spool placed on top	Drive gear with Idler gear with Resistive heater	PTFE Tube
Design range	90-100%	80-100%/ 90-100%	80-100%
Expectation	Does not affect accuracy	Low Weight and High Performance	Does not affect accuracy
System Range	95-100%	85-95% / 95- 100%	85-95%
Reason	Standard Configuration	Simple mechanism and High Efficiency	Standard Configuration
Sequence	**DP611**	**DP612**	**DP621**
Selection	Linear rails	Snap action limit switch	Linear rails
Design range	-0.05 to + 0.05 mm	5-10 micron	+/- 0.01 - +/- 0.05
Expectation	High Performance	High Performance	High Performance
System Range	- 0.03 to + 0.03 mm/ 5 m/s	5-10 micron	- 0.03 to + 0.03 mm
Reason	Highest accuracy	Fast Response Time And Low cost	Highest accuracy
Sequence	**DP622**	**DP631**	**DP632**
Selection	Snap action limit switch	Rack and Pinion	Stopper limit switch
Design range	5-10 micron	+/- 0.01 - +/- 0.05	5-10 micron
Expectation	High Performance	Does not affect accuracy	Does not affect accuracy
System Range	5-10 micron	0.0254 mm	10 micron
Reason	Fast Response Time And Low cost	Standard Configuration	Standard Configuration
Sequence	**DP54**	**DP613**	**DP623**
Selection	Screwable Nozzle	Servo Motor	Servo Motor
Design range	200 - 400 deg C	1000-2000 rpm	1000-2000 rpm
Expectation	Low Weight and High Performance	Accuracy at High Speed	Accuracy at High Speed
System Range	280 deg C	0.33 degree/ 1000- 5000 rpm	0.33 degree/ 1000- 5000 rpm
Reason	Low Weight and High Efficiency	Highest accuracy and High Speed	Highest accuracy

IOP Conf. Series: Materials Science and Engineering 1174 (2021) 012006 doi:10.1088/1757-899X/1174/1/012006

4.3.4. Partial Improvement

In the final step, DP Alternatives would be selected from the Alternative Matrix. If the system range of the Alternative falls within the design range of the DP, the alternative is selected. If it is outside the design range, either another alternative is chosen or change the ES of the U-Set. The alternatives (summarized below) selected satisfied the given design range of DP.

Table 8. Selected Alternatives of the FFF 3-D Printer Hardware System

Sequence	DP	Alternative	Design Range	System range
41	Material Storage Space	Spool covered with a box on four sides with open lid on top	90-100%	90-100%
{51, 52}	Filament Extruder / Heater	Drive gear with idler gear with Resistive heater	80-100%/ 90-100%	85-95% / 95- 100%
42	Material Conveyor	PTFE Tube	80-100%	85-95%
611	X Positioning Drive	High accuracy Linear rails	-0.05 to + 0.05 mm	- 0.03 to + 0.03 mm/ 5 m/s
612	X Limit Switch	Snap action limit switch	5-10 micron	5-10 micron/ Easily available and lowest cost
621	Y Positioning Drive	High accuracy Linear rails	+/- 0.01 - +/- 0.05	- 0.03 to + 0.03 mm
622	Y Limit Switch	Snap action limit switch	5-10 micron	10 micron/ Easily available and lowest cost
631	Z Positioning Drive	Ball screw	+/- 0.01 - +/- 0.05	±0.02mm
632	Z Limit Switch	Snap action limit switch	5-10 micron	10 micron/ Easily available and lowest cost
54	Nozzle Head	Screwable small Nozzles	200 - 400 deg C	280 deg C/ Lightweight
613	X Actuator	Stepper Motor	1000-2000 rpm	1.8 degree and no constant feedback/ 1000-1500 rpm
623	Y Actuator	Stepper Motor	1000-2000 rpm	1.8 degree and no constant feedback/ 1000-1500 rpm
633	Z Actuator	Stepper Motor	1000-2000 rpm	1.8 degree and no constant feedback/ 1000-1500 rpm

5. Discussion

The DCS Sequence helps a designer identify the scope for improvements sequentially such that they do not affect other parameters. This helps to reduce reiteration of complete design, and to & fro iterations between DPs. Once the modified DM was obtained, the next step is to assemble a list of design parameters which would solve the functional requirements. Each FR can be satisfied using different DPs. The acumen of a designer would enable them to choose an alternative which would perfectly satisfy the need. However, a designer is limited by their knowledge of the alternatives and thus sometimes they may overlook selecting a DP because they simply do not have enough knowledge about it. Thus, we can see from the modified DM that once we improve DP41 and move to {DP51, DP52}, we need to think of alternatives and solutions that are compatible with the selected DP41 and then move on further to DP42 and so on. Therefore, this method helps to sequentially improve the design of the product. To improve the design process, we use selection methods which would enable designers without considerable knowledge about a field to make and evaluate design selection.

This major observations from the case study are -

1. From the ES, we can clearly identify that Filament Module, Extruder Module and Control Module have higher precedence and thus a larger design space/ flexibility. Therefore, the designer should develop these modules first, followed by other modules.
2. We rearrange the design matrix according to the execution sequence. From the design matrix, we can clearly identify that there is a central module comprising of filament, extruder and control module. Also, there is an auxiliary module comprised of power, base plate and display module. Between the two modules, we can clearly see that there are relations/ couplings.
3. From the DM in Figure 9, we can see that the central module has the highest number of couplings and thus, most of the improvements will come from this section. Whereas the auxiliary module is a decoupled design and has less scope for improvements.

The two major improvements which can be implemented from the case study are -

1. For {DP51, DP52} if we consider it as a module and select the best alternative then we can eliminate DP42, which is a better solution from downstream perspective. But if we solve this coupling with TRIZ then it suggests Segmentation principle through which we can separate the extruder and

heater (Bowden extruder) which helps to eliminate couplings between DP51, DP52 and Module 6, and thus, is a better solution from upstream perspective.
2. If we can eliminate the C-Set by removing the couplings between DP72 and DP73, then we can have a more ideal solution.

Yet some improvements are not feasible, and we need to stick to the original design because they require change in the controller module which was fixed earlier to all the parameters.

6. Conclusion and Future Work

This paper outlines a method to design, analyze and enhance a hardware system. The first step demonstrates the process of information collection to build a functional model. The information collecting step outlines the relevant data which would be required for construction of a system model. This was followed by sketching out the steps to make a Design Matrix (DM) to visualize the system. The second step, Design Coupling Sequence (DCS) methodology was used to generate an Execution Module (EM), a hierarchical structure of the functional model. The DCS methodology helps the designer determine the best sequence of functional couplings to reduce the complexity as much as possible. It also allows the designer to uncover improvement opportunities in the system. The third step, we outline improvement strategy to partially improve the design of the system. This begins with understanding the current drawbacks in the product/system and mapping them to the causal DPs. Once the improvement DPs have been highlighted, the EM would be used to define the improvement strategy. This strategy would suggest all the DPs which need to be modified to improve the performance of the product. In the next step, a procedure for comparing different alternatives of DPs is outlined and method for implementing the modified DPs. Finally, this methodology was explained using a case study on an FFF 3D printer.

From the case study, a precedence in designing sub-systems was setup, with highest precedence given to the Filament Module, Extruder Module and Control Module. They should have access to largest design space and be developed first, followed by other modules. This precedence should allow designers to partially improve modules for each generation of product without the necessity to redesign the product.

The design matrix showed that a 3D Printer can be separated into two modules - a central module (comprising of filament, extruder and control module) and an auxiliary module (comprised of power, base plate and display module). This should help companies design platforms for their products and reduce their development cycle times and manufacturing costs.

Finally, presence of large number of couplings within Control Module suggests that the current generation of products have high complexity and in future, largest number of improvements will come from this section.

The future works should focus on application of DCS on integrated systems. The current set of methodology is designed to work with homogenous modules like physical or non-physical components. A comprehensive methodology which would allow the designer to compare all the elements of a system – hardware and software and that would enable them to make better decisions.

References

[1] Gassmann, O. & Von Zedtwitz, M. Organization of industrial R&D on a global scale. R D Manag. 28, 147–161 (1998).
[2] Suh, N. P. Engineering Design Axiomatic Design Theory for Systems. Res. Eng. Des. 10, 189–209 (1998).
[3] Van Hoek, R. I., Vos, B. & Commandeur, H. R. Restructuring European Supply Chains by Implementing Postponement Strategies. Long Range Plann. 32, 505–518 (1999).
[4] Gassmann, O. & Von Zedtwitz, M. New concepts and trends in international R&D organization. Res. Policy 28, 231–250 (1999).
[5] Fine, C. & Whitney, D. Is the make-buy decision process a core competence? MIT Cent. Technol.

Policy Ind. Dev. 1–31 (2002).
[6] Farid, A. & Suh, N. Axiomatic Design in Large Systems. Axiomatic Design in Large Systems (2016). doi:10.1007/978-3-319-32388-6.
[7] Jang, B. S., Yang, Y. S., Song, Y. S., Yeun, Y. S. & Do, S. H. Axiomatic design approach for marine design problems. Mar. Struct. 15, 35–56 (2002).
[8] Zhu, A., He, S., He, D. & Liu, Y. Conceptual Design of Customized Lower Limb Exoskeleton Rehabilitation Robot Based on Axiomatic Design. Procedia CIRP 53, 219–224 (2016).
[9] Suh, N. P. Complexity in engineering. CIRP Ann. - Manuf. Technol. 54, 46–63 (2005).
[10] Cheng, Q., Zhang, G., Gu, P. & Shao, X. A product module identification approach based on axiomatic design and design structure matrix. Concurr. Eng. Res. Appl. 20, 185–194 (2012).
[11] Bai, Z. H., Zhang, S., Ding, M. & Sun, J. G. Research on product innovation design of modularization based on theory of TRIZ and axiomatic design. Adv. Mech. Eng. 10, 1–15 (2018).
[12] Thielman, J. & Ge, P. Applying axiomatic design theory to the evaluation and optimization of large-scale engineering systems. J. Eng. Des. 17, 1–16 (2006).
[13] Wang, C. Y. & Lu, S. C. Y. Managing Functional Coupling Sequence to Decrease Complexity and Increase Modularity in Conceptual Design. MATEC Web Conf. 223, 1–9 (2018).
[14] Wang, C. & Lu, S. C. Managing Functional Coupling Sequences To Reduce Design Complexity During Concept Improvements. ASME 2015 Int. Des. Eng. Tech. Conf. Comput. Inf. Eng. Conf. 1–7 (2017).
[15] Fan, L. X., Cai, M. Y., Lin, Y., & Zhang, W. J. (2015). Axiomatic design theory: Further notes and its guideline to applications. International Journal of Materials and Product Technology, 51(4), 359–374
[16] Zhang, W. J., Li, J. W., & Zettl, B. (2012). Classification of design theories and methodologies for effective industrial applications. Proceedings of the 2012 7th IEEE Conference on Industrial Electronics and Applications, ICIEA 2012, (1), 1255–1260.
[17] Zhang, R., Cha, J., & Lu, Y. (2007). A conceptual design model using axiomatic design, functional basis and TRIZ. IEEM 2007: 2007 IEEE International Conference on Industrial Engineering and Engineering Management, 1807–1810.

IOP Conf. Series: Materials Science and Engineering 1174 (2021) 012007 doi:10.1088/1757-899X/1174/1/012007

Axiomatic design of a man-machine interface for Alzheimer's patient care

Sergio A. Navarro-Tuch[1,2], Jack Gammack[1], David Kang[1], Sang-Gook Kim[1]

[1] Park Center for Complex Systems, Department of Mechanical Engineering, Massachusetts Institute of Technology, Cambridge, MA, USA

[2] School of Engineering and Sciences, Tecnológico de Monterrey, Mexico City, Mexico

E-mail: sangkim@mit.edu

Abstract. Axiomatic thinking framework is used to develop a novel concept of "Care-Toy," which helps caregivers of Alzheimer's and dementia patients by providing much needed respite for themselves without compromising patients' safety. "Care-Toy" interacts with a patient via stimulating images and sounds for touch interactions, actively keeping the patient's attention for a needed duration as well as monitoring the patient's status or alerting the caregiver if necessary. Systems solution for this concept could be effectively elaborated with the Axiomatic Design framework, leading to the development of a key module intended for measuring and maintaining the attention of the patient during the respite of caregivers. Gaze detection is the key module based on deep learning-based facial recognition system to monitor whether "Care-Toy" attracts the patient's interest and maintains attention, otherwise to alert the caregiver. The module-junction structure could be derived from the decomposed functional requirement (FR)-design parameter (DP) tree structure and design matrices of them, which effectively defines the software system structure for this product.

1. Introduction

As part of the current global demographic transition, we have seen a rise in the average age of the population in almost every part of the globe.The global population aged 60 years or older numbered 962 million in 2017, more than twice as large as in 1980, and is expected to double again by 2050 to reach nearly 2.1 billion [1]. Being that age is the strongest known factor to the onset of degenerative mental diseases such as dementia [2], [3], and since 60 to 70% of dementia causes are related to Alzheimer's [4], it is very important to develop strategies and tools for the care of such individuals.

As the individuals' disease progresses, the dependence towards the caregiver, mostly family members of the patients, increases, as does the hardship on the whole family. Severe symptoms can frequently hinder the caregivers' capability to manage necessary daily tasks, such as food preparation, supply acquisition, and cleaning, as losing sight of the patient could result in the patient wandering off and getting hurt in such a time period. It is essential to provide a much needed respite for a caregiver so that minimum quality of living can be maintained.

In order to aid the caregivers and provide them with a kind of short respite, about 30 minutes to 1 hour, the idea of "Care-Toy" was conceived, which is similar to the concept of providing toys to toddlers in order to make a short respite for parents doing essential tasks. As a toy must attract toddlers' interest for a duration of respite while not harming them at all, Care-Toy should be so. In addition to these basic functions, Care-Toy can bring more functions for the goodness of patients enabled by the widely available smart tablets. Based on this initial concept, Axiomatic Design thinking framework has been applied [5] to identify and define the FRs and DPs needed to detail the design of the system. Once the key modules are determined from the upper-level FR-DP tree structure, the main focus of this development becomes the "Attention Module" which is further explored in the section following the general design concept development.

2. Conceptual Design

An overview of the highest level functional requirements (FR) and design parameters (DP) are first presented to address the needs of the customer. An outline and brief discussion of the full Axiomatic decomposition of the system follows. Lastly, some of the most important modules are expanded upon and implemented.

2.1. High-Level System Design

Solution development requires definition of the customer needs in order to understand the high-level FRs and the corresponding DPs for "Care-Toy". In order to provide the caregiver with a brief respite, the caregiver must be certain that the patient will not lose interest in the device, resulting in the patient wandering off and potentially endangering themselves. Additionally, due to the nature of the disease, patients may be extremely sensitive to unpleasant stimuli or physical restraint, limiting options for maintaining attention and preventing movement of the patient. Patients also often have impaired motor skills and cognitive abilities, meaning that any toy must be easy and comfortable to use.

The task of maintaining the attention of the patient requires both capturing their initial interest and maintaining that interest after it is captured. As a result of the symptoms of the disease, it is difficult to complete these tasks, and a different DP may be necessary to capture the interest than the DP used to retain the attention. For example, when patients lose interest in the toy and become interested in something else, an immediate external stimulus could be provided to break the patients' diminished attention span to recapture their interest. This led to the designers creating independent modules to clarify the need for separate DPs to address the multiple FRs present.

The top-level FRs and DPs for "Care-Toy" are defined as follows. The main requirements of "Care-Toy" to provide the respite are to maintain the attention of the patient (FR1), to notify the caregiver in the case of an emergency (FR2), and to allow the patient to easily use the system (FR3). The designers conceive of a device that provides audio-visual stimulation to capture the initial interest and to hold the long-term attention of the patient (DP1). The device will have a system for detecting emergencies and notifying the caregiver (DP2), and a simple and durable physical design for ease of use (DP3). Figure 1 shows the defined high-level FRs and DPs of "Care-Toy", as well as the relevant design matrix.

FR Provide caregiver with brief respite (30 to 60 min)
- FR1 Maintain attention of patient
- FR2 Notify caregiver in case of emergency
- FR3 Allow patient to easily use/play the toy

DP "Care-Toy" for patient to interact with
- DP1 Device provides audio-visual stimulation
- DP2 System for detecting an emergency and notifying
- DP3 Simple and durable physical design

	DP1	DP2	DP3
FR1	X	0	0
FR2	X	X	0
FR3	0	0	X

Figure 1. Top level FRs, DPs, and design matrix of "Care-Toy".

2.2. Decomposition and Module-Junction Structure

An overview of the decomposed FRs and DPs is presented in this section. Table 1 shows the first 3 levels of decomposition representing the system structure of the "Care-Toy". A module-junction structure based on the design matrices for Module 1 is presented in Figure 2, representing the decision order and information flow in which key modules are defined to be developed for FR1, maintaining the

IOP Conf. Series: Materials Science and Engineering 1174 (2021) 012007 doi:10.1088/1757-899X/1174/1/012007

attention of the patient. The module-junction structure shows the gaze detection module is the key for this system, which is described in detail in a later section.

Table 1. Full decomposition of FRs and DPs for "Care-Toy".

FR0	Provide brief respite to caregiver	DP0	Toy for patient to interact with
FR1	Maintain attention	DP1	Device that provides audio-visual stimulation
FR1.1	Display images	DP1.1	LED Screen
FR1.2	Play sounds	DP1.2	Speakers
FR1.3	Capture short term interest	DP1.3	Attention-grabbing flashing and sounds.
FR1.4	Keep long term attention	DP1.4	Show familiar images and sounds
FR1.4.1	Allow upload of media by caretaker	DP1.4.1	Upload module
FR1.4.2	Allow selection of uploaded media	DP1.4.2	Selection module
FR1.5	Measure attention	DP1.5	System uses sensors and camera to detect patient
FR1.5.1	Detect face of patient	DP1.5.1	Face Detection Module
FR1.5.2	Detect gaze of patient	DP1.5.2	Gaze Detection Module
FR1.5.3	Detect inputs of patient	DP1.5.3	Input Detection Module
FR1.6	Recover attention if lost	DP1.6	Update images and sounds appropriately
FR2	Get help in case of emergency	DP2	System for detecting an emergency and contacting caregiver
FR2.1	Detect abnormalities	DP2.1	System for comparing current and baseline conditions
FR2.1.1	Collect baseline conditions	DP2.1.1	Data log from calibration period
FR2.1.2	Collect current conditions	DP2.1.2	Data log from Attention Module
FR2.1.3	Compare baseline and current conditions	DP2.1.3	Module for comparing conditions
FR2.2	Alert caregiver	DP2.2	Page or notification directly to caregiver
FR2.3	Collect and provide relevant data	DP2.3	Output report on information collected from patient
FR3	Allow patient to easily use toy	DP3	Simple and durable physical design
FR3.1	Survive drops and minor misuse	DP3.1	Material and structural properties
FR3.2	Allow to rest in position on table	DP3.2	Stand on back of device
FR3.3	Allow to be comfortably holdable	DP3.3	Thin and lightweight
FR3.4	Allow UI to be easily accessible	DP3.4	Proximity of holding position to buttons
FR3.5	Allow freedom of movement	DP3.5	Size and weight

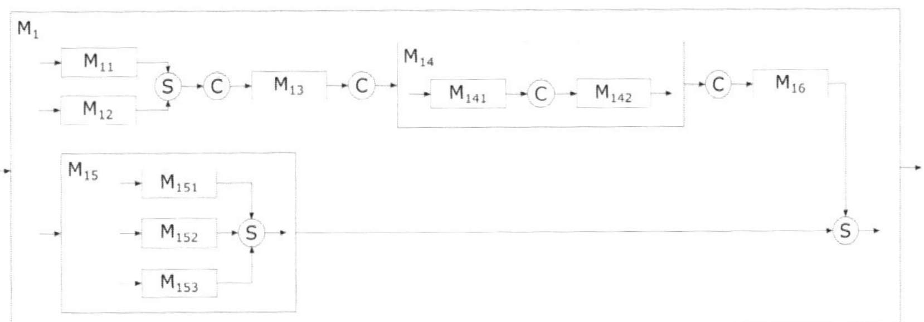

Figure 2. Module-junction structure for Module 1 of "Care-Toy".

To maintain the attention of the patient through a device that provides audio-visual man-machine interface, the target device must display images to the user (FR1.1) and play sounds to the user (FR1.2) that can get the patient's attention. As patients are unlikely or unwilling to pay attention to something when they are told to, the patient must first be attracted to the device to capture short-term interest of the patient (FR1.3), which can be done though attention-grabbing flashing with interesting sounds. The device must then keep long-term (at least 30 min) interest of the patient (FR1.4) by showing familiar or interesting images and sounds to the patient. To ensure that the patient is paying attention, the system measures the attention of the patient (FR1.5) and must attempt to regain the attention of the patient if it is going to lose attention (FR1.6).

To allow the toy to use familiar images and sounds, a personalized content module needs to be created with relevant images and sounds such as family photos and popular songs that are still vivid in patients' long-term memory (FR1.4.1). The caregiver must also be able to collect, build, and select different images and sounds to be played in different scenarios via cloud sourcing from the family (FR1.4.2).

The Attention Module, which measures the attention of the patient, will use sensors and a camera to detect if the patient is in front of the device, looking at the device, or interacting with the device. A camera is used to determine whether the user is in front of the camera through facial recognition (FR1.5.1). The camera is also used to detect the gaze of the user (FR1.5.2). Lastly, the device will determine if the user is interacting by pressing buttons or moving the device around (FR1.5.3).

In the case that the patient's attention cannot be recovered when lost, or if there is an emergency, the system must be able to notify the caregiver (FR2). The system must be able to detect an abnormality during use (FR2.1), alert the caregiver (FR2.2), and provide the caregiver with relevant data related to the abnormality (FR2.3).

To detect abnormalities, the system will compare the current conditions with a set of baseline conditions. To do so, the system must collect the baseline conditions during a calibration period (FR2.1.1), collect current conditions from the camera and sensors during use (FR2.1.2), and determine if the current conditions are outside of what is acceptable (FR2.1.3). In order to alert the caregiver that there is an abnormality, the device can send a page or message directly to a device held by the caregiver. From the collected data, the device can output a time-stamped report.

The last main functional requirement of the device is to be easily usable by the patient (FR3), who may have impaired motor skills. The device should have a simple and durable design such that it is able to survive drops and minor misuse (FR3.1), and it must be usable in a variety of different positions, such as being able to rest on a table during use as well as able to hold in one's hands during use (FR3.2 & FR3.3). In any of these positions, the user must be able to easily interact with the device by pushing buttons or touching the screen (FR3.4). The device must also be portable so that the user can use it in multiple different locations (FR3.5). The size, shape, material properties, and proximity of the User Interface (UI) to the holding position all affect these FRs. Figure 3 shows the coupled design matrix for FR3 (specifically for FRs 3.1, 3.3, 3.5), illustrating the need for feedback in the process of ensuring the device is strong enough to withstand abuse without being overly cumbersome to use.

	DP3.2	DP3.4	DP3.1	DP3.3	DP3.5
FR3.2	X	0	0	0	0
FR3.4	0	X	0	0	0
FR3.1	0	0	X	X	X
FR3.3	0	0	X	X	X
FR3.5	0	0	X	X	X

Figure 3. "Care-Toy" design matrix for FR3.

2.3. Attention Measurement and Recovery Exploration

From the module-junction diagram at the 1st level FR-DP tree, the main focus in this article for the desired system becomes the task of maintaining the attention of the user. To address this, we elaborate this module on how to attract the initial attention and how to make sure that the patient is still focused (immersed) on the content of the toy. We also elaborate on the capability of the module to adapt to the severity of the affliction and the personalized behaviors of the specific patient.

2.3.1. Measure Attention. Multiple studies have been developed on the attention and engagement measurement of attention, especially with the current progress of computational power and online learning. As for unobtrusive methods, some of the technologies that have been explored include computer vision for facial expression interpretation [6], [7], pose and user gaze tracking and speech recognition [8], [9], and even remote pulse measurement [10]. These methods can be used to detect, measure and determine the user´s attention and engagement [11]–[13] while interacting with virtual systems [14], [15]. From the literature review, an approach combining facial analysis and gaze tracking was selected. Further explanation of the module development as the core of this project is presented in Section 3.

2.3.2. Recover Attention. The basis of the development works under the assumption that the attention span of the user is quite irregular and thus can easily lead to changes and loss of focus due to either external distractions or the evolution of the affliction, as previously stated. Under this assumption, it is essential to handle a wide range of elements that stimulate the patient, thus inviting them to continue the interaction, while continually feeding back the changes on the state to undertake the appropriate steps. Achieving the task requires to take into consideration and combine two parts, which can be divided into passive and active components. The passive components are those that provide a continuity and structure that allow the patient to maintain a level of familiarity and control.

2.3.3. Content. Guaranteeing the engagement of a patient requires that the system is capable of presenting instantly attractive material or stimuli to the user. The literature exploration led to the finding that the recommended content intends to stimulate the memory of the user, while also aiding in the use and honing of their capabilities. Among the studies, we can find recommendations of Sensory stimulation [16], [17] and Multi-Sensory Stimulation (MSS) [18]. The content selection would later affect the characteristics of the required interface and the ease of use. Given that the patients' abilities keep deteriorating as a result of the affliction, an interface with high level of complexity and abstraction is undesirable. Thus the initial content selection was limited to visual stimuli from the patient's life for memory stimulation, and low-level, task oriented games.

2.3.4. Accessibility. Previously it was mentioned that the patient's circumstances and situation may vary, depending not only on the initial condition of the user, but also the evolution of the affliction. Thus, it is important that the system is capable of adapting to the specific scenario, however, as previous studies have found, it is very important for the elder population that the feeling of control over the system is maintained [19]. To allow the adaptation of the system to the user, the proposed solution is to allow the manual input of material by the caregiver, building the library of memories to be presented, and allow interface personalization.

3. Attention Module

Instant generation and maintaining the attention of an Alzheimer's patient for a period of 30-60 min is the key to the success of "Care-Toy". The selected approach is the use of computer vision for gaze tracking and user face detection. The monitoring capabilities are at the core of the "Care-Toy" system to drive audio-visual human-machine interface to get instant initial attraction, provide various content to maintain attention and notify the caregiver in case that the user's safety may be compromised.

3.1. Module Logic.

In this subsection, we present the logic and corresponding flow diagram of the Attention Module. The description of the components for the working of the module is presented in the subsection afterwards.

The Attention Module takes care of the monitoring of the interaction between the user and the system. To ensure the continuity of such interaction, the system must complete two main tasks: the sensing of the user and the exposure to the corresponding stimuli. The stimuli are intended to be modified depending on the user and by the active input of the caregiver or recommendation of the physician. Thus, the remaining task is the monitoring or attention tracking, which falls completely on the system. In Figure 4, we can appreciate the working of such module. The first task of the module is ensuring that the user is present for the activity to start. To do this the user should be in front of the camera integrated in the system, otherwise the use of displayed and sounded alerts are used to grasp their curiosity. Once the interaction is secured, the system allows the implementation of the activities, which may be either passive or active. One addition that the activities received after early designs was the implementation of a calibration activity for the gaze tracking.

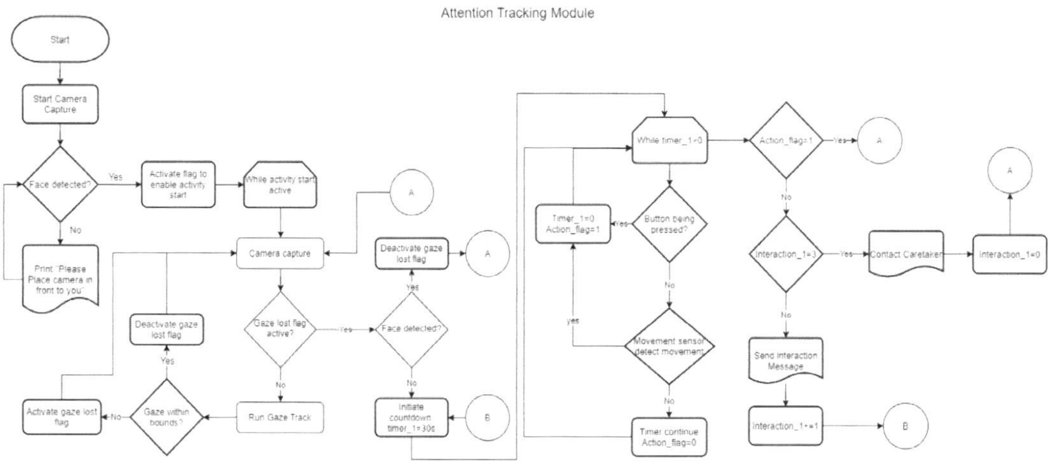

Figure 4. Attention Module Flow Diagram.

During the activities, the gaze of the user is monitored to determine if it falls within the predetermined area of interest. The monitoring of the instances when the gaze is lost may allow for the feedback to the caregiver to eliminate or substitute activities, and it may be used to monitor the progress of the patient. If the patient's gaze is lost, the next confirming instance is the face detection. If the face is also out of range or can't be detected, the next confirmation element is the activity sensing, which is comprised of two parts, the button pressing and the movement sensing. These two components rely on the integrated accelerometers and touch screens or keyboard of the system.

If the attention of the patient is unable to be found after a predetermined time, the system will attempt to communicate with the patient through alerts and messages or calls. If a certain number of calls remain unanswered, the system will determine a state of abandonment, thus will attempt to contact the caregiver through the corresponding strategy. If the patient resumes the interaction, the system will resume the normal monitoring procedures.

3.2. Module Components.

The prototype uses the *dlib* and *OpenCV Python* libraries for the face detection and tracking of the face position. The facial tracking is combined with the gaze tracking. To test the initial logic implementation, the gaze tracking in early stages was performed by a third party software named GazePointer [20]. The intention for this interaction is to guarantee that the user is not only facing the system, but is also looking at the desired area. The interaction of both systems allows to implement different kinds of alerts or responses depending if the user gaze is lost, but the user is still in the vision field of the system, or if the subject has abandoned it completely.

The logic testing with the use of the GazePointer tool and the facial tracking provided an acceptable response and ease to achieve the desired task, thus proving that it was possible to work with a monocular eye tracking instead of needing to implement an additional hardware piece for the viability of the system. This led to the task of developing a gaze tracking component independent of a third party. The proposed approach is the following.

Even though the final position of the gaze is delivered by a learning model, the possible approaches of which kind of data and model to use are multiple. After some analysis of the options, the designers opted to approach the problem by building on previous modules to gather the data for the training of the model. First, the eye positions were extracted by implementing the *dlib* libraries and defining as an area of interest the rectangle containing the points that delimit each eye. It was decided to extract each eye independently instead of using both as a single piece, as it would not only allow to minimize the additional burdening of the system in case of face pitching, but it also allowed to test the idea of developing a system capable of tracking an estimated gaze even if the obstruction of one of the eyes may occur. Once the area containing the eye is defined, the contour of the eye is delimited using border detection to locate the pupil, and the content in such delimited area is processed by analyzing the Gaussian concentration of pigment. The response is used to generate a black and white blob whose borders are then brushed to eliminate the area of transition. With the shape acquired, the center of the area is calculated by finding the mass center of the shape. The center is then returned as the coordinates to be used in the tracking model, being referenced with the relative and global coordinates of the shape.

The training database for this model is still being built, however, the processing of the eyes has been showing promising results when evaluating the use with various light conditions and angles, although there haven't been enough tests made with the use of eyeglasses.

4. Summary and Future Work

The current design of assistance system is the result of the consideration of the needs of the patients and caregivers, while also focusing on the available technology. By implementing the Axiomatic Design methodology, the correct identification of the functional requirements allowed for an optimized detection and prevention of potential couplings. This approach allowed the team to identify the most relevant tools needed to complete the task without sacrifying efficiency.

The currently achieved prototype is intended to be tested by potential users, which include caregivers, patients and physicians, to evaluate additional adjustments that may be needed to maximize the quality of the User Experience (UX). In addition to the interface changes that may result from such evaluation, the team is also working on optimizing and further improving detection tools. Another instance that will require further work is the expansion of the menu for available activities for the patients.

Acknowledgements
This research has been partially funded by the Tec de Monterrey and MIT Nanotechnology Program.

References

[1] P. D. United Nations, Department of Economic and Social Affairs, "World Population Ageing 2017 - Highlights (ST/ESA/SER.A/397)," 2017.
[2] World Health Organization, "Global action plan on the public health response to dementia 2017 - 2025," 2017.
[3] W. A. Kukull et al., "Dementia and Alzheimer Disease Incidence," *Arch. Neurol.*, vol. 59, no. 11, p. 1737, Nov. 2002.
[4] Alzheimer Association, "2019 Alzheimer's disease facts and figures," 2019.
[5] N. P. Suh, *Axiomatic Design: Advances and Applications*. 2001.
[6] A. Kumar and A. Agarwal, "Emotion recognition using anatomical information in facial expressions," in *2014 9th International Conference on Industrial and Information Systems (ICIIS)*, 2014, pp. 1–6.
[7] S. Stöckli, M. Schulte-Mecklenbeck, S. Borer, and A. C. Samson, "Facial expression analysis with AFFDEX and FACET: A validation study," *Behav. Res. Methods*, vol. 50, no. 4, pp. 1446–1460, Aug. 2018.
[8] M. Aparna and P. Wanare, "Human Emotion recognition from speech," *Int. J. Emerg. Technol. Adv. Eng*, vol. 4, no. 7, pp. 74–78, 2014.
[9] R. A. Calvo and S. D'Mello, "Affect Detection: An Interdisciplinary Review of Models, Methods, and Their Applications," *IEEE Trans. Affect. Comput.*, vol. 1, no. 1, pp. 18–37, Jan. 2010.
[10] M.-Z. Poh, D. J. McDuff, and R. W. Picard, "Non-contact, automated cardiac pulse measurements using video imaging and blind source separation," *Opt. Express*, vol. 18, no. 10, p. 10762, May 2010.
[11] H. Monkaresi, N. Bosch, R. A. Calvo, and S. K. D'Mello, "Automated Detection of Engagement Using Video-Based Estimation of Facial Expressions and Heart Rate," *IEEE Trans. Affect. Comput.*, vol. 8, no. 1, pp. 15–28, Jan. 2017.
[12] S. Dhamija, "Learning based visual engagement and self-efficacy," in *2017 Seventh International*

Conference on Affective Computing and Intelligent Interaction (ACII), 2017, vol. 61, pp. 581–585.

[13] B. M. Booth, A. M. Ali, S. S. Narayanan, I. Bennett, and A. A. Farag, "Toward active and unobtrusive engagement assessment of distance learners," in *2017 Seventh International Conference on Affective Computing and Intelligent Interaction (ACII)*, 2017, vol. 2018-Janua, pp. 470–476.

[14] J. Hernandez, Z. Liu, G. Hulten, D. Debarr, K. Krum, and Z. Zhang, "Measuring the engagement level of TV viewers," *2013 10th IEEE Int. Conf. Work. Autom. Face Gesture Recognition, FG 2013*, 2013.

[15] J. Whitehill, Z. Serpell, Y. C. Lin, A. Foster, and J. R. Movellan, "The faces of engagement: Automatic recognition of student engagement from facial expressions," *IEEE Trans. Affect. Comput.*, vol. 5, no. 1, pp. 86–98, 2014.

[16] B. S. Strøm, S. Ytrehus, and E. K. Grov, "Sensory stimulation for persons with dementia: A review of the literature," *J. Clin. Nurs.*, vol. 25, no. 13–14, pp. 1805–1834, 2016.

[17] J. M. Witucki and R. S. Twibell, "The effect of sensory stimulation activities on the psychological well being of patients with advanced alzheimer's disease," *Am. J. Alzheimers. Dis. Other Demen.*, vol. 12, no. 1, pp. 10–15, 1997.

[18] L. A. Wareing et al., "A randomized controlled trial of the effects of multi-sensory stimulation (MSS) for people with dementia," *Br. J. Clin. Psychol.*, vol. 40, no. 1, pp. 81–96, 2001.

[19] M. Zallio, D. Berry, and L. J. Leifer, "Meaningful Age-Friendly Design. Case Studies on Enabling Assistive Technology," 2020, pp. 779–790.

[20] S. Deja, "GazePointer." GazeRecorder, 2019.

Supervised machine learning applied to gas leak detection in air conditioner cooling system

S. F. Moreira[1], V. Shah[2], M. L. R. Varela[1], A. C. Monteiro[3], G. D. Putnik[1]

[1]Department of Production and Systems, Universidade do Minho, Guimarães, Portugal
[2]Department of Information Systems, Universidade do Minho, Guimarães, Portugal
[3]Department of Mechanical Engineering, Universidade do Minho, Guimarães, Portugal

E-mail: leonilde@dps.uminho.pt

Abstract This paper aims to present a concept test for an alternative refrigerant gas leak detection method, to be used in air conditioning manufacturing processes, in order to increase confidence in retaining products with gas leak in the factory, minimizing human interference in the test. To analyse the proposed solution, experimentation cycles were conducted, involving variables of industrial environment, product, and a thermographic camera with infrared technology, responsible for collecting the thermal image of the leak study area. Supervised machine learning method was used to train algorithms on temperature dataset to classify an area either as "Gas leakage" or "Normal". The regression logistic algorithm had the best performance in the predictions, showing that it is possible to detect "Gas leakage" area in automatic decision-making in an industrial environmental.

1. Introduction

Since its invention in 1902 by Willis Carrier, air conditioner appliance, for domestic and industrial use, designed to control the temperature and humidity of a closed environment, has undergone constant technological developments, making its domestic, industrial, commercial, and automotive use feasible, bringing thermal comfort to the environment, in a sustainable manner. At the end of 2016, from 1.6 billion residential air conditioning units installed in the world, China has the largest quantity, the United States in second position and Japan in third position, with the first 2 being responsible for a little more than 50% of the total; Brazil occupied the 9th position [1]. This increase in demand over the years adds to the factories' challenge of ensuring product quality in their large-scale manufacturing processes. One of the main product quality problem with great impact on the consumer's home is the failure "Doesn't Cool" or "Doesn't Heat", where the main reason is related to the lack of refrigerant gas, which escaped through some micro hole present in the refrigeration system. In the air conditioning manufacturing process, tests to detect the gas leakage of the product are applied, but one of the difficulties encountered is related to the invisibility of the refrigerant gas. Among the techniques and methods usually used to gas leak detection, we can mention [2]: pressure drop tests, gas detectors, bubble test in water tanks, hermetically closed cabin, such tests have significant dependence on equipment and human processing and therefore, the vulnerability of defect products goes to the final consumer. It is expected this dependence can be minimized using image or data and machine learning techniques.

Machine learning (ML) is one of the areas of computer science that has been growing lately, with application in automatic detection of significant patterns in a database [3]. It has the benefit of improving analysis and problem solving, due its ability to work with a lot of data, different attributes, learn and adapt using appropriated algorithms.

In the Industry 4.0 (I4.0) context, one of the main characteristics focuses on digitization of physical assets and integration of digital systems, where digital interfaces have the main communication role. One of the digital enablers allows converting physical elements into digital information for subsequent treatment [4]. In this sense, it is expected through this project to obtain the benefit advance provided by I4.0, to use the image captured by a thermographic camera to learning about the presence of gas leakage in the air conditioner. Thermography is described as a technique for a body thermal analysis or system, without physical contact due to its Infrared technology [5], started to be used for military purposes, space investigation and in the 1950s this procedure was already used in medicine and veterinary medicine as diagnostic tool.

The proposal solution is to use a thermal camera for diagnosing products with gas leakage in industrial air conditioning processes, using image and temperature data for training algorithms, to get an automatic test, reducing human interference in the current test.

This work describes the methodology used to test the concept of the solution, discussion of the results, conclusions, and future work.

2. Methodology

To study the test concept, 6 sigma tools were used to assist in conducting experiments and ML techniques for developing intelligence's test.

2.1. Experimentation strategies and concept testing

The 6 sigma methodology, developed by Motorola in 1987, is based on a set of statistical and quality tools used for continuous improvement of products and processes. To conduct the concept test for solution, some tools were used to organize the study, plan, carry out and analyse the experiments.

The PDSA (Plan, Do, Study, Act) cycle, also known as the Deming cycle, is the tool to guide detailed activities at each stage to implement an improvement. It is considered some elements of the scientific method within PDSA cycle and the execution of several cycles, allows the iterative process of knowledge under construction, moving from low knowledge on the subject to converging high knowledge for research in progress [6], illustrated by figure 1.

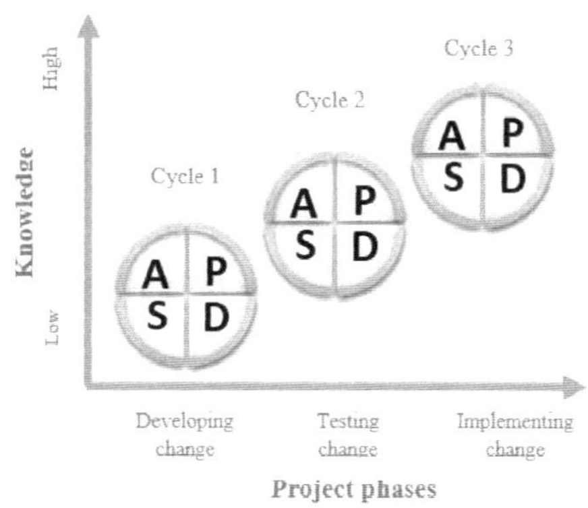

Figure 1. Sequential Learning: Multiples Deming Cycles [6].

Four cycles of PDSAs were conducted to acquire knowledge on the use of the thermal camera to detect gas leak patterns in air conditioners.

Design of Experiments (DOE) is one of the main tools for 6 sigma, used to test a proposal solution. It is a planned factorial experiment, where a change is made in a product or process, in a controlled manner with purpose of learning [6]. The experiments for this study were conducted within each PDSA cycle, considering variables (x's) that may impact Y, mapped in the cause-and-effect diagram in figure 2.

IOP Conf. Series: Materials Science and Engineering 1174 (2021) 012008 doi:10.1088/1757-899X/1174/1/012008

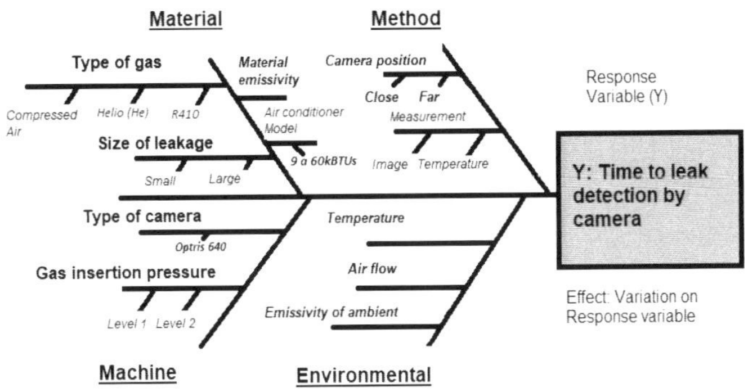

Figure 2. Cause & Effect Diagram: Mapping variables.

The Cause and Effect diagram helps to formalize in a visual way, the statements of what people already know about the current process and proposed solution, based on theories, past experience or things people do routinely, an important step when applying Axiomatic Design [7,8]. It helps planning the experimentation to confirm, or reject, the theories before implementation of new process design, in this case, in industrial environment.,

Another way to present and discuss with expert matters about the project design is a decomposition between FRs (Functional Requirements) and DP (Design Parameters) considering the first axiom of Axiomatic Design – the independence of functional elements [8]. It is presented in the table 1. After definition about the main FRs ad DPs, 24 tests were generated to learn about the impact of variables on Y: Time to detect leakage, which must meet the process time limit, a constrain of the project.

Table 1 – Relationship between Functional Requirements (FRs) and Design Parameters (DPs).

Functional Requirements		Design Parameters				
		DP1	*DP2*	*DP3*	*DP4*	*DP5*
FR1	Identify different size of leakage (hole size)	0,12mm² to 0,22mm²				
FR2	Camera distance from the product in test (Assembly line)		10cm to 50cm			
FR3	2 different kind of gas available for test inside the product			He and R410		
FR4	Environment noisy (Emissivity)				Blanket use	
FR5	Size of the product (Biggest production volume)					12.000 BTUs

2.2. Machine Learning for leak detection

Machine learning is a branch of artificial intelligence that allows machines to perform their jobs with skill, using intelligent software [9]. It uses data that represents real-world characteristics, which are processed by algorithms or models with the ability to perform forecasting tasks, such as Classification (Identifies which class/category/label belongs to), Estimation (The record is identified by a numeric and not a categorical value) and Prediction (It aims to predict the future value of an attribute). The model/algorithm can be considered as an approximation of the process that we want machines imitate, which use learning techniques: Supervised (uses classified data), Unsupervised (uses Unclassified data), Semi-supervised (Uses classified and Unclassified data) and reinforcement learning [9].

Supervised machine learning algorithms are trained based on the attributes of entry and classification of the instances with interest output. In this study, the classification is "Leakage" and "Normal "regions. Data for training and testing the algorithms were extracted from the last experiment (DOE#4) mapping the input attributes such as Product, Size of leak, Temperature over time ($t_{0,2}$), ($t_{0,4}$), ($t_{0,6}$) ... (t_{20}). Table 2 shows part of the dataset for training – 32 instances are the total used.

Table 2. Dataset sampling for algorithm training.

	Product	(A) Size of leakage	(B) Use Blanket	(C) Camera distance	Classification	Meas. Position	Temp. (t 0)	Temp. (t 0,2)	Temp. (t 0,4)
INSTANCES	B	0,12mm2	No Blanket	10cm	Leakage	P1	27,4	27,4	27,3
	B	0,12mm2	No Blanket	10cm	Normal	P2	27,9	28	27,8
	B	0,12mm2	No Blanket	10cm	Normal	P3	28,2	28,2	28,2
	B	0,12mm2	No Blanket	10cm	Normal	P4	28,1	28,1	27,9
	B	0,12mm2	No Blanket	10cm	Normal	P5	27,9	27,9	27,8
	B	0,12mm2	No Blanket	10cm	Normal	P6	30	30	30,1
	B	0,12mm2	No Blanket	10cm	Normal	P7	27,4	27,3	27,2
	B	0,12mm2	No Blanket	10cm	Normal	P8	28,9	28,9	28,9
	A	0,22mm2	With Blanket	10cm	Leakage	P1	28,8	28,9	28,7

The training and testing datasets are summarized in the table 3. It is possible to notice unbalanced quantity of data between "Normal" and "Leakage" areas.

Table 3. Total Instances for Algorithm Training& Testing.

	Training data		Testing data		
	Normal	Leakage	Normal	Leakage	Total
Instances	28	4	28	4	64

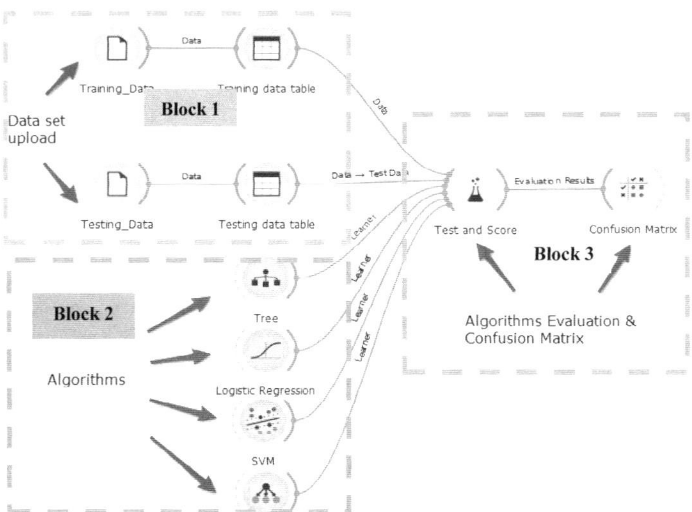

Figure 3 - Machine learning project to identify "Leakage".

The figure 3 presents the ML study based on 3 blocks: Block 1, data up loading; Block 2, learning algorithms; Block 3, algorithms performance evaluation based on the input data (training & testing). It was used "Orange", an open source software that works with machine learning and data visualization [10].

Four algorithms were used for training and testing (Decision Tree, Logistic Regression, Support Machine Vector and Naïve Bayes) to Supervised Machine Learning techniques, using confusion matrix and performance indicators to evaluate the ability to classify and predict "Leakage" area as True Positive (TP) and "Normal" area as True Negative (TN).

3. Results & Discussion

In this section, the results and discussion about sequential experiments carried out to learn about the variables involved in the concept testing design, assessing the ability to detect gas leakage in air conditioner using thermal camera are presented. Furthermore, Machine Learning algorithms performance applied on temperature dataset to identify regions with leakage or no leakage is discussed.

3.1. Sequential Experiment Results & Discussion

After 3 PDSAs for concept testing design, running 24 product testing with different industrial conditions (size of leakage, gas insertion pressure, camera position, thermal blanket, type of gas), the last experiment block (DOE#4) results in less time to leak detection by camera. The main factor to get less time was the type of gas, R410, when compared to Helium (He) gas in the first 2 blocks. The time to detect was less than process limit (Figure 4).

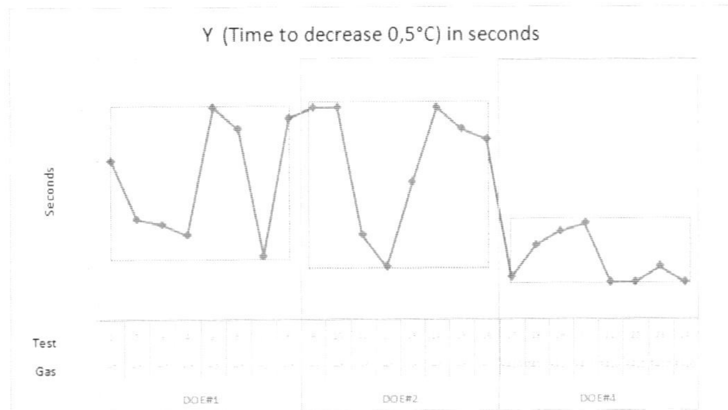

Figure 4. PDSA's summary – Y time to leak detection by camera.

3.2. Algorithms Performance Results & Discussion

Table 3 show the performance results comparison for algorithms used to training on data set for supervised learning, the most of algorithms can be found in [3]. Here there are most used indicators to assess the ability on classification task.

Table 4. Algorithm's assessment on classification.

	Higher is better			Lower is better	
Algorithms	Accuracy	Recall	Precision	% FP	% FN
Tree	0,84	0,00	0,00	0,03	0,13
SVM	0,91	0,25	1,00	0,00	0,09
Log Regression	0,94	0,50	1,00	0,00	0,06
Naive Bayes	0,44	0,50	0,11	0,50	0,06
Assessment	% defects and no defects classifyed correctly	% real defects	% defects correct classifyed	% False Positive	% False Negative
Metric	$\frac{(TP+TN)}{(TP+TN+FP+FN)}$	$\frac{(TP)}{(TP+FN)}$	$\frac{(TP)}{(TP+FP)}$	$\frac{(FP)}{(TP+TN+FP+FN)}$	$\frac{(FN)}{(TP+TN+FP+FN)}$

- The 4 algorithms' comparison shows the Decision Tree had the worst result in the Precision and Recall indices with low performance, followed by Naive Bayes with a high FP index, indicating that the product is "Normal" but it has "Leak".
- Logistic Regression showed better results: 100% Accuracy in TP classification, Zero FP classification. There is an opportunity for improvement in the Recall indicator, with 50% of correct real defects (TP), affected by the FN classification (the algorithm says there is no defect "Normal" but there is defect "Leakage").

- To increase confidence in the forecasts, more air-conditioners manufacturing process data would be needed, mainly from "Leakage", to improve the accuracy assessment using more balanced TP and TN instances.

4. Conclusions and future works

As the work progressed, it was possible to observe the evolution of knowledge on the use of an IR thermal camera to identify gas leaks in an air conditioning cooling system, based on the constrains of process time limit and the composition of FR and DP, resulting in more important Design Parameter using R410 gas to reduce the time to detect leakage.

The product's surface temperature data were used for supervised machine learning, with instances classified by region with "leakage" or "normal" region. With the aid of software, it was possible to train algorithms and evaluate their performance, where Logistic Regression achieved a better result in the classification, with opportunities for improvement with the collection of more data to increase confidence.

Thus, this project provides subsidies for the validation of a new gas leak test concept design, providing knowledge for the company move forward with the software, hardware, and camera development phase for production process, with the purpose of transferring the judgment of a "Leakage" and "Normal" product for an automatic detection system.

Acknowledgement

This work has been supported by FCT – Fundação para a Ciência e Tecnologia within the R&D Units Project Scope: UIDB/00319/2020.

References

[1] International Energy Agency. The Future of Cooling: Opportunities for energy-efficient air conditioning. 2018; Available from: www.iea.org
[2] Rottländer H, Umrath W, Voss G. Fundamentals of leak detection [Internet]. Leybold; 2016. p. 48. Available from: www.leybold.com
[3] Akinsola JET, Osisanwo FY, Awodele O, Hinmikaiye JO, Olakanmi O, Akinjobi J. Supervised Machine Learning Algorithms: Classification and Comparison. Int J Comput Trends Technol [Internet]. 2017 [cited 2021 Mar 24];48. Available from: http://www.ijcttjournal.org
[4] Ribeiro JM. O conceito da indústria 4.0 na confeção: análise e implementação. 2017;1–99. Available from: http://repositorium.sdum.uminho.pt/handle/1822/49413
[5] Filho O, Araujo S, Florio A. Detecção automática do aquecimento em componentes de um sistema industrial , baseada no reconhecimento de imagens térmicas . 2006;1–9.
[6] Moen R, Nolan T, Provost L. Quality Improvement through Planned Experimentation, 3rd edition. 3rd ed. United States: Mc Graw Hill; 2012. 476 p.
[7] Suh NP. The Principles of Design. Oxford University Press; 1990.
[8] Suh NP. Engineering Design Axiomatic Design Theory for Systems. Res Eng Des. 1998;10(4):189–209.
[9] Mohammed M, Khan MB, Bashie EBM. Machine learning: Algorithms and applications. Machine Learning: Algorithms and Applications. 2016. 1–204 p.
[10] LARHUD. Orange [Internet]. LARHUD Laboratório em rede de humanidade digital. 2018 [cited 2020 Dec 7]. Available from: http://www.larhud.ibict.br/index.php?title=Orange

IOP Conf. Series: Materials Science and Engineering 1174 (2021) 012009 doi:10.1088/1757-899X/1174/1/012009

Using the principles of axiomatic design in the development of bushing manufacturing technology

L Slatineanu[1], O Dodun[2], G Nagîţ[3], P Duşa[4], M Coteaţă[5], A Hriţuc[6], D Luca[7], and I Carp[8]

[1, 2, 3, 4, 7]Professor, "Gheorghe Asachi" Technical University of Iaşi, RO
[5]Lecturer, "Gheorghe Asachi" Technical University of Iaşi, RO
[6]Ph.D.student, "Gheorghe Asachi" Technical University of Iaşi, RO
[8]Ph.D. Engineer, "Gheorghe Asachi" Technical University of Iaşi, RO

E-mail: hrituc.adelina3295@yahoo.com

Abstract. There are different types of bushings used in mechanical equipment and they can be used for different purposes. In the case of bronze bushings used to ensure rotation conditions of shaft-type parts, there are some difficulties in addressing the problem of their manufacture. In essence, in order to meet the machining accuracy requirements specific to the outer and inner revolution surfaces, appropriate manufacturing technologies must be established. Some distinct technological paths can be considered for machining and checking the outer and inner cylindrical surfaces of bushings. In solving certain problems specific to the development of bushing manufacturing technologies, the principles of axiomatic design were used. Thus, the axiom of the independence of the functional requirements was used preferentially in establishing the machining operations to be applied to the workpieceto obtain a bushing. The second axiom was applied to identify the best solutions for the machining of the inner revolution surface. In this way, more convenient manufacturing technology for bronze bushings was identified.

1. Introduction

The bushings are mechanical parts that have coaxial inner and outer revolutionsurfaces. These revolutions surfaces could be essential cylindrical and conical, but also there are other types of surfaces (toroidal, spherical, etc.).Various criteria could be used to classify the bushings [1]. Thus, if the outer surface shape is analyzed, there are smooth bushings and step bushings. The steps could have diameters decreasing on one side or decreasing on both sides.

The bushings are used to support theshafts or other parts found in rotational motion (pinions, cams, etc.). Two surfaces of the bushings have an essential role. Thus, an inner revolution surface takes contact with the journals of the part found in rotation motion. The outer revolution surface is used to fix the bearing bush in a housing. The two surfaces mentioned above are generally characterized by a high accuracy and a low roughness. Usually, the outer surface used in fixing the bearing bushing in the housing has an accuracy that corresponds to the tolerance classes7-8 and a surface average roughness Ra=1.6 μm. On the other hand, the inner revolution surfaces used to support the shaft journal must be achieved in the tolerance classes 6-8 and has a surface average roughness Ra=0.8 μm.

As materials for bearing bushings, the bronzes, cast irons, steels, plastics could be used. In the machine manufacturing, the most used material for bearing bushings seems to be certain types of bronzes that are characterized by low values of the friction coefficient.

An investigation by Morozov et al. showed that the presence of lead in the material of bronze bushings with tin, zinc and lead results in such an arrangement of lead on the movable contact surface that the pressure is reduced (by increasing the contact surface) and the coefficient of friction decreases [2]. In the end, this means an increase in the durability of the mobile joint.

The manufacture and modelling of the operating behavior of hybrid aluminum and steel bushings by successive plastic deformation processes were investigated by Behrens et al. [3-5].

An experimental research on the behavior of a zirconium ceramic bushing was performed on a universal friction machine by Alisin [6]. This research allowed the development of a procedure for the evaluation of the reliability and service life of the bushings made of indicated material.

The objective of the research presented in this paper was to find some solutions for particular aspects of the manufacturing of bearing bushings using the axiomatic design principles.

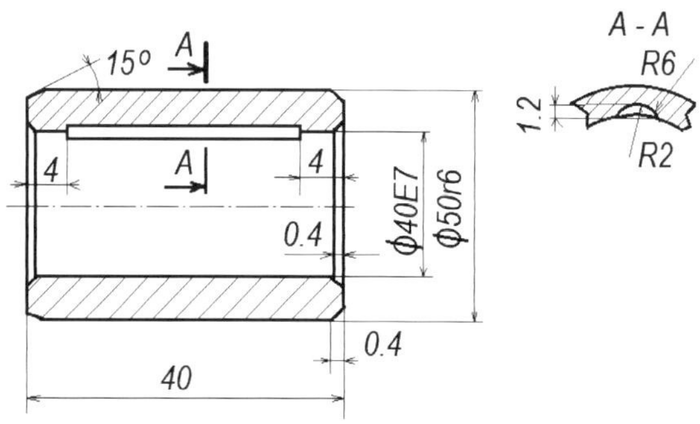

Figure 1. Mechanical drawing of the bronze bushing to be obtained (adapted from [7]).

2. Analysis of information provided by the mechanical drawing of the bronze bushing and the technological solutions that can be used to obtain it

An example of mechanical drawing of a bronze bearing bushing can be seen in figure 1. The part had to be made by the students during the practical activities developed in a school workshop that worked in the Faculty of Mechanics from the "Gheorghe Asachi" University of Iaşi – Romania. The part material was an antifriction bronze. As workpieces, tubes achieved by casting were used.

Analyzing the existing information in the mechanical design of the bearing bushing, it was concluded that they were sufficient to proceed to the development of a bushing manufacturing technology, by taking into account the machine tools, jigs, fixtures and tools available, the number of parts to be manufactured, the level of qualification of the operators (students coordinated by foremen) and other conditions specific to the workshop in which these bushings were to be manufactured.

After the analysis of the initial conditions that correspond to the bearing bushing proposed to be manufactured, it also concluded that the main groups of manufacturing phases could be:

1. Machining of the outer surfaces in two distinct phases (roughing and finishing);
2. Machining of the inner surfaces also in two distinct phases, roughing and finishing;
3. Machining of the bushing frontal flat surfaces;
4. Machining of the grease groove.

3. The use of the first axiom to define the technologies for the manufacture of bushings

To apply the first axiom from the axiomatic design, a normal sequence of manufacturing stages has been completed.

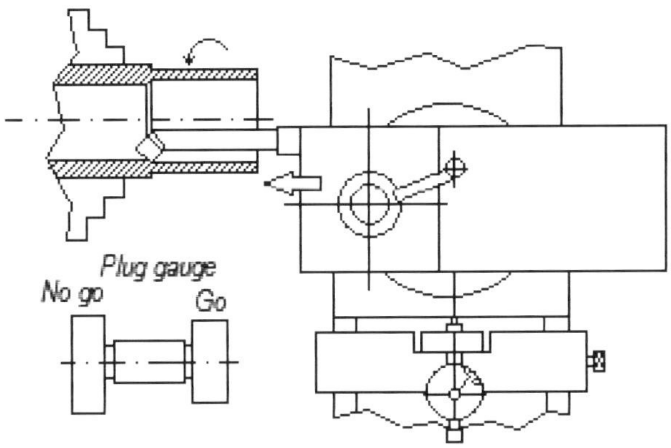

Figure 2. Turning of the inner cylindrical surface using a dial comparator device and a go/no-go plug gauge.

If we take into consideration only the process of designing the technology of manufacturing the bearing bushings, as first customer the representatives of an office for organizing the manufacturing process could be considered. These representatives must ensure the providing of the workpieces, jigs, fixtures, and devices, transmit the technological documentation to the manufacturing workshop coordinators etc. As *customer need* directed to the office of designers of manufacturing processes, it can consider:

CN: develop the manufacturing technology to obtain 1000 bearing bushings, in accordance with the information found in the mechanical drawing and the available processing conditions.

In this way, *the functional requirement of zero order* could be:

*FR*0: design the technological process of manufacturing 1000 bearing bushings following the mechanical drawing and the endowment of the manufacturing workshop.

The functional requirements of first order could have the following shape:

*FR*1: Analyze the mechanical drawing of the bearing bush;

*FR*2: Verify the part manufacturability;

*FR*3: Establish the type of workpiece for manufacturing the bushings;

*FR*4: Establish the technological itinerary (the succession of operations and phases);

*FR*5: Determine the values of the machining allowances and intermediate dimensions;

*FR*6: Establish the values of the operating parameters;

*FR*7: Calculate the operation time;

*FR*8: Evaluate the manufacturing process efficiency;

*FR*9: Complete the technological documentation.

Because the mechanical design was developed in a design office where engineers with extended experience were working and there were inclusively standards referring to the design of bearing bushings [6,7], it has considered that the mechanical drawing is complete and ensures the information necessary to pass to the next stage. On the other hand, similar arguments were considered to evaluate the bush manufacturability. In fact, as it has noticed later, some less convenient aspects concerning the bush manufacturability were observed when the problem of establishing the technological itinerary was approached.

As mentioned above, a tubular workpiece obtained by casting was taken into consideration. The selection was evaluated as convenient since the chips obtained during the machining process and eventually wrong parts could be recasted just in a metallurgical workshop existing also in the university.

In this way, the fourth functional requirement was approached.

As *functional requirements of second order*, it was considered:

*FR*4.1: Establishhow to get the revolution and flat surfaces;

IOP Conf. Series: Materials Science and Engineering 1174 (2021) 012009 doi:10.1088/1757-899X/1174/1/012009

*FR*4.2: Establish how to get the grease groove;

*FR*4.3: Establish how to control the part from the technical point of view;

*FR*4.4: Establish how to prepare the machined parts for transport to the beneficiary;

The considered *functional requirements of third order* could be the following:

*FR*4.1.1: Establish how to get the flat surface;

*FR*4.1.2: Establish how to get a rough outer cylindrical surface;

*FR*4.1.3: Establish how to get a finish outer cylindrical surface;

*FR*4.1.4: Establish how to get the conical surfaces of the outer cylindrical surface, at the right end of the tube;

*FR*4.1.5: Establish how to get the rough inner cylindrical surface;

*FR*4.1.6: Establish how to get the finish inner cylindrical surface;

*FR*4.1.7: Establish how to get the conical surface at the inner cylindrical surface, at the right end of the tube;

*FR*4.1.8: Establish how to cut the part from the tube type workpiece;

*FR*4.1.9: Establish how to get the flat surface at the left end of the detached bushing;

*FR*4.1.10: Establish how to get the conical surfaces of the outer cylindrical surface, at the left end of the part;

*FR*4.1.11: Establish how to get the conical surfaces of the inner cylindrical surface, at the left end of the part;

*FR*4.2: Establish how to get the longitudinal grease groove;

*FR*4.3: Establish how to control of the machined part from technical point of view;

*FR*4.4. Establish how prepare the machined parts for transport to the beneficiary.

Using the zigzag process, design parameters appropriate to the functional requirements were identified.

Thus, the general design matrix of the bearing bushes manufacturing process is presented in table 1. It can be observed a decoupled design, the result being a diagonal matrix.

Some particular solutions of identifying the design parameters in the case of functional requirement *FR*4 can be seen in table 2. In the decomposition and zigzagging process, it was observed that turning the flat surface at the right end of the bush and the straight turning of the outer cylindrical surface can be made using the same lathe tool. Also, the same tool can be used both for finishing the outer cylindrical surface and for chamfering the outer cylindrical surface at the right end of the bearing bush. The lathe tool used for boring contributes both to the finishing turning of the inner cylindrical surface and to the chamfering of the inner cylindrical surface at the right end.

After cutting the part from tube type workpiece, the tool used for flat turning also provides conditions for chamfering the outer cylindrical surface at the left end of the bearing bush in the mechanical drawing.

Under the above conditions, it is found that a triangular design matrix is obtained, as some lathe tools are used to meet several functional requirements.

It was necessary to solve some technological difficulties for the realization of grease groove. The initial variants of machining processes considered either did not ensure a sufficiently high productivity (e.g., use of electrical discharge machining) or it was not possible to use them under normal conditions (slotting or broaching). A solution was identified based on the use of an end mill, which does not ensure the realization of flat front surfaces at both ends of the grease groove. However, given the acceptable productivity of the process and with the consent of the beneficiary, it was preferred to end mill processing located, during machining process, in an inclined position relative to the symmetry axis of the bearing bush.

Table 1. General design matrix.

Line no. 1	**Design parameters**		**Design parameters**								
2			**Design parameters of zero order**								
3			*DP*0: Checks performed and documentation completed								
4			**Design parameters of the second order**								
5	**Functional requirements**		*DP*1: Verified mechanical drawing	*DP*2: Verified part manufacturability	*DP*3: Determined workpiece	*DP*4: Established technological itinerary	*DP*5: Determined values of the machining allowances and intermediate dimensions	*DP*6: Determined values for the operating parameters	*DP*7: Calculated operation time	*DP*8: Evaluated manufacturing process efficiency	*DP*9: Completed technological documentation
6 Column no. 1	2	3	4	5	6	7	8	9	10	11	12
7	Functional requirement of the zero order	Functional requirements of the first level	Design parameter *DP* that corresponds to each functional requirement								
8		*FR*1: Analyze the mechanical drawing of the bearing bush	X								
9		*FR*2: Verify the part manufacturability		X							
10		*FR*3: Establish the type of workpiece for manufacturing the bushings			X						
11	*FR*0: Design the technological process of manufacturing 1000 bearing bushings following the mechanical drawing and the endowment of the manufacturing workshop	*FR*4 : Establish the technological itinerary				X					
12		*FR*5: Determine the values of the machining allowances and intermediate dimensions					X				
13		*FR*6: Establish the values of the operating parameters						X			
14		*FR*7: Calculate the operation time							X		
15		FR8: Evaluate the manufacturing process efficiency								X	
16		*FR*9: Complete the technological documentation									X

IOP Conf. Series: Materials Science and Engineering 1174 (2021) 012009 doi:10.1088/1757-899X/1174/1/012009

Table 2. Design matrix for the functional requirement *FR*4.

Line no. 1	**Design parameters**			**Design parameters**									
2				**Design parameters of second order for the functional requirement FR4**									
3				*DP*4.1: Lathe							*DP* 4.2: End milling tool	*DP* 4.3: Tools for checking dimensions, shape and roughness	*DP* 4.4: Oil impregnation tools
4				**Designparameters of third order**									
5	**Functional requirements**			*DP*4.1.1: Lathe tool for facing and straight rough turning	*DP*4.1.3: Straight finish turning tool	*DP*4.1.5: Rough boring tool	*DP*4.1.6: Finish boring tool	DP4.1.8: Cutoff tool	*DP*4.1.9: Facing tool	DP4.1.10 Finish boring tool			
6 Column no. 1	2	3	4	5	6	7	8	9	10	11	12	13	14
7	Functional requirement of the first order	Functional requirements of the second order	Functional requirements of the third order	Design parameter *DP* that corresponds to each functional requirement									
8			FR4.1.1: Establish how to get the flat surface	X									
9			FR4.1.2: Establish how to get a rough outer cylindrical surface	X									
10	*FR*4: Establish the technological itinerary	*FR*4.1: Establish how to get the revolution and flat surfaces	FR4.1.3: Establish how to get a finish outer cylindrical surface		X								
11			FR4.1.4: Establish how to get the conical surfaces of the outer cylindrical surface, at the right end of the tube		X								
12			FR4.1.5: Establish how to get the rough inner cylindrical surface			X							

13	FR4.1.6: Establish how to get the finish inner cylindrical surface	X						
14	FR4.1.7: Establish how to get the conical surface at the inner cylindrical surface, at the right end of the tube	X						
15	FR4.1.8: Establish how to cut the part from the tube type workpiece		X					
16	*FR*4.1.9: Establish how to get the flat surface at the left end of the detached bushing			X				
17	*FR*4.1.10: Establish how to get the conical surfaces of the outer cylindrical surface, at the left end of the part			X				
18	*FR*4.1.11.Establish how to get the conical surfaces of the inner cylindrical surface, at the left end of the part				X			
19	*FR*4.2: Establish how to get the grease groove					X		
20	*FR*4.3: Establish how to control the part from the technical point of view						X	
21	*FR*4.4: Establish how to prepare the							X

IOP Conf. Series: Materials Science and Engineering 1174 (2021) 012009 doi:10.1088/1757-899X/1174/1/012009

	machined parts for transport to the beneficiary

4. Using the second axiom to select the alternative usable for finishing the inner cylindrical surface

According to the information found in the part mechanical drawing, the inner surface must have a diameter $\emptyset 40\ E7^{+0.075}_{+0.050}$. This means that the maximum size will be 40.075, the minimum 40.050 mm, and the size corresponding to the middle of the tolerance field will be 40.0625 mm.

A grinding operation cannot be used, due to the risk of clogging the abrasive tool with chips detached from the workpiece material (bronze). The following versions of finishing the inner surface of the bearing bush were taken into consideration:

1. Finishing turning in the usual conditions offered by the universal lathe, so without the use of work devices that ensure the increase of the accuracy of adjusting the position of the lathe tool tip to the working dimension. The machining will be performed by the method of test chips, which means a succession of measurements and successive machining sequences, until obtaining a diameter of the processed surface in the tolerance field. The size of the division that can be observed and used by the operator to adjust the radial position of the lathe tool tip when moving the transverse slide is 0.05 mm. The various corrections of the position of the tool tip will be affected also by the subjectivity of the lathe operator in the appreciation of some fractions of the distance between two divisions that correspond to an advancement of the transverse slide by 0.05 mm. Usually, the lathe operator shows a tendency to overlap the adjustment position of the lathe tool tip with the middle of the prescribed tolerance field[1,9]. It is expected, as such, that the distribution of the diameters of the inner surfaces of parts is carried out in accordance with a Gaussian distribution, but with a fairly large distribution field, due to the use of the scale with the division of 0.05 mm. Such a distribution corresponds, in fact, to tolerance class 9 (class usually appreciated as achievable by finishing turning). Taking into account a dispersion field corresponding to this class (T = 0.039 mm) and the value corresponding to the middle of the tolerance field prescribed on the mechanical drawing (valid for tolerance class E7), we find that the actual dimensions should be between a minimum value of

$$40.0625-(1/2)\cdot 0.062=40.0625-0.031=40.0315 \text{ mm} \tag{3.1}$$

and a maximum value

$$40.0625+(1/2)\cdot 0.062=40.0625+0.031=40.0935 \text{ mm}. \tag{3.2}$$

The Gaussian distribution function (which provides information on obtaining a certain dimension between certain values, so on the probability of having a dimension in a predetermined range) is as follows:

$$P = \frac{1}{\sigma\sqrt{2\pi}} \int_{x_1}^{x_2} e^{-\frac{x-M(x)}{2\sigma^2}} dx, \tag{3.3}$$

where x is the diameter of the processed surface, σ- the standard deviation, and $M(x)$ - the arithmetic mean of the measured values.

Taking into account the values valid in the analyzed case, we will have:

$$P = \frac{1}{\sigma\sqrt{2\pi}} \int_{40.0315}^{40.0935} e^{-\frac{x-40.0625}{2\sigma^2}} dx. \tag{3.4}$$

Determining with the help of statistical calculation tables the proportion of the number of diameters achievable by ordinary finishing turning (i.e., with a tolerance field corresponding to tolerance class 9), but within a dimensional range corresponding to class 7, we find that the probability is 77.2% according to what will be shown later.

2. The second version of finishing turning could consider the use of a device using a dial gauge with a value between two divisions of 0.001 mm to adjust the position of the lathe tool tip at the finishing passes necessary to obtain the inner surface by the trial chip method. This device can be mounted on the guide of the transverse slide of the lathe (figure 2), in order to be able to obtain more precise information regarding the movement of the transverse slide, together with the tip of lathe tool, when adjusting the position of tip of the inner turning lathe tool.

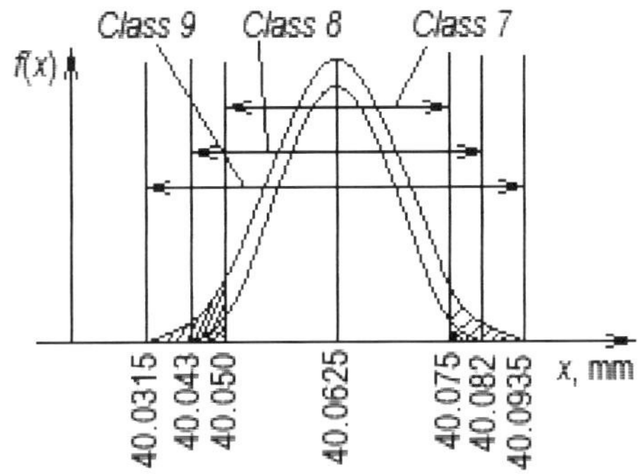

Figure 3. Gauss distributions valid for certain tolerance classes.

It can be appreciated that the use of the dial gauge with a division of 0.001 mm will result in an increase in machining accuracy, and it is possible to achieve tolerances corresponding to tolerance class 8. If this tolerance class 8 corresponds to a tolerance field$T$ = 0.039 mm arranged symmetrically in relation to the mean size to be obtained according to the mechanical drawing (ϕ40.0625), it means that the minimum size of the new distribution field could be:

$$40.0625\text{-}\ (1/2)\cdot 0.039 = 40.0625\text{-}0.0195 = 40.0430\ \text{mm} \tag{3.5}$$

and the maximum size could be:

$$40.0625+ (1/2)\cdot 0.039) = 40.0625 + 0.0195) = 40.0820\ \text{mm}. \tag{3.6}$$

Determining the probability of having dimensions within the prescribed tolerance field(figure 3) will lead to:

$$P_8 = \frac{1}{\sigma\sqrt{2\pi}}\int_{40.0430}^{40.0820} e^{-\frac{x-40.0625}{2\sigma^2}}\,dx. \tag{3.7}$$

Assuming that we could obtain an accuracy of the dimensions corresponding to class 8 (which means a dispersion field of 0.039 mm, but arranged around the average size ϕ40.0625) and using the tables that provide the values of the Laplace function, we arrive at:

$$P_8 = 2\oint_0^{40.082}(u) = 2\cdot 0.472 = 0.944, \tag{3.8}$$

dealing with a value u = 1.923, for a standard deviation:

$$\sigma = 0.039/6 = 0.065. \tag{3.9}$$

This means that there is a 94.4% probability of obtaining machined surface diameters within the prescribed tolerance field or that about 5.6% of the diameters will not fall within the prescribed tolerance field.

In reality, however, even when using the devices to increase the machining accuracy by finish turning on the available universal lathe, it is difficult to obtain a tolerance field corresponding to tolerance class 8 and the actual dispersion is expected to correspond to class 9 (which will correspond to a tolerance/distribution field of 0.062 mm, i.e., between the dimensions 40.0315 mm and 40.0935 mm). In this way, we will reach:

$$P_9 = 2\oint_0^{40.0935}(u) = 2 \cdot 0.386 = 0.772, \quad (3.10)$$

in the calculation using a value $u = 1.21$, for a standard deviation

$$\sigma = 0.062/6 = 0.0103 \text{ mm}. \quad (3.11)$$

This means a probability of obtaining diameters within the tolerance field of 77.2% or a risk that 22.8% of the diameters will not fall within the prescribed tolerance range.

3. *The third version* could involve the use of both the device which includes the dial gauge and a go plug gauge to detect the moment when, by successively removing layers of material from the tubular workpiece, the diameter of the machined surface becomes slightly larger than the prescribed minimum diameter. It is obvious that the new dispersion field will be framed between the minimum size corresponding to the prescription on the mechanical drawing (ϕ 40.050) and a maximum size that takes into account a tolerance field specific to tolerance class 8 (due to the use of the dial gauge: 40.050+ 0.39 = 40.089 mm). According to the information in the literature [1], due to the use of plug gauge go and the tendency of the lathe operator to stop machining as soon as it has obtained a size larger than the minimum diameter, there is a probability that the surface diameter will fall in the prescribed tolerance field and we will no longer have a dispersion corresponding to the model constituted by the Gaussian curve. Now, we will have to consider a Maxwell-Boltzmann distribution (figure 4), which is characterized by a certain degree of asymmetry. The mathematical relation corresponding to a distribution function (probability) of the Maxwell-Boltzmann distribution has the form:

$$P = \frac{2}{\sqrt{\pi}}\int_0^{\frac{x}{\sqrt{2}a}} e^{-x^2}dx - \sqrt{\frac{2}{\pi}} \cdot \frac{xe^{-x^2}/(2a^2)}{a}, \quad (3.12)$$

where$a$=1 or$a$=2.

As can be seen in figure 4, outside the prescribed tolerance field (40.050 mm – 40.075 mm) will remain the parts in which the diameters of the inner surfaces are in the field (40.075 mm – 40.089 mm). Accepting a value$a$ = 2, and using a method of graphically approximating the size of the surface between the curve corresponding to the Maxwell-Boltzmann distribution and the abscissa axis, it was found that the interval (40.075-40.089) will correspond to a percentage of 7.9% of the entire surface located between the curve and the abscissa axis in the field (40.050-40.089), which means that 7.9% of the parts will have dimensions outside the prescribed tolerance field.

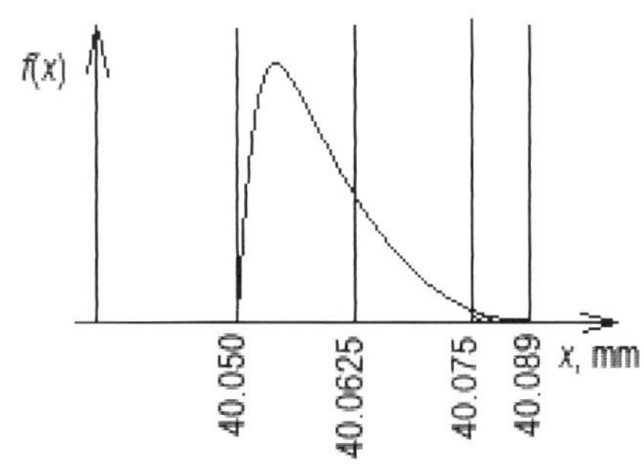

Figure 4. Maxwell-Boltzmann distribution valid when using a smooth plug gauge go.

Summarizing the above considerations, by using the device that includes a dial gauge to increase machining accuracy, we will have a 77.2% probability of obtaining diameters of machined surfaces within the prescribed tolerance range. However, if the device including the dial gauge and the check of machined surfaces with that plug gauge go is used, the probability of obtaining diameters within the prescribed tolerance range reaches 92.1%.

Referring to the second axiom of axiomatic design, it is obvious that for the finishing of the inner surface in the bushing will be preferred the version that provides the use of the device with dial gauge and a plug gauge go to check the acceptability of the bushing if the tolerance field of the hole diameter is taken into account.

5. Conclusions

Bearing bushings are parts used in industrial practice to support rotating shafts. These bushings can be obtained by applying various manufacturing processes. Such a process may be, for example, based in principle on turning pipe-shaped blanks. To establish the sequence of processing operations, axiom I of axiomatic design was used. In this respect, different design parameters were taken into account to meet the functional requirements. On the other hand, it has been found that even for obtaining a certain surface by turning, different processes can be used, if the needs of reaching the machining accuracy established by the indications on the mechanical drawing of a bearing bush are taken into account. To select the most convenient procedure, the second axiom of axiomatic design was used. As such, a machining process has been adopted that offers the highest probability of obtaining hole diameters within the prescribed tolerance field. In this way, the process that is most convenient for machining the surface of the bushing hole was identified, by taking into account the propermachining conditions. The selected procedure is based on the use of a device incorporating a dial gauge to increase the accuracy of adjusting the position of the tool tip to the working size and a plug gauge go for testing the diameter of the machined hole. In the future, it is intended to analyze the possibilities of using the second axiom for selecting machining processes for other bushing surfaces, when several such processes can be used to obtain a certain surface.

References

[1] Picoş C 1974 *Machine manufacturing technology* (in Romanian) (Bucharest: Editura Didactică şi Pedagogică)
[2] Morozov A, Fedotov G and Kundrotas K 2020 *IOP Conf. Ser. Mater. Sci. Eng.* **709** 3 033058
[3] Behrens B A, Maier H J, Poll G, Wriggers P, Aldakheel F, Klose C, Nürnberger F, Pape F, Böhm C, Chugreeva A, Coors T, Duran D, Thürer S E, Herbst S, Hwang J I, Matthias T, Heimes N and Uhe J 2020 *Prod. Eng. Res. Devel.* **14** 569–581
[4] Behrens B A, Sokolinskaja V, Chugreeva A, Diefenbach J, Thürer S and Bohr D 2019 *AIP Conference Proceedings***2113** 040028
[5] Behrens B A, Uhe J, Petersen T, Klose C, Thürer S E, Diefenbach J, Chugreeva A 2021 *Materials***14** 4 803
[6] Alisin V 2020 *IOP Conf. Series: Earth and Environmental Science* 548 052019
[7] STAS 772-6 1967 *Sliding bearings. Bushes. Typification* (Romanian standard; in Romanian)
[8] Popinceanu N G, Răileanu A, Gafiţanu M, Iliescu G, Merticaru V, Hostiuc L, Hantelman M, Diaconescu E, Rădăuceanu D. 1969 *The accuracy of machine elements. Design Handbook* (in Romanian) (Iaşi: Institutul Politehnic)
[9] Komissarov V I, Leontiev V I., Starostin V G 1968 *Dimensional adjustment of universal cutting machine tool* (in Russian) (Moscow: Mashinostroenie)

Technical Design Approach for Autonomous Outdoor Transport Systems based on an Extension of Axiomatic Design using Metrics

D Clauer[1], J Fottner[1], E Rauch[2], G Roman[2] and T Irrenhauser[3]

[1] Technical University Munich, Department of Mechanical Engineering, Chair of Material Handling, Material Flow, Logistics, 85748 Garching, Germany
[2] Free University Bozen-Bolzano, Faculty of Science and Technology, Industrial Engineering and Automation (IEA), 39100 Bolzano, Italy
[3] Bayerische Motoren Werke (BMW) AG, 80788 Munich, Germany

E-mail: dana.clauer@tum.de

Abstract. This contribution focuses on the technical design of autonomous outdoor transport systems operating on factory premises. Outdoor applications present a high degree of complexity, which is reflected in the technical design of these vehicles. The requirements they must fulfill are higher than for indoor applications. Hence, the results of the technical design are crucial to enable the correct use of such systems. Axiomatic Design hereby offers a methodological basis for the systematic derivation of solution parameters for the technical design of autonomous outdoor transport systems. Functional metrics are introduced in addition to the conventional Axiomatic Design method to support the formulation of functional requirements and to reduce the subjectivity in the design process. The technical design approach that was developed is applied in the form of a framework.

1. Introduction

The term automated guided vehicle (AGV) is widely used in both literature and practice [1]. Conventional AGVs perform logistics tasks, such as transporting and handling materials that are assigned to them in an automated manner. They do so by following a pre-defined route within a defined action area [2]. Extensions of these vehicles are often described in literature as autonomous mobile robots (AMRs), autonomous intelligent vehicles (AIVs) or autonomous transport systems [3]. They represent a smarter and more intelligent version of an AGV [4]. This means that they have a decentralized planning, execution and control and help optimize internal material and information flows by cooperating and interacting with other systems and with humans [5]. However, since they not only drive indoors but also outdoors on factory premises, and since they also act autonomously, they have to satisfy significantly higher requirements, which are reflected in the complexity of their technical design.

Adaptable, comprehensive and intelligent transport solutions are required to meet the increasing demands on the flexibility of material supply chains within logistics [6]. Autonomous outdoor transport systems are one possible answer to this. However, the technical design of these systems is highly complex [7]. Even if single requirements are known, it is still not clear how these should be organized and which technological solutions are the best for certain requirements. Axiomatic Design (AD) provides a systematic research method to tackle this issue, generating a structured technical design approach for autonomous outdoor transport systems. Based on the collection of customer needs (CNs) and the derivation of functional requirements (FRs), the research team also introduces and adopts functional metrics (FMs) to complement FRs. Thus, it is more likely that the chosen design solutions will successfully satisfy the CNs [8]. Furthermore, FMs allow the most suitable design parameters (DP) to be selected for a certain FR. According to the decomposition process (zig-zagging and mapping), FMs are assigned at each hierarchy level to the respective FRs, for which DPs are then generated. The aim of this paper is thus to present a technical design approach for autonomous outdoor transport systems in the form of a framework based on an extension of AD with the aid of metrics.

2. Methodology

In addition to the traditional use of AD according to Suh [9], FMs were also implemented in this research work. Several authors have already dealt with the use of metrics within AD. These include Brown and Henley, Cochran et al. and Melnyk et al., who refer to metrics as an important link between strategy, execution and eventual value creation [8, 10-13]. FMs are used for quality assurance and suitability verification of the individual FRs and the respective DPs. Metrics ensure that if multiple DPs are selected, the most appropriate DP for a given FR will be chosen. Moreover, they support the formulation of the FRs and consolidate their informative value by also strengthening the independence axiom [8]. Thus, we are able to reduce subjectivity in the design process and provide a basis for comparison.

The methodology for determining FMs in this paper is based on the approach developed by Henley and Brown [11]. The authors derive an FM for each FR-DP pair at each hierarchy level of the decomposition. However, before we can proceed with the gathering of CNs as well as the mapping and decomposition of DPs and FRs, we first present the methodology needed to derive the technical design framework of autonomous outdoor transport systems using metrics (figure 1).

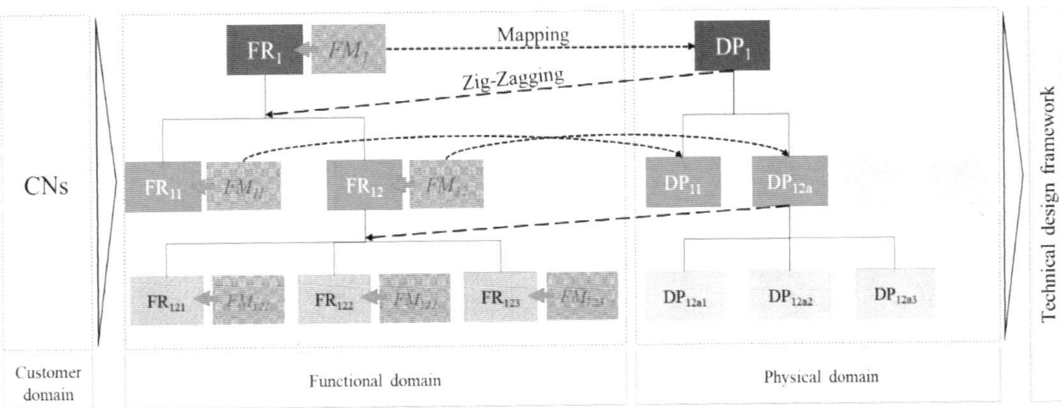

Figure 1. Derivation of the design approach using Axiomatic Design and metrics.

As implied by the AD approach, the two axioms (independence and information) serve to verify FRs and avoid coupled designs [14]. To reinforce the independence axiom, we additionally attribute an FM to each FR on each hierarchical level. On this basis, the research team implements the following five steps: determination of CNs (i), transformation of CNs into FRs at the top-level (ii), application of FMs at the top-level (iii), mapping of the DPs to the top-level FRs (iv) and finally decomposition (zig-zagging) and mapping of lower FRs using FMs according to the top-down principle (v).

2.1. Determination of CNs

The CNs used in this study are adapted from the previous work of Clauer et al. [15]. A systematic literature review and semi-structured interviews about the implementation of outdoor AMRs were used to identify these CNs. However, since this paper focuses primarily on the technical design of autonomous outdoor transport systems, and not on the overall implementation process, the following attribute is selected: provide solutions for autonomous material handling on factory premises.

2.2. Transformation of CNs into FRs at the top-level
Based on the representative CN designated in the previous subsection, it is possible to define the FRs at the top-level that represent the key requirements. Considering the objective of this work, the top-level key requirement (FR_0) for the technical design of an autonomous outdoor transport system is defined as: "Handle and transport material autonomously on the factory premises". To satisfy this requirement, FR_0 is assigned a solution (DP_0) in the physical design domain: "Technical design framework for autonomous outdoor transport systems on the factory premises". The proposed DP_0 is formulated in a very general and abstract way. Therefore, it has to be decomposed into further FRs to prompt concrete parameters and solutions for the design framework:

FR_1 Certify sturdy hardware components for outdoor use
FR_2 Ensure robust and sustainable navigation software for the outdoor environment
FR_3 Interact independently in a dynamic environment

2.3. Application of FMs at the top-level
According to the approach shown in Figure 1, we first assign FMs to the top-level FRs before formulating their respective DPs. The assigned FMs aim at identifying the most appropriate DP for a given FR. The correctness and meaningfulness of the metrics are indispensable in order to achieve this [13]. For this purpose, the following FMs are assigned to the top-level FRs:

FM_1 Mean time between failures (MTBF)
FM_2 Availability in the operative environment
FM_3 Availability in a highly complex environment

To ensure that the individual hardware components of the vehicle are robust enough for outdoor use and are thus able to withstand different weather conditions (snow, ice, rain or fog) as well as road conditions (bumps, curbs or ruts), the MTBF is identified as an appropriate metric. The higher the MTBF, the more reliable and robust the components will be. The second FR focuses on the software part of the transport systems. The higher the availability in the operating environment, the more robust and sustainable the navigation software. In addition to hardware and software requirements, independent interaction in a dynamic environment characterizes the autonomy of the vehicle and thus represents an elementary pillar of the technical design framework. The availability in a highly complex environment beyond the standard operational setting is a suitable metric for FR_3 to measure this requirement.

2.4. Mapping DPs to the top-level FRs
According to Suh, a central feature of AD during the mapping and decomposition process of DPs is a solution-neutral formulation of FRs. The goal is to think creatively and freely about possible solutions without being influenced by existing or seemingly obvious solutions, thus restricting the design space [16]. Top priority is always given to compliance with the independence axiom and the information axiom as well as the consideration of the collectively exhaustive and mutually exclusive approach (CEME) according to Thompson [17]. Starting from the top-level key requirement FR_0 and its design solution DP_0, as well as the previously derived top-level FRs including their FMs, we are now able to derive DPs at the first hierarchy level. Figure 2 illustrates the FR-DP tree at the top-level.

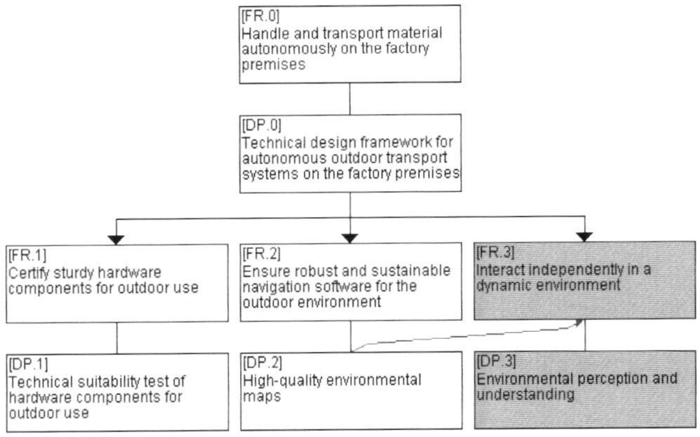

Figure 2. Graphical representation of the FR-DP tree at top-level using Acclaro DFSS.

To verify whether the independence axiom is fulfilled, the top-level FRs and DPs are expressed in a matricial form inside a design matrix (DM) [18]. This DM, presented in Equation 1, is characterized by a decoupled design in the form of a triangular matrix.

$$\begin{Bmatrix} FR_1 \\ FR_2 \\ FR_3 \end{Bmatrix} = \begin{bmatrix} X & 0 & 0 \\ 0 & X & 0 \\ 0 & X & X \end{bmatrix} \begin{Bmatrix} DP_1 \\ DP_2 \\ DP_3 \end{Bmatrix} \quad (1)$$

The resulting decoupled triangular matrix shown in equation 1 illustrates the connection between DPs and FRs. The technical suitability testing of hardware components (DP_1) fulfills FR_1 and can be functionally verified using the MTBF (FM_1). DP_1 has no relation to any other subsequent FR, since it refers exclusively to the frame, chassis and housing of the transport system. High-quality environmental maps with a robust and accurate localization (DP_2) have an impact on the independent interaction in a dynamic environment (FR_3). The more accurate the environmental maps, the more independently a transport system can operate within a dynamic environment. Moreover, a closer look shows that the dependencies existing between the FRs and DPs create manageable feedback loops if the DPs are implemented in the correct order from left to right [19-20].

2.5. Decomposition and mapping of lower FRs using FMs

The proposed solutions ($DP_{1\text{-}3}$) are still too abstract for a concrete design framework. Therefore, they are decomposed into further sub-requirements. Starting from DP_1, the following lower level FR-DP-FM trios for determining technical suitability testing (DP_1) are derived (table 1).

Table 1. Decomposition of DP_1 – technical suitability test of hardware components for outdoor use.

FR		DP		FM	
FR_{11}	Protect against the intrusion of particles and water	DP_{11}	Protection rating and protection class	FM_{11}	IP-Level
FR_{12}	Test mechanical stability and durability	DP_{12}	Virtual mock-up tests of designed components	FM_{12}	Performance limits of components
FR_{13}	Employ reliable mechatronic components	DP_{13}	Endurance tests	FM_{13}	Failure rate

High-quality environmental maps with a robust and accurate localization (DP_2) are also decomposed into lower FRs. Table 2 shows the FR-DP-FM trios for DP_2.

Table 2. Decomposition of DP_2 – high-quality environmental maps.

FR		DP		FM	
FR_{21}	Send/receive information about the current location	DP_{21}	Localization technologies	FM_{21}	Precision and accuracy of the data
FR_{22}	Determine own position on a map at any time	DP_{22}	Detailed and synchronized map	FM_{22}	Deviation compared to real position (+/- 1 m)
FR_{23}	Reduce sensor noise	DP_{23}	Suitable software and hardware solutions	FM_{23}	Signal-to-Noise-Ratio (SNR)

Localization technologies, especially for outdoor environments, have a great impact on the robust and accurate localization of an autonomous transport system. DP_{21} therefore needs to be further decomposed to a third hierarchical level so as to obtain accurate and concrete DPs (table 3).

Table 3. Decomposition of DP_{21} – localization technologies.

FR		DP		FM	
FR_{211}	Determine the position in a fixed reference system	DP_{211}	GNSS	FM_{211}	Deviation from the real value (+/- 1 m)
FR_{212}	Determine the distance covered	DP_{212}	IMU and odometry	FM_{212}	Precision and accuracy of the calibration
FR_{213}	Determine the position relative to a known position	DP_{213}	Combination of visual sensors with environmental objects	FM_{213}	Robust recognition

Detailed and synchronized maps (DP_{22}) include further sub-requirements, which are shown in table 4.

Table 4. Decomposition of DP_{22} – detailed and synchronized map.

FR		DP		FM	
FR_{221}	Calibrate robot position	DP_{221}	Definition of reset positions (start/finish/wait position)	FM_{221}	Accuracy from known positions
FR_{222}	Update map continuously	DP_{222}	SLAM algorithms	FM_{222}	Update frequency
FR_{223}	Define possible driving areas	DP_{223}	Definition of driving and restricted areas on the map	FM_{223}	Clarity of boundaries

DP_2 has now been broken down to a level of detail suitable for the technical design framework. The last decomposition within the technical design field of an autonomous outdoor transport system concerns the environmental perception and understanding, which represents the final pillar of the framework. It is further decomposed into three sub-requirements to once again obtain feasible physical solutions.

Table 5. Decomposition of DP_3 – environmental perception and understanding.

FR		DP		FM	
FR_{31}	Obtain sufficient information from sensors	DP_{31}	Data from cameras and laser scanners (2D/3D)	FM_{31}	Density of the point clouds
FR_{32}	Process sensor data	DP_{32}	Sensor fusion	FM_{32}	Quality of results

FR_{33}	Respond to different scenarios	DP_{33}	Definition of response scenarios based on the environment	FM_{33}	Responsiveness in time measurement unit

3. Technical design framework

The proposed technical design framework for autonomous outdoor transport systems results from the extracted DPs, which in turn can be traced back to the corresponding FRs and FMs. The framework is based on three main pillars: first, technical suitability testing of hardware components for outdoor use; second, high-quality environmental maps; and third, environmental perception and understanding through sensor technology with corresponding response scenarios. Each pillar includes different solutions that must be considered in the technical design of an autonomous outdoor transport system. It should be noted that the order of the pillars also plays an important role during the implementation process. Due to the connection between DP_2 and FR_3, failure to follow the predefined order can lead to confusing circular reasoning, characterized by the violation of the independence axiom.

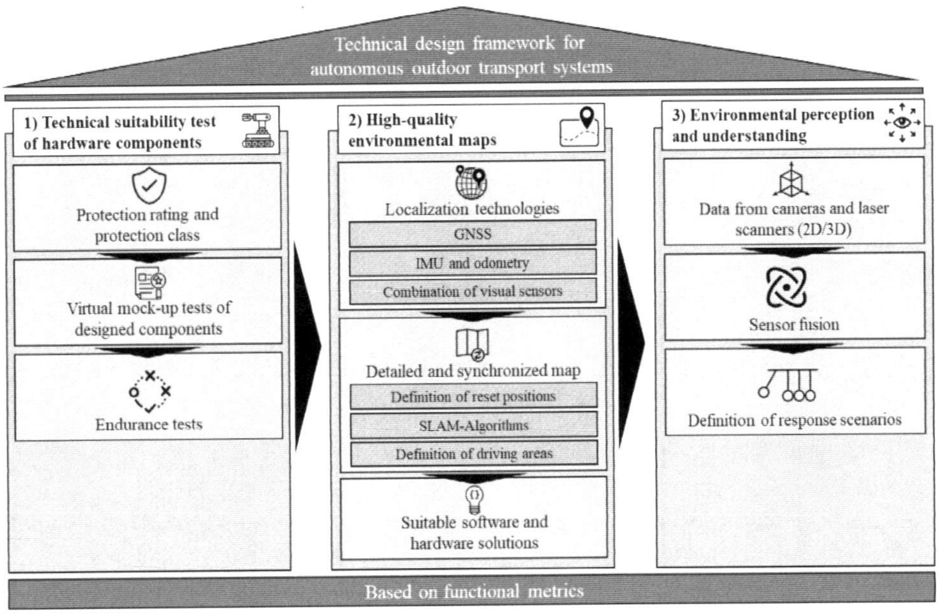

Figure 3. Proposed technical design framework for autonomous outdoor transport systems.

4. Conclusion

In connection with the numerous challenges and difficulties faced by autonomous transport systems in an outdoor environment, demands on the technical design of these vehicles have grown substantially. To counter this increasing degree of complexity, in particular during the design phase, the present work presents an extension of the AD method with the use of FMs. Through their implementation in the approach, FMs support the mapping process of DPs and reinforce the importance of FRs. This ensures an optimal choice of DPs with regard to CNs. This systematic approach enables the identification and classification of three main technological requirements that characterize the technical design of autonomous outdoor transport systems. On the basis of these findings, the research team built and presented a framework based on three pillars for the technical design of autonomous outdoor transport systems, laying the foundation for further work in this field.

5. Acknowledgement

This research has been funded by the Bayerische Motoren Werke (BMW) AG (F37917J). We extend our sincerest thanks to the BMW AG for the comprehensive support for this work.

References

[1] Ullrich G 2015 *Automated Guided Vehicles Systems* (Berlin Heidelberg: Springer Berlin Heidelberg) pp 4–14
[2] Fedorko G, Honus S and Salai R 2017 Comparison of the traditional and autonomous AGV systems *MATEC Web of Conf.* **134** p 13
[3] Rubio F, Valero F and Llopis-Albert C 2019 A review of mobile robots: concepts, methods, theoretical framework, and applications *Int. J. of Advanced Robotic Systems* **15** pp 1–2
[4] Fragapane G, Ivanov D, Peron M, Sgarbossa F and Strandhagen J O 2020 Increasing flexibility and productivity in industry 4.0 production networks with autonomous mobile robots and smart intralogistics *Annals of Operations Research* **18** p 15
[5] Fottner J et al. 2021 Autonomous systems in intralogistics – state of the art and future research challenges *J. Logistics Research* p 3
[6] Pantano M, Regulin D, Lutz B and Lee D 2020 A human-cyber-physical system approach to lean automation using an industrie 4.0 reference architecture *Proc. Manufacturing* **51** pp 1082–90
[7] Nassiraei A A F and Ishii K 2007 Concept of intelligent mechanical design for autonomous mobile robots *J. of Bionic Engineering* **4** pp 217–26
[8] Brown C A and Henley R 2016 Metrics for developing functional requirements and selecting design parameters in axiomatic design *Proc. CIRP* **53** pp 113–18
[9] Suh N P 1990 *The Principles of Design* (New York: Oxford University)
[10] Henley R 2015 Using functional metrics to facilitate designing collectively exhaustive mutually exclusive systems in the context of managing return on investment *Proc CIRP* **34** pp 31–36
[11] Henley R and Brown C A 2016 Axiomatic design applied to play calling in american football *Proc. CIRP* **53** pp 206–12
[12] Cochran D S, Hendricks S, Barnes J and Bi Z 2016 Extension of manufacturing system design decomposition to implement manufacturing systems that are sustainable *J. of Manufacturing Scinece and Engineering* **138** 11 p 249
[13] Melnyk S A, Stewart D M and Swink M 2004 Metrics and performance measurement in operations management: dealing with the metrics maze *J. of Operations Management* **22** pp 209–18
[14] Suh N P 1995 Axiomatic design of mechanical systems *J. of Mechanical Design* **117** chapter B pp 2–10
[15] Clauer D, Fottner J, Rauch E and Prueglmeier M 2021 Usage of autonomous mobile robots outdoors – an axiomatic design approach *Proc. CIRP* **96** pp 242–47
[16] Suh N P 2001 *Axiomatic Design – Advances and Applications* (New York: Oxford University)
[17] Thompson M K 2013 *ICAD 2013 Proc. of the 7th Int. Conf. on Axiomatic Design* (Worcester, Massachusetts, USA)
[18] Suh N P 2005 *Complexity – Theory and Applications* (Oxford: Oxford University Press)
[19] Rauch E, Spena P R and Matt D T 2019 Axiomatic design guidelines for the design of flexible and agile manufacturing and assembly systems for SMEs *Int. J. on Interactive Design and Manufacturing* **13** pp 1–22
[20] Lee T and Jeziorek P N 2006 Understanding the value of eliminating an off-diagonal term in a design matrix *4th Int. Conf. on Axiomatic Design* (Firenze)

Axiomatic Design applied to CRUD matrix in Information Systems

Luís Cavique[1,2,5], **Miguel Cavique**[3, 4]

[1] Universidade Aberta, Rua da Escola Politécnica, 147, 1269-011 Lisboa, PORTUGAL
[2] LASIGE, Faculdade Ciências de Lisboa, Campo Grande, 1749-016 Lisboa, PORTUGAL
[3] R&D Unit for Mechanical and Industrial Engineering - UNIDEMI, NOVA School of Science and Technology, Campus de Caparica, 2829-516 Caparica, PORTUGAL
[4] Naval Academy, Base Naval de Lisboa, Alfeite, 2810-001 Almada, PORTUGAL
[5] Corresponding author, mailto: luis.cavique@uab.pt

Abstract. In Information Systems (IS), the search for a good design is still a relevant issue. In IS, two sub-systems coexist, applications and data. To articulate these sub-systems, some authors opt for the CRUD matrix. Similarly, Axiomatic Design (AD) theory studies how functional requirements are related to design parameters using a design matrix. In this paper, the application sub-system corresponds to the functional requirements, the data sub-system to the design parameters, and the CRUD matrix to the design matrix. The goal of the CRUD matrix is to maintain the independence of its items to minimize the information content of the design. Similarly, AD uses the design matrix to define a good design. This work aims to develop a theory to create object-oriented elements based on the CRUD matrix aligned with the business strategy.

1. Introduction

Axiomatic Design (AD) theory is a powerful systems design methodology based on matrix methods that analyze the pipeline from customer needs to process variables. For a thorough explanation see [11], [12], [13], [14].

Information Systems (IS) can be seen as a simple input/output mechanism that collects, stores, and distributes information. However, new legislation, new entities, and new interactions create new challenges in information management. A significant effort is given to align the business strategy with the technology, i.e., the process in which a business organization uses information technology to achieve business objectives [8].

Despite the high number of methodologies [16] in IS, the search for a good design is still a relevant issue, and the concept of a good design is still a subject of controversy. However, the essential SI elements remain constant. BSP (Business Systems Planning) [7] presented four essential elements for Information Systems Planning, coined as the Iron Cross: organization (people), applications (software), data, and technological systems (hardware).

To articulate the elements, applications, and data, we opt for the CRUD matrix. CRUD, achromic of <create, read, update, delete>, means the four basic operations possible on a data value. In SQL it corresponds to <insert, select, update, delete>. CRUD matrix was developed by IBM [7] and popularized by James Martin [9] in his book 'Managing the Database Environment'.

The CRUD matrix is an essential tool to perform diagnostics in software. Unfortunately, there are several practical examples with the CRUD matrix that corresponds to coupled designs. In [1] the authors use the CRUD matrix to know the current state (as-is) and to model the improved future state (to-be). In the current state, the CRUD matrix shows a poor design. After swapping rows and columns, cells are aggregated into macro-cells in a decoupled matrix in the improved state.

The CRUD matrix is still beneficial when integrated with more recent tools, like UML (Unified Modeling Language) [5]. UML in analysis and design allows the fulfillment of the system requirements with object-oriented design and relational databases models.

This work aims to add to CRUD the scientific support given by the Axiomatic Design theory. The design matrix brings value to the project implementation in IS. This work contributes to developing a theory to create object-oriented elements based on the CRUD matrix and aligned with the business strategy.

In a similar approach in enterprise architectures [2], the authors use data as FRs and applications as DPs. In this work, we follow the approach presented in chapter 5, Suh [13], where FRs correspond to applications and DPs to data.

The paper is organized into three additional sections. Section 2 introduces the CRUD matrix in information systems. Section 3 details the proposed CRUD design. Finally, in Section 4, conclusions are drawn.

2. CRUD Matrix

Enterprise Architecture is usually defined by the three-layers model [8]: business/process architecture, information system architecture, and technological architecture, as shown in Figure 1.

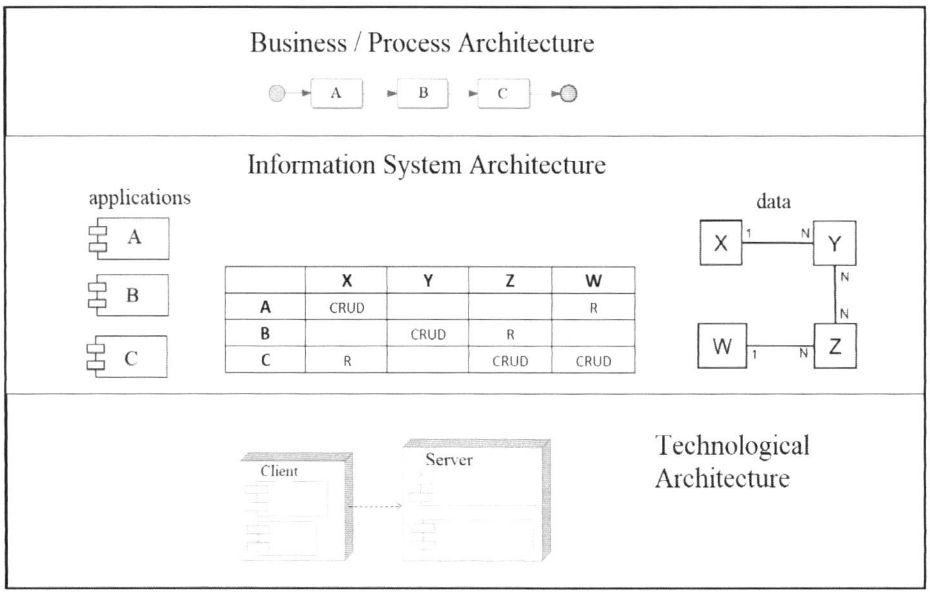

Figure 1. Three-layers model of Enterprise Architecture

In Enterprise Architecture, many synonyms depend on the layer, methodology, framework, or tool used, so it is crucial to find a synthesis with a reduced number of names.

In Cavique et al. [3], only three essential elements were chosen: the actors, the activities, and the data. Some of the synonyms are as follows: (i) actor is a synonym of lines-of-responsibility, (ii) activities is about applications, tasks, uses-cases, or operational-processes, (iii) data is a synonym of classes or informational entities.

In the following examples, we will use actors (α, β), activities (A, B, C), and data (X, Y, Z, W). This version of Enterprise Architecture with three layers includes:

- Business / Processes Architecture: where processes are made up of activities (A, B, C) and managed by human actors (α, β);
- Information System Architecture: with two different software groups, the data (X, Y, Z, W) and the applications (A, B, C);
- Technological Infrastructure Architecture: consists of hardware components and essential software (operating systems and database management systems).

The CRUD matrix plays a vital role in the Information System Architecture by combining data between applications (A, B, C) and data (X, Y, Z, W). In a relational data model approach, data (X, Y, Z, W) correspond to tables of a database and the applications (A, B, C) to operations that maintain the data sets.

To develop straightforward projects, the authors Pereira and Sousa [10] and Vasconcelos [17] presented a set of three heuristic rules that guarantee the alignment of the information system. The heuristic rules can be summarized as follows:

#1 Each processing activity of the business process is automated by a single application;

#2 The data items must support the processes activities of the business processes;

#3 Each data item is managed (CUD) by a single application in the CRUD matrix.

In this work, rule #3 emphasizes developing a more managed CRUD matrix since data items updated by different applications are harder to manage.

applications vs data	X	Y	Z	W
A	CRUD			R
B		CRUD	R	
C	R		CRUD	CRUD
CRUD counters	1211	1111	1211	1211

Figure 2. CRUD matrix with counters

To ensure rule #3, Figure 2 shows the CRUD matrix with CRUD counters that validate possible inconsistencies [4]. Counter 1211 indicates that there are 1 Create, 2 Reads, 1 Update, and 1 Delete. There must be a single application that performs Create, Update and Delete, (CUD), and there may be multiple applications with the Read operator.

Preferably, the CRUD counters should be 1N11, i.e., a unique Create, Update and Delete and multiple Read operators. As previously stated, Read operations correspond to the SQL Select and usually involve many tables to obtain new information. The multiple Reads are allowed and do not produce concurrence; therefore, they will not be considered in the rest of our analysis.

The CRUD approach intends to obtain a close view of the system to avoid coupled sub-systems by finding a matrix with the main diagonal filled with the least number of elements.

With the three-layer model, Enterprise Architecture presents similarities with Axiomatic Design, which uses three mapping zones. Firstly, Axiomatic Design studies the transformation of the customer attributes into functional requirements, which corresponds to the transformation of activities into applications. Then, Axiomatic Design maps the functional requirements with a set of design parameters using the design matrix, corresponding to the CRUD matrix matching, between applications and data.

3. Proposed CRUD design

This section focuses on how Axiomatic Design can improve and better justify the CRUD structure in information systems projects. After clarifying the design matrix, the transformation to objected-oriented design is shown. Finally, we discuss the similarities between AD and IS rules.

3.1. Axiomatic Design approach

This section presents a practical example of the concepts presented in chapter 5, Axiomatic Design of Software [13].

In AD, Functional Domain and Physical Domain are associated as follows: functional requirements {FRs} and design parameters {DPs}. The most common designs maps between FRs and DPs, expressed by the design equation {FRs} =[A].{DPs}. Each DP must fulfill an FR, so the design matrix [A] is square.

The {FRs} correspond to 'what' are the goals to achieve, and {DPs} match with 'how' to reach them. In this work:

- functional requirements {FRs} correspond to the applications {A, B, C}, that are aligned with the business architecture;
- the design parameters {DPs} match the data items {X, Y, Z, W}, that should support the business architecture;
- the design matrix [A], which maps FRs and DPs, coincides with the CRUD matrix.

In other words, {FRs} = [A]. {DPs} corresponds to {Apps}= [CRUD]. {Data}. DPs can also be seen as Design Patterns, as referred in Thomas and Mantri [15]. In the design matrix, the cells with operators CRUD are replaced by an X. Such as the axioms of AD, the goal of the CRUD matrix is to maintain the independence of the functional requirements, which usually correspond to minimize the information content of the design.

When there are more DPs than FRs, the design is either a redundant design or a coupled design, as stated by Theorem 3 [11]. Gonçalves-Coelho et al. [6] claim that redundancy designs belong to particular kinds of designs and propose seven theorems. Figure 2 shows a redundant design, as application C operates on tables Z and W. According to AD, the design equation is by Equation 1:

$$\begin{Bmatrix} FR\_A \\ FR\_B \\ FR\_C \end{Bmatrix} = \begin{bmatrix} X & & & \\ & X & & \\ & & X & X \end{bmatrix} . \begin{Bmatrix} DP\_X \\ DP\_Y \\ DP\_Z \\ DP\_W \end{Bmatrix} \tag{1}$$

However, an application makes a change on a table and then on another table. Therefore, the redundant design of application C and data (Z, W) is equivalent to two different states of the same FR, that can be fulfilled by *DP_Z* or *DP_W*. Equation 2 shows the mentioned redundant design:

$$\{FR\_C\} = [X \quad X] . \begin{Bmatrix} DP\_Z \\ DP\_W \end{Bmatrix} \tag{2}$$

Equation 2 shows the first state of operation on Equation 3.

$$\{FR\_C1\} = [X \quad \ ] . \begin{Bmatrix} DP\_Z \\ DP\_W \end{Bmatrix} \tag{3}$$

And the other state by Equation 4.

$$\{FR\_C2\} = [\ \quad X] . \begin{Bmatrix} DP\_Z \\ DP\_W \end{Bmatrix} \tag{4}$$

Application C can split in *C1* when addressing table *Z* and *C2* when addressing table *W*. It makes all applications belong to the same level of decomposing. Therefore, the design matrix can turn into a square matrix by the decomposition of FR-C as shown in Figure 3, using a square sub-matrix, where FR-C is decomposed into FR-C1 and FR-C2.

IOP Conf. Series: Materials Science and Engineering 1174 (2021) 012011 doi:10.1088/1757-899X/1174/1/012011

		DP-X	DP-Y	DP-Z	DP-W
FR-A		X			
FR-B			X		
FR-C	FR-C1			X	
	FR-C2				X

Figure 3. Decomposition of FR-C

3.2. Object-oriented design

Object-oriented (OO) techniques are some of the most potent paradigms widely used in computer science in software design. An object is a bundle of data associated with a set of operations that act on that data. The object-oriented paradigm supports four crucial features: abstraction, encapsulation, inheritance, and modularity.

Object-oriented techniques encapsulate data and the operations that manipulate that data in a specific object. OO paradigm allows a higher level of abstraction, focusing only on those characteristics available by the object's interface and omitting the details of data and operations. The system's modularity is also achieved by decomposing each problem into loosely coupled intra-objects so that internal changes to one object do not affect another one.

In this work, the data items can be enriched with the OO approach. Four objects {A.X, B.Y, C1.Z, C2.W} can be found in the design matrix. Each cross ('X') in the design matrix [A], in Figure 3, corresponds to an object. Figure 4 shows the object {A.X} that associates the data DF-X with the operations/application FR-A.

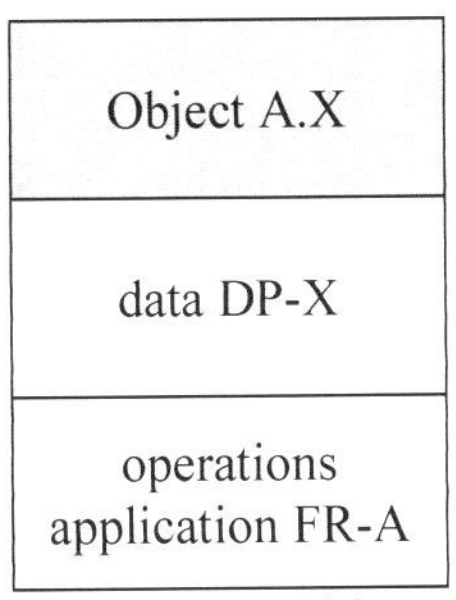

Figure 4. Object A.X

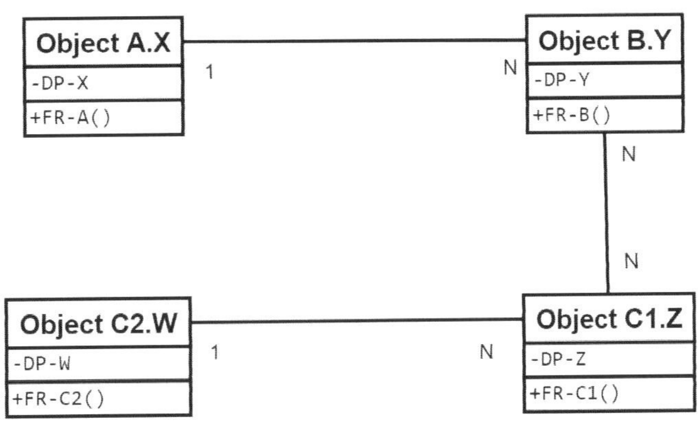

Figure 5. Object-oriented design supported by AD

IOP Conf. Series: Materials Science and Engineering 1174 (2021) 012011 doi:10.1088/1757-899X/1174/1/012011

The data schema presents in Figure 1 can be enriched with the functional requirements. The result is presented in Figure 5, where relations link the four objects. The object-oriented design supported by AD reinforces the OO paradigm to incorporate the advantages of modularity and reuse, providing savings in the software project development.

3.3. Discussion of the two approaches

Since there are similarities between IS and AD, some relationships between the axioms of Axiomatic Design and the heuristic rules in Information Systems are detailed in Table 1. Using AD theorems, a new rule, number 4, is proposed.

Table 1. Heuristic rules versus Axiomatic Design

Heuristic rules in IS	Axiomatic Design
#1 a single application automates each process activity of the business processes	applications ≅ {FRs}
#2 the data items must support the processes activities of the business processes	data items ≅ {DPs}
#3 each data item is managed by a single application in the CRUD matrix	design matrix [A] ensuring the independence of FRs
#4 (new) the number of applications is equal to the number of data items/tables	Theorem 4 in Axiomatic Design: #(FRs) = #(DPs)

Rule #1, a single application that automates each process activity, also corresponds to the mapping of Customer domain (process activities) and Functional domain (applications).

Rule #2, the data items must support the process activities of the business processes, or in order words, data items must support the applications.

Rule #3, each data item is managed by a single application in the CRUD matrix, as already seen, almost corresponds to the design matrix. In the transformation, two additional issues should be mentioned.

On the one hand, AD imposes a square design matrix, according to Theorem 4 [11]. Theorem 4 states that the number of DPs is equal to the number of FRs in an ideal design, so the CRUD should be square. To overcome this issue, we add rule #4, where the number of applications is equal to the number of data items.

On the other hand, using the CRUD counters, the CRUD matrix only allows uncoupled designs, showing decoupled matrices are not adequate for IS. This statement agrees with Theorem Soft 1 [13] that uncoupled software or hardware systems can be operated without precise knowledge of the design elements (i.e., modules) if the design is uncoupled.

Heuristic rules #1 and #2 ensure the vertical alignment with the business processes, and rule #3 ensure the horizontal alignment between applications and data.

4. Conclusions

Business strategy alignment with Information Systems and Information Technology is a longstanding problem because of the increasing complexity of the designs. In Information Systems (IS), the horizontal alignment of applications and data items is possible with the CRUD matrix.

However, regarding IS, the concept of a good design is still a subject of controversy. Adopting Axiomatic Design theory validates what a good design can be in information systems. AD helps to justify and simplify the design of the CRUD matrix of IS projects. We propose to adopt rule #4 that makes the number of applications equal to the number of data items, to create object-oriented modules.

By applying heuristic rule #3, the CRUD matrix can return a redundant matrix with more DPs than FRs. The re-design of the project using Axiomatic Design, by adding rule #4, allows creating object-oriented elements, improving the data design, and ensuring alignment with the business.

References

[1] Alves P, Vasconcelos A, Silva A 2011 Arquitecturas de sistemas de informação de referência para atendimento multicanal na administração pública [Reference information system architectures for multichannel service in public administration], *Proceedings of the Portuguese Association for Information Systems Conference*, Portugal

[2] Behrouza F, Fathollahb M 2016 A systematic approach to enterprise architecture using axiomatic design, *the 10th International Conference on Axiomatic Design, ICAD, Procedia*, Elsevier, **53** pp 158-165

[3] Cavique L, Cavique M, Mendes AB 2021 Integration of UML Diagrams from the Perspective of Enterprise Architecture, In: Rocha Á., Adeli H., Dzemyda G., Moreira F., Ramalho Correia A.M. (eds) Trends and Applications in Information Systems and Technologies, *WorldCIST 2021, Advances in Intelligent Systems and Computing*, **1366**, Springer, Cham. https://doi.org/10.1007/978-3-030-72651-5_44

[4] Cavique, L 2020 Modelação de Sistemas de Informação: Elementos essenciais na visão integrada das ferramentas do UML com a matriz CRUD [Information Systems Modeling: Essential elements in the integrated view of UML tools with the CRUD matrix,], Recursos Educativos, Universidade Aberta, Portugal

[5] Fowler, M 2003 *UML Distilled: A Brief Guide to the Standard Object Modeling Language*, Addison-Wesley Professional, 3rd Edition, ISBN: 978-032-119-368-1

[6] Gonçalves-Coelho AM, Neştian G, Cavique M, Mourão A 2012 Tackling with redundant design solutions through axiomatic design, *International Journal of Precision Engineering and Manufacturing*, **13**, pp 1837–1843, https://doi.org/10.1007/s12541-012-0241-x

[7] IBM Corporation 1978 *Business System Planning Information - System Planning Guide*, International Business Machines Corporation, 2nd edition, New York

[8] Lankhorst, M 2013 *Enterprise Architecture at Work: Modelling, Communication and Analysis*, 3rd Editions, Springer, ISBN: 978-364-229-650-5

[9] Martin, J 1983 *Managing the Database Environment*, Prentice-Hall, Englewood Cliffs, New Jersey, ISBN: 013-550-582-8

[10] Pereira CM, Sousa P 2005 Enterprise architecture: business and IT alignment, In H. Haddad, L.M. Liebrock, A. Omicini, R.L. Wainwright, eds., SAC, pp 1344–1345, ACM

[11] Suh NP 1990 *The Principles of Design* (New York: Oxford University Press)

[12] Suh NP 1995 Design and operation of large systems, *Journal Manufacturing Systems*, 14 (3)

[13] Suh NP 2001 *Axiomatic Design: Advances and Applications* (New York: Oxford University Press)

[14] Suh NP 2005 *Complexity: Theory and Applications* (New York: Oxford University Press)

[15] Thomas J, Mantri P 2015 Axiomatic Design/Design Patterns Mashup: Part 1 (Theory), *the 9th International Conference on Axiomatic Design, ICAD, Procedia*, Elsevier, **34**, pp 269-275

[16] Van-Bon J, Verheijen T 2006 *Frameworks for IT Management: an Introduction*, ITSM Library, Publisher:Van Haren, ISBN-13 : 978-9077212905

[17] Vasconcelos, A 2017 Slides Aulas Arquiteturas Tecnológicas Empresariais [Slides of the Classes in Business Technological Architectures], DEI, Instituto Superior Técnico, Universidade Lisboa, Portugal

IOP Conf. Series: Materials Science and Engineering 1174 (2021) 012012 doi:10.1088/1757-899X/1174/1/012012

Identifying Poor Designs with Conflicts in Design Parameters

Kenji Iino, Masayuki Nakao

The University of Tokyo, 7-3-1 Hongo, Bunkyo-ku, Tokyo 113-0033, Japan

kiino@esc.u-tokyo.ac.jp

Abstract. Mechanical design is an act of defining machines that produce values to people, however, a poor design sometimes causes mechanisms to fail. Such failures can lead to harms to people, the environment, or both and in the worst case, it can even cause people's death. Axiomatic design is a design tool that helps the designer come up with more stable and robust designs by pointing out troubling elements in a design by identifying interference among parts. Taking a design through axiomatic analysis only at an abstract level, however, can miss finding interferences within. A product, through its lifetime, can also experience disturbances that the designer overlooked at the time of design. This paper discusses a trouble that arose with bicycles with a new safety feature of locking the front wheel when parked. The locking mechanism faced an unexpected part failure that caused the handle to lock while riding. Another topic is an inherent interference that existed from the very beginning of a design that the designers could not identify probably due to the extreme complexity of the system, the Fukushima Nuclear Power Plant-1, Unit-1. Analysis for evaluating how well a design will meet its functional requirement without trouble should go into the detail level, especially for all modes of operation, to avoid oversight of hard-to-find interferences and conflicts. This paper also suggests carrying out design analysis after production if phenomena, unexpected or unforeseen at the time of design, arise after a product is placed into service.

1. Introduction

Mechanical design is an intriguing act of defining shapes of new parts, how they assemble with machine elements and instrumentation to form new products, how the sub-assemblies interact with other subassemblies, or even defining complex systems like automobiles with tens of thousands of parts [1]. In this paper, we call the subject of design, a machine, regardless of its size or complexity, from a single part shape to a complex system.

The designer aims at providing value to the users, societies, or even the entire globe, however, responsibility comes with such development of machines to make sure they do not cause harm to the target users or the environment.

As complexity of our design grew with modern industrialization that started in the 18th century, assuring safety with design turned more challenging and since the latter half of the 20th century, people devised many analytical tools to help the designer keep the design trouble free. Such tools include finite element method for stress analysis, computational fluid dynamics to predict fluid flow behaviors, design review to take advantage of knowledge of more people, or quality control primarily to keep eyes on the manufacturing processes.

Design theorists, around this time, pointed out that the product cost and whether it will succeed in the market or not are almost determined in the early design stage of conceptual design, e.g., [2], and they came up with tools for helping the designer with conceptual design. Among such tools, this paper discusses Design Record Graph [3] and Axiomatic Design [4], especially, how they can detect design problems in the conceptual design stage.

The design problems this paper discusses are two; one with a recent recall of a bicycle, and the other, the 2011 Fukushima Nuclear Power Plant (NPP) accident.

2. Design Record Graph and Axiomatic Design

This section briefly describes the two design tools that designers often use in their conceptual design.

2.1. Design Record Graph

Figure 1 shows a design record graph (DRG). The use of design record graph is twofold; in producing a new design or analyzing an existing design. For creative design of coming up with a new product that did not exist before, the diagram starts from the left end node, the product functional requirement, and generally proceeds to the right. The designer iteratively divides the functions on the left to collections of smaller functions to the right until the designer reaches a set of functional elements (FEs). An FE is one that is not practical to divide into smaller functions.

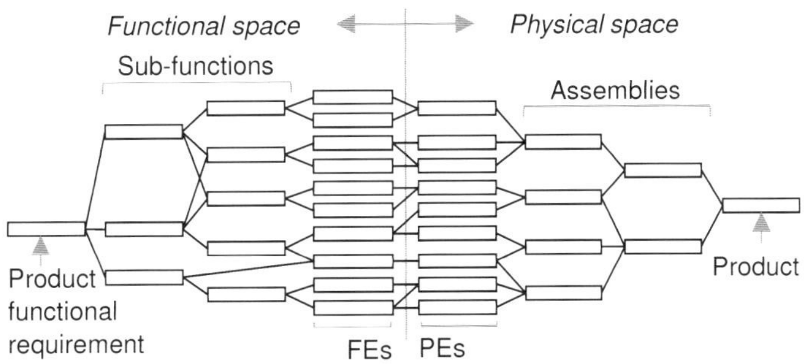

Figure 1 Design Record Graph

The designer then maps each FE to one or more physical elements (PEs) which are typically machine parts that produce the corresponding function. This time, the designer is moving into the physical space from the functional space. The mapping can take any form of one to one, one to many, or many to one. Once in the physical space, the designer can start defining assemblies that collect multiple parts, then higher-level assemblies or modules consisting of multiple assemblies until the designer reaches the destination of a product; this goal, the product meats the product functional requirement defined at the starting point. This design does not necessarily proceed strictly from the left to right; the designer can jump back and forth over the border of functional space and physical space in a zig-zagging manner.

A DRG is also useful in analyzing existing products. In this case, the product functional requirement at the left end and the product at the right end are known, as well as all the PEs. The analysis proceeds by figuring out FEs for the PEs and how the two types of elements combine into higher level functions and assemblies, in other words, the analysis identifies the arcs in the DRG.

2.2. Axiomatic Design

Axiomatic Design (AD) calls FEs and PEs in DRG, functional requirements (FRs) and design parameters (DPs), respectively. It shows the correspondence among FRs and DPs with a matrix named design matrix (DM). By placing the FRs and DPs in the form of vectors ***FR*** and ***DP***, AD expresses the design with the following matrix equation.

$$\boldsymbol{FR} = \mathbf{A}\,\boldsymbol{DP} \tag{1}$$

where **A** is the DM. If we number the 10 FRs in figure 1, FR1 to FR10 and the 9 DPs, DP1 to DP9, AD expresses this design with the following matrix.

$$\begin{Bmatrix} FR1 \\ FR2 \\ FR3 \\ FR4 \\ FR5 \\ FR6 \\ FR7 \\ FR8 \\ FR9 \\ FR10 \end{Bmatrix} = \begin{pmatrix} X & 0 & 0 & 0 & 0 & 0 & 0 & 0 & 0 \\ X & 0 & 0 & 0 & 0 & 0 & 0 & 0 & 0 \\ 0 & X & X & 0 & 0 & 0 & 0 & 0 & 0 \\ 0 & 0 & X & 0 & 0 & 0 & 0 & 0 & 0 \\ 0 & 0 & 0 & X & 0 & 0 & 0 & 0 & 0 \\ 0 & 0 & 0 & X & 0 & 0 & 0 & 0 & 0 \\ 0 & 0 & 0 & 0 & X & X & 0 & 0 & 0 \\ 0 & 0 & 0 & 0 & 0 & 0 & X & 0 & 0 \\ 0 & 0 & 0 & 0 & 0 & 0 & 0 & X & 0 \\ 0 & 0 & 0 & 0 & 0 & 0 & 0 & X & X \end{pmatrix} \begin{Bmatrix} DP1 \\ DP2 \\ DP3 \\ DP4 \\ DP5 \\ DP6 \\ DP7 \\ DP8 \\ DP9 \end{Bmatrix} \quad (2)$$

The DM in Equation (2) consists of elements 0 and X, where 0 indicates that the corresponding DP has no effect on the FR and X that the DP affects the FR to some level without quantitative indication.

When a design is uncoupled in AD, its DM is diagonal. When this condition holds for a design, its DRG has all its arcs across the functional space and physical space show one to one correspondence and they form a ladder like structure.

3. Two accident cases

This section discusses a consumer product, a bicycle recall case that caused major injuries but luckily no fatalities, and part of a catastrophic accident that still has residual effects on us, the Fukushima Nuclear Power Plant (NPP) accident.

3.1. Bicycle recall case

First is a design problem with our first subject, a bicycle. A major bicycle manufacturer in Japan introduced a new model in 2003 with a device to lock the rotation of the steering shaft when the user parks the bike and locks the rear wheel. The company named the mechanism "single action double lock" to have it stand out in the market where its competitors had models with separate steering locks that required the riders to engage them separately from the actions of locking the rear wheels. Figure 2 shows a photo of the recalled model. The red circles identify the rear wheel and steering shaft locks.

Figure 2 Recalled bicycle

The purpose of the rear wheel lock is to stop the rear wheel from rotating while the bicycle is parked, that is to prevent theft. Locking the steering shaft is a recent development for bicycles. One of the purposes is to prevent it from toppling over when parked on sloped surfaces, and another function, favored by parental riders, is to keep a stable position while seating a child in a child seat.

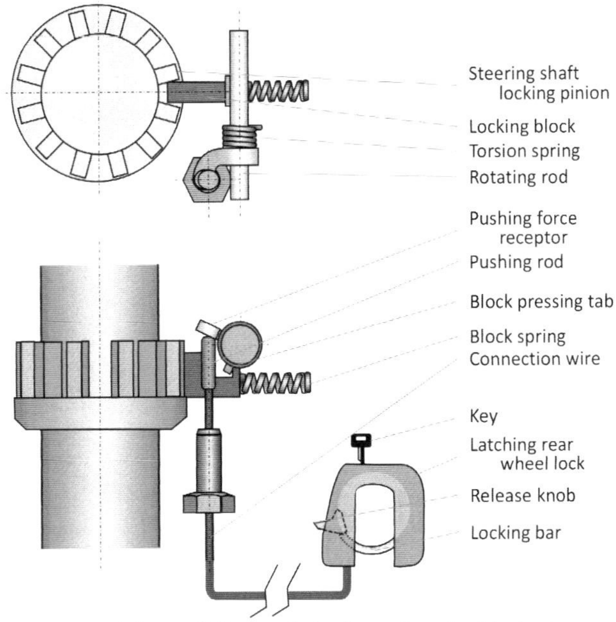

Figure 3 Parts of single action double lock

Figure 3 shows parts of the single action double lock mechanism [5]. The steering shaft lock is housed inside in a diecast case, seen as an oversized gray box in the right-hand upper photo in figure 2, with a window to show the status of the lock.

Figure 4 shows how the single action double lock works. The figure on the left shows when the lock is disengaged, and the one on the right when engaged. The movements numbered 1 through 5 in the right figure happen at the same time; the numbers are for easier understanding of the mechanism.

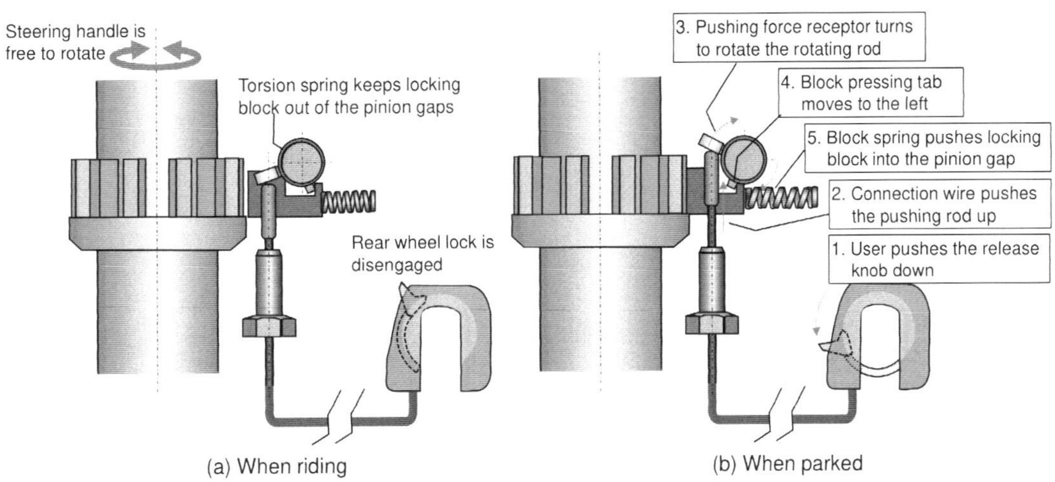

Figure 4 How single action double lock works

Figure 5 is the DRG for this bike locking system. All the arcs that cross the boundary between the functional space and physical space have one to one mapping indicating that the design is uncoupled in AD.

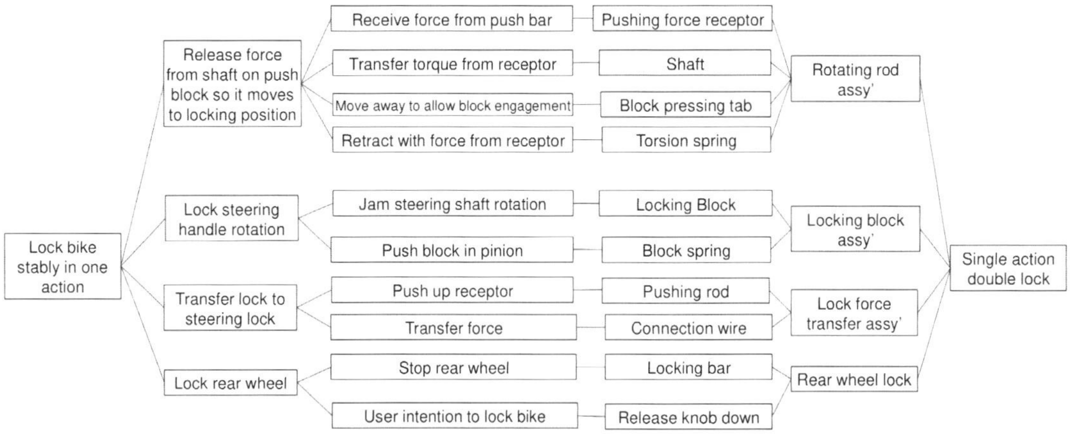

Figure 5 DRG of "Single action double lock" for a bicycle in parked mode

So far, we discussed how "single action double lock" works in parking a bike. The primary function of a bike, however, is not in parking it, but rather in riding it. Figure 6 next shows a typical DRG for the primary function of riding a bike without the detail of the braking system for simplicity. The two nodes with heavy lines and pink background show where the FE-PE relation failed when the locking system failed as we describe later.

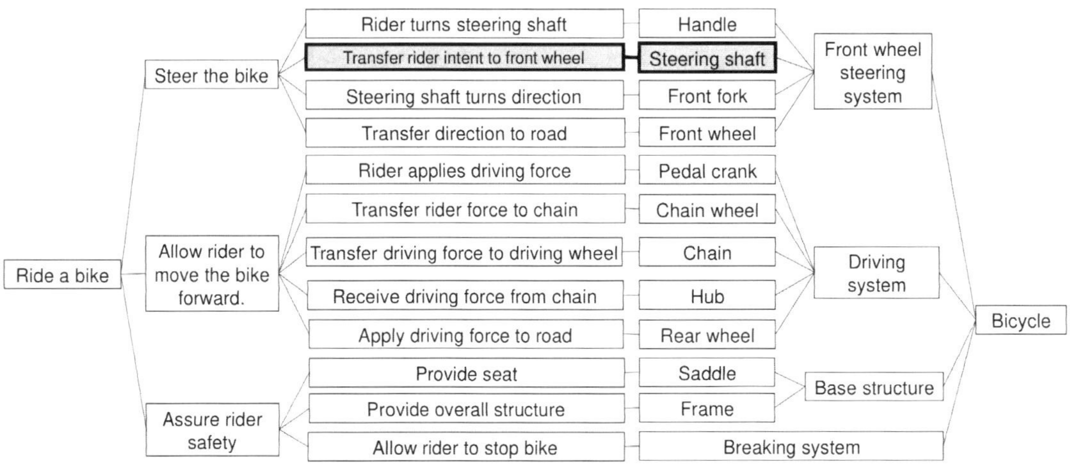

Figure 6 DRG of a bicycle in riding mode

Note that the designer is not very concerned about the locking system. One may add the function of "keep the bike locked when parked" with a corresponding mechanism of "locking system," or for a deliberate functional analysis the designer may add the contents of figure 5 into figure 6.

Knowing that this "single action double lock" led to the recall, we will take a closer look at the locking system when the user is riding the bike. Figure 7 shows the DRG in this mode.

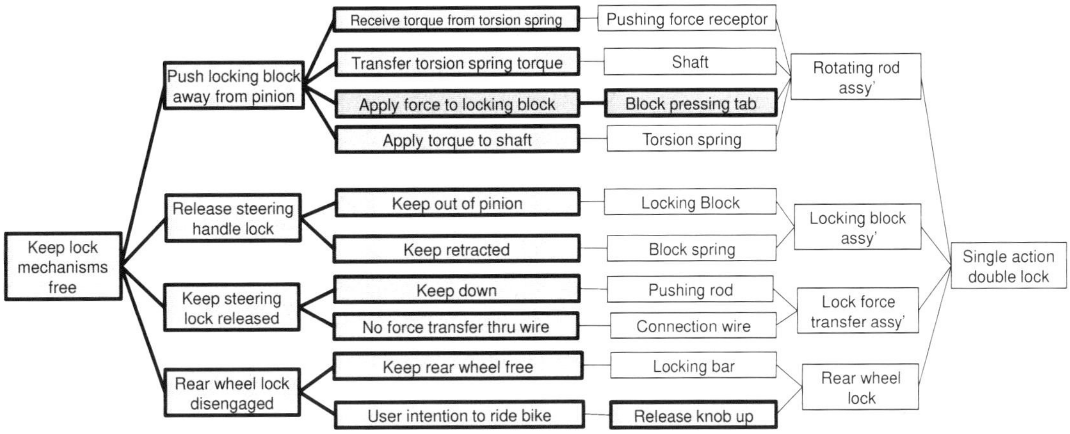

Figure 7 DRG of "Single action double lock" for a bicycle in riding mode

Note that with the mechanism working in the other direction, the parts change their functions shown with colored nodes in Fig. 7. What causes the locking block to engage into (parked mode) and retract from (riding mode) a pinion gap is the balance among the torsion spring, block spring, and force from the pushing rod. In riding mode, the pushing rod does not push up the receptor, and spring force from the torsion spring overcomes the block spring force. On the other hand, in parked mode, the pushing rod gains upward force that overcomes torque from the torsion spring and block pressing tab moves to the left to allow the block spring to push the locking block into a gap of the pinion.

The locking block in riding mode has the function of retracting itself from the pinion gap it engaged in parked mode. The torsion spring in riding mode, in the meantime, must constantly keep its spring force on the pushing force receptor to keep the locking block retracted.

The recall was due to cracking of the diecast case of the steering handle lock assembly. Ministry of Economy, Trade and Industry (METI) had a news release [5] showing photographs of a normal and a cracked case. Figure 8 shows the illustration that accompanied the photos and comparing it with the photo of the steering shaft lock in figure 2 clearly shows the problem.

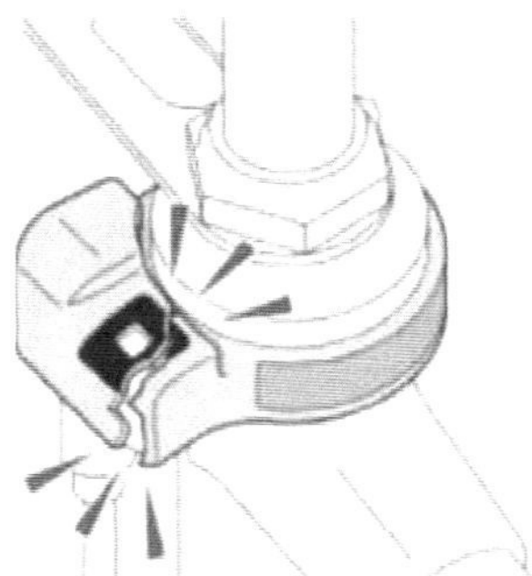

Figure 8 Illustration of a cracked steering shaft lock case

When the case is cracked, parts inside it lose tight positioning and the block pressing tab can lose its contact with the locking block. Figure 7 shows this origin of failure with the pink colored FE-PE mapping. The locking block and block spring also failed to meet their corresponding FEs. Without force from the torsion spring, the block spring pushes the locking block into a pinion gap. Consequently, the steering handle locks while the user is riding the bike and that is a totally unexpected behavior that can cause the rider to fall on the road with the bike.

The recall covered about 3.4 million bicycles, and at the time of the METI press release on June 24, 2019 [6], there had been 6 serious injury cases. Consumer Affairs Agency (CAA) of the Government of

Japan later reported that the number of accidents during April of 2019 to March of 2020 totaled 42 cases, and 11 during April to August 18, 2020 [7].

3.2. Accident of Unit 1 of Fukushima 1 NPP

The 2011 off the Pacific coast of Tohoku Earthquake broke out on March 11, 2011. It caused tsunami waves that attacked Fukushima Nuclear Power Plant-1 (Fukushima 1 NPP) about an hour later. The wave heights exceeded limits that Fukushima 1 NPPs were built for. The seawater flooded the basements of all six units at Fukushima 1 NPP, where most of all switchboards for units 1 to 4 were located. The four units were forced into station blackout, i.e., loss of all AC power [8]. Units 1 and 2 also lost all their DC power as well.

The earthquake caused all the units, running at the time, units 1, 2, and 3, to go through emergency insertion of their control rods. At the same time, all isolation valves on the main steam lines to the turbines and feedwater lines back from the condensers closed. The 3 reactors entered the state of core isolation. Since all the control rods succeeded in full insertion, main power of the cores stopped generating most of their powers. Nuclear fuel, however, even when they stop generating heat from nuclear chain reaction, continue to generate heat called decay heat.

Decay heat is only a few percent of full power (6.4% for unit 1 at the time of shutdown [9]), however, without removing this heat, the reactor will quickly boil the water inside and uncover the fuel. Once uncovered, the fuel will start to melt. Unit-1 had 78 minutes after station blackout to recover its lost core cooling function [9].

Unit 1, the oldest of all units that had started operation in March of 1971, had two Isolation Condenser (IC) systems, A and B, designed to remove the decay heat at time of core isolations. Figure 9 shows one of the systems at the time of core isolation.

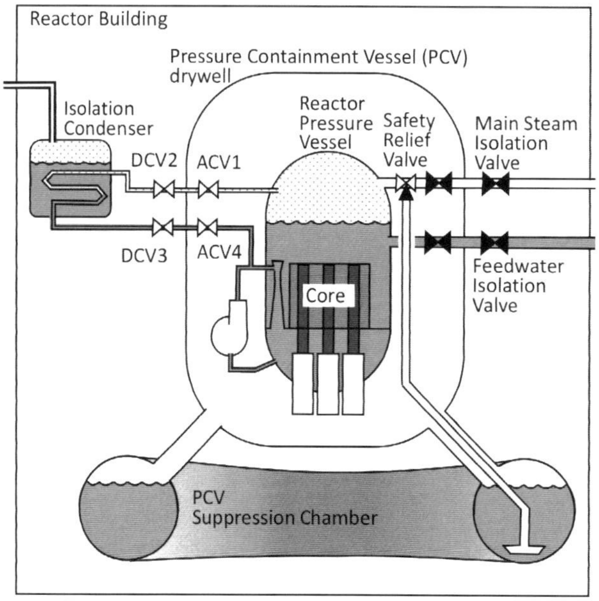

Figure 9 Fukushima 1 NPP Unit-1, IC system upon core isolation

IC is a simple condenser with a condensing chamber filled with cooling water and steam inlet pipes that connect to a high elevation of the reactor pressure vessel and water return lines to the recirculation piping that connects to a low elevation. This piping has 4 isolation valves; 2 AC valves inside the pressure containment vessel and 2 DC valves outside. During operation, the operator turns just the DC valve DC3 on and off to control temperature drop inside the pressure vessel and the remaining three valves are open all the time. Figure 10 shows the DRG for this IC system.

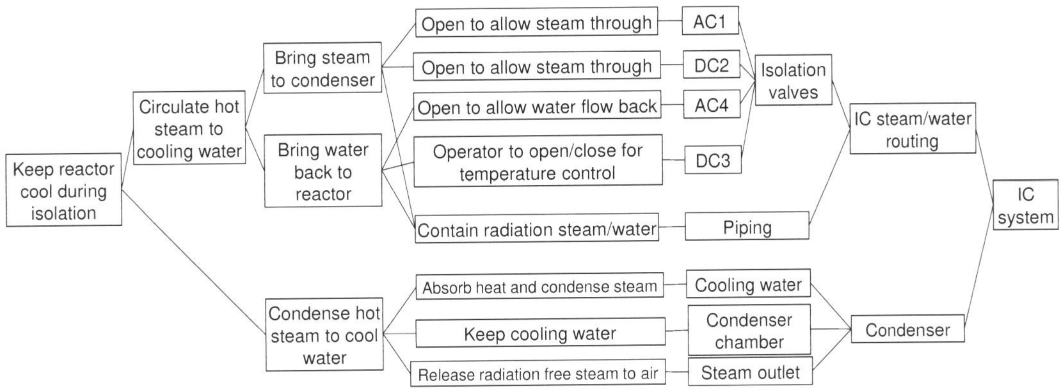

Figure 10 DRG of Fukushima 1 NPP Unit-1 IC System in operation

This IC system, however, had another mode of operation for safety. The piping that routes the steam through the cooling water to condense steam into water had a pressure sensor on it. The purpose of this steam sensor was to monitor any possible cracking on the piping. From figure 9, one can tell if the piping had a crack, steam with radiation can leak into the surrounding air by escaping through the crack to the condenser chamber and eventually through the steam outlet to the atmosphere.

This was IC piping failure mode that closes all the four IC isolation valves to avoid any radiation leakage from the IC. It is not very difficult to realize that this DC power operated instrument was designed to send out a leakage detection signal when it loses its DC power. It was a fail-safe design to assume leakage when the sensor lost its means of detecting the reality. Figure 11 shows the DRG of this safety feature.

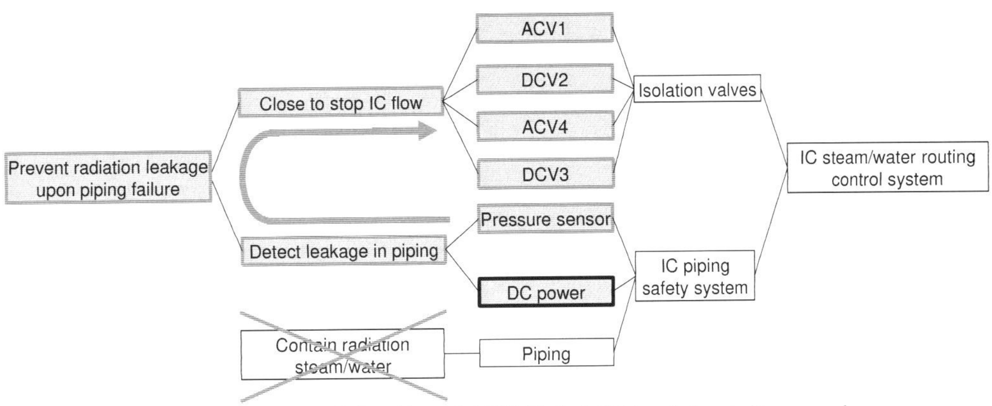

Figure 11 DRG of Fukushima 1 NPP Unit-1 IC in piping failure mode

It was ironic that at the time of station blackout, DC power (shown with pink background in Fig. 11) to this pressure sensor was first lost and the system sent out signals to close all the four IC isolation valves. In other words, the IC system entered a false piping failure mode. The operator was controlling the reactor temperature by opening and closing DCV3 on System A, thus, IC System B had its DCV3 fully closed. For this reason, the statuses of the other three valves on System B made no difference.

The operator, without recognizing this safety feature, kept hitting the open and close switch of DCV3 on IC System A. He even went outside to observe steam from the two steam outlets outside the reactor building. He only saw a small amount of steam rising from the exhaust pipe but still was not sure if the IC was working. It was later pointed out that the plant had no experience of training about loss of DC power and the IC systems were tested only during plant startup, but never during the 40 years of operation that followed.

The consequence was meltdown of the core of Fukushima 1 NPP Unit-1 and the process generated hydrogen gas that leaked up inside the reactor building. The reactor building blew up in a hydrogen explosion a day after the station blackout.

4. Discussion

The previous section reviewed two accidents, one with a consumer product that caused a number of injuries to users and eventually was recalled, and the other about a major catastrophic accident of an NPP. Both accidents share the same difficulty in design that parts of the design had to operate in different manners for different modes of the products.

4.1. Bike recall case

The two modes for the bicycle were parking and riding. When we think of safety, the riding mode obviously has higher priority over the parking mode. Imaginable failure of the parking mode is theft or falling over. Theft is a definite loss to the user and falling may cause damage to the product, however, there is no way of getting injured with these failures. Users have voiced the convenience of steering shaft lock when they need to lift their children into the child seats, however, it is just "easier" and users used to do so without the steering shaft lock with added carefulness for seating their children.

In fact, the bike manufacturers no longer have the single action double lock features in their new products and instead, placed an independent steering shaft lock that the user must engage manually.

The designer in applying AD or DRG to design tends to write out part names for DPs. That can easily lead to oversight if the design requires conflicting operations of the part in different modes. We have to analyze all modes of a product and pay special attention if the same part has different functions or different configurations in different modes. For the case of the bike, trouble with the part integrity let the "Locking block" slide into a different posture while in a mode where safety was more important.

4.2. Nuclear plant case

The purpose of our analysis in this paper with regards to Fukushima 1 NPP Unit 1 was not in pointing out responsibilities, but rather it was for revealing how easy it is for the designer to overlook conflicts in detail design.

It is, in fact, easy to point out design deficiencies after the trouble. However, in the case of Fukushima, among many reasons that can lead to core isolation, station blackout was definitely one that needed consideration. Moreover, before the Fukushima accident, even the Japanese design guideline stated that station blackouts would be short, and the practice was to assume 30 minutes or less [10]. Probably this minor concern with loss of power led to overlooking that the IC systems would shut down upon a station blackout closing all the IC valves, while they needed to be kept open to keep the shutdown core from melting.

5. Conclusion

We looked into two accidents where parts of the product needed to behave differently in different modes. When we apply AD or DRG in analyzing our design, we need to write out how the design expects the parts to behave. If we just write down the part names, it is easy to overlook conflicting behaviors of the parts in different modes. We further see the need to execute design analysis of products in all modes of their operations and find out what triggers the products from shifting from one mode to another.

In case of the bicycle, the two modes were "parked" and "ridden." The "locking block" designed to lock the steering shaft turned loose and interfered with the function of safely riding the bike by allowing the rider to steer the handle into the desired direction. It was an interference of a secondary function with one of the crucial ones for safety.

In case of Fukushima 1 NPP Unit 1, the two modes were "IC piping failure" and "core isolation" and the designers overlooked consequences of the serious event of a station blackout. When this event happened, instrumentation first lost its DC power supply and enforced the safety function of closing all isolation valves on the IC systems. It was at the time of "core isolation" when cooling was needed for the isolated core and the reactor without means for releasing decay heat of the core headed into core meltdown.

Designers must review all modes of operation of products and acknowledge the priorities among different modes. In case of the bike, "safe riding" had priority over "stable parking" and in case of the NPP, "cooling during core isolation" had priority over "preventing possible radiation leakage." The designer must be careful in adding features that can possibly damage FRs of modes with higher priority.

References

[1] "How many parts is each car made of?" Toyota Motor Corporation, 1995 [https://www.toyota.co.jp/en/kids/faq/d/01/04/]
[2] Ullman D G *The Mechanical Design Process Fourth Edition*, McGraw-Hill, 2010, New York, NY
[3] Iino K, Nakao M, Design Record Graph and Axiomatic Design for creative design education, PROCIR-D-16-00377, *Proceedings of ICAD 2016*
[4] Suh N P, *Axiomatic Design, Advances and Applications*, Oxford University Press, 2001
[5] Accident talks: Design to notify trouble triggers accident/Recall of 3.43 million bikes with "Single action double lock" *NIKKEI MONOZUKURI*, Dec. 2019 (in Japanese).
[6] Recall of Bicycles and Electric Power Assisted Bicycles with Built-in Locking Mechanism Simultaneously Working Handlebar Lock and Circle Lock (Free Product Inspection and Repair), METI News Release, June 24, 2019 (https://www.meti.go.jp/english/press/2019/0624_003.html)
[7] Announcing serious product caused accidents by a consumer product, CAA News Release, August 18, 2020 (https://www.caa.go.jp/notice/assets/consumer_safety_cms202_200818_01.pdf in Japanese)
[8] Hatamura Y, Abe S, Fuchigami M, Kasahara N., *The 2011 Fukushima Nuclear Power Plant Accident*, Woodhead Publishing, 2015
[9] Iino K., Yoshioka R, Fuchigami M, Nakao M, Precautions at Fukushima That Would Have Suppressed the Accident Severity, *Journal of Nuclear Engineering and Radiation Science*, ASME, February 2018
[10] Report: A comparison of U.S. and Japanese regulatory requirements in effect at the time of the Fukushima accident, NRC, November 2013 (https://www.nrc.gov/docs/ML1332/ML13326A991.pdf)

IOP Conf. Series: Materials Science and Engineering 1174 (2021) 012013 doi:10.1088/1757-899X/1174/1/012013

Design of an Assembly Workplace for Aging Workforce and Worker with Disabilities

Benedikt G. Mark[1,*], Erwin Rauch[1], Christopher A. Brown[2], Dominik T. Matt[1,3]

[1] Free University of Bozen-Bolzano, Faculty of Science and Technology, Industrial Engineering and Automation (IEA), 39100 Bolzano, Italy
[2] Worcester Polytechnic Institute, Department of Mechanical Engineering, 01609 Worcester, MA, USA
[3] Innovation Engineering Center (IEC), Fraunhofer Research Italia s.c.a.r.l., A.-Volta-Straße 13 A, 39100 Bolzano, Italy

[*]Corresponding author: Tel: +39-0471-017111, E-mail address: Benediktgregor.mark@unibz.it

Abstract. With the introduction of the Fourth Industrial Revolution (also known as Industry 4.0 or I4.0) the use of worker assistance systems gets more and more important. These assistance systems should support the worker to reduce both mental as well as physical stress and at the same time increase ergonomics. Besides the before mentioned positive aspects of assistance systems they can also support the inclusion of workers with limitations in the daily industrial production environment. In this paper, we apply Axiomatic design (AD) to systematically design a future operator's workplace for older workers and workers with disabilities. Therefore, we conducted interviews with employers and physical and mental disabled people to identify the customer needs (CNs) and set constraints (Cs). In a next step, we defined the first level functional requirements (FRs) and the deduced corresponding design parameters (DPs). Following, the decomposition and mapping process and considering the independence and information axiom we designed a workplace equipped with worker assistance systems for better integrating elderly people and people with disabilities in manufacturing.

1. Introduction

The production environment has been changing fundamentally over the last years. This is mostly due to technologies and concepts based on "Industry 4.0", which is also known as the fourth industrial revolution [1]. On the one hand, there is a high trend towards automatization and repeated activities and on the other hand the products are made in many variations, which leads to a smaller lot size [2]. This diversity, consequently, results in a growing total of product adjustments for the operator working in the production environment [3]. It can be said that the industry we are currently part of changes from "mass production" towards "mass customization" together with a lower product life cycle. In order to cope with the new challenges, the operator needs to be flexible and trained but first and foremost there must be enough workers available on the market.

Having this background in mind a worrying trend is noticeable. The population development in Europe varies from country to country. Overall, however, the continent's share of the world's population will continue to decline. The average age of people in Europe is the highest of all regions of the world, and the trend is rising. In fact, this century Europe will be marked by enormous demographic changes [4]. It is not surprising that Europe is desperately looking for workers. Europe is heading towards the demographic trap. More and more people are leaving the labor market due to age, while the number of new entrants is not enough to close this gap [5]. In order to counteract this certainty, elderly people and people with disabilities could be involved in the labor market. Therefore, useful and appropriate worker assistance systems have to be found. Thus, these groups of people might turn out as a useful asset for the production industry and consequently get included into the daily life better.

Due to the lack of assistance systems especially for elderly people and people with disabilities and the high prices that had to be paid for such technology in the past there is a high potential for inventing a workplace equipped with aid systems. With this paper, we want to contribute to the research on worker assistance systems, as tools for a more inclusive manufacturing and to relax the extreme lack of workers due to the demographic change.

2. AD Based Design Approach

The aim of this research article is to develop a framework for the successful design of an assistant workplace for elderly people and people with disabilities to help the human in production 4.0.

This section presents a procedure to receive a framework and design concept for such an assistant workplace. The research team used AD as a research methodology to deduce a list of design solutions. In the following, this technique is shown in detail.

2.1. Axiomatic Design Methodology

Axiomatic Design has become one of the more commonly applied engineering theories regarding design in industrial practice and academic literature [6], since 1978 when it was first mentioned in a publication [7]. It was founded by Nam P. Suh in order to develop a generalized, systematic and scientific procedure for design. AD uses a concept of four different domains of the design world, which can be summarised as follows: 1) the customer domain with Customer Needs (CNs) where the customer wishes are shown, 2) the functional domain with Functional Requirements (FRs) where the translation of CNs into FRs takes place, 3) the physical domain with Design Parameters (DPs) where the design parameters are derived, and 4) the process domain with Physical Variables (PVs) where the DPs are translated into real PVs [8]. The two following axioms must be respected when performing AD [9]:

I. Axiom 1: The "Independence Axiom" to maintain the independence of the functional requirements (FRs),
II. Axiom 2: The "Information Axiom" to minimize the information content of the design.

Parallel physical and functional parts are constructed in order to apply these axioms. The physical hierarchy contains the final physical design solutions. Through AD the designer learns to keep an overview and how to construct and structure large design hierarchies quickly. Hence, there is more time for managing applications [10]. CNs were gathered and based on this, FRs and Cs could be defined, and design parameters were derived in an AD mapping process and top-down decomposition.

2.2. Collection of Customer Needs

Based on discussions and interviews with the different stakeholders specified in chapter three, nine customer needs (CN) could be collected. These are the following:

CN_1: Assistance system must be as flexible as possible.
CN_2: System must ensure the quality of the product.
CN_3: It must be possible to subdivide the working tasks in parts.
CN_4: Safety of the worker must be guaranteed during the tasks.

CN_5: System must overcome the counting difficulties of the workers.
CN_6: System must overcome the reading and understanding problems of the workers.
CN_7: System must ensure/maintain the attention of the worker
CN_8: System should overcome the memorization problems of the worker.
CN_9: There must be always a structure in what the workers need to do.
CN_{10}: System should be able to support the operator with handling of heavy weight.

These mentioned customer needs as gathered from the interviews with caregiver of elderly people and people with disabilities, elderly people and people with disabilities themselves and companies give a starting point for the later discussed decomposition.

3. Decomposition and Mapping Process

3.1. Definition of Top-Level Functional Requirements and Design Parameters
The following FR_0 and DP_0 were defined based on the previously identified CNs:

FR_0 *Enable the inclusion of elderly people and people with disabilities in production*
DP_0 *Workplace equipped with assistance systems to support elderly people and people with disabilities in production*

In the following, the top-level FRs and DPs are listed which were defined by the research team:

FR_1 *Enable people with mental health issues the inclusion in daily work life*
DP_1 *Workplace to assist people with cognitive or memory issues*

This upper level FR-DP pair indicates one part of the target group, namely people with mental health issues, that must be included in daily work life. Further decomposition is required (see section 4.4.1).

FR_2 *Enable people with physical health issues the inclusion in daily work life*
DP_2 *Workplace to assist people with physical health issues*

This upper level FR-DP pair indicates the other part of the target group, namely people with physical health issues, that must be included in daily work life. Further decomposition is required (see section 4.4.2).

When looking at the interdependencies in the design matrix (see Eq. 1) a decoupled design can be seen. This displays the existence of an established order of the design solutions implementation concerning the compliance of the first axiom.

$$\begin{Bmatrix} FR1 \\ FR2 \end{Bmatrix} = \begin{bmatrix} X & 0 \\ 0 & X \end{bmatrix} \begin{Bmatrix} DP1 \\ DP2 \end{Bmatrix} \tag{1}$$

3.2. FR1-DP1 – Enabling Inclusion of People with Mental Health Issues in Daily Work Life
As mentioned before, the top-level FR-DP pairs need to be further decomposed on more levels in order to get more palpable solutions for design and creation. The following sections 4.4.1 and 4.4.2 show the mapping process and decomposition results.
The following lower level FR-DP pairs were derived by the research team based on the top-level FRs and DPs in order to determine mental health issues of elderly people and people with disabilities in daily work life.

Table 1. Decomposition of FR_1-DP_1 pair.

	FR			DP		
Level 1	$FR_{1.}$	$FR_{1.1.}$	Overcome cognitive or memory issues.	$DP_{1.}$	$DP_{1.1.}$	Solutions to overcome cognitive or memory issues.
		$FR_{1.2.}$	Overcome communication difficulties.		$DP_{1.2.}$	Solutions to overcome communication difficulties.
		$FR_{1.3.}$	Overcome attention issues.		$DP_{1.3.}$	Solutions to overcome attention issues.
		$FR_{1.4.}$	Overcome reading, linguistic and verbal comprehension issues.		$DP_{1.4.}$	Pictures and videos in stand-alone or combined with multilanguage text.
		$FR_{1.5.}$	Message and inform the caregiver/supervisor.		$DP_{1.5.}$	Andon button (physical or virtual).
		$FR_{1.6.}$	Overcome quantity control issues.		$DP_{1.6.}$	Solutions for quantity control.
		$FR_{1.7.}$	Overcome quality control issues.		$DP_{1.7.}$	Solutions for quality control.
		$FR_{1.8.}$	Increase attention to important subjects.		$DP_{1.8.}$	System that ensures the clear visual understanding.
Level 2	$FR_{1.1.}$	$FR_{1.1.1.}$	Provide work steps.	$DP_{1.1.}$	$DP_{1.1.1.}$	Dynamic instruction manual.
		$FR_{1.1.2.}$	Support the worker with information about what has to be done and what has already been done.		$DP_{1.1.2.}$	Dynamic checklist.
		$FR_{1.1.3.}$	Structure the assembly process in several tasks.		$DP_{1.1.3.}$	Assembly sequence workflow.
		$FR_{1.1.4.}$	Guarantee the easy accessibility of the work instructions.		$DP_{1.1.4.}$	Easily accessible hard- and software.
	$FR_{1.2.}$	$FR_{1.2.1.}$	Calling up individual content that is difficult to remember.	$DP_{1.2.}$	$DP_{1.2.1.}$	Customizable note/memo page.
		$FR_{1.2.2.}$	Send and visualize general company/product relevant content.		$DP_{1.2.2.}$	Chat/news function.
	$FR_{1.3.}$	$FR_{1.3.1.}$	Provide motivational information.	$DP_{1.3.}$	$DP_{1.3.1.}$	System that congratulates after each XX part.
		$FR_{1.3.2.}$	Bridge the time until the next break.		$DP_{1.3.2.}$	Visualization of the timeline until the break.
	$FR_{1.6.}$	$FR_{1.6.1.}$	Manual counting of parts.	$DP_{1.6.}$	$DP_{1.6.1.}$	Counting gauge.
		$FR_{1.6.2.}$	Automatic counting of parts.		$DP_{1.6.2.}$	Scale combined with human interface.
	$FR_{1.7.}$	$FR_{1.7.1.}$	Check presence of parts.	$DP_{1.7.}$	$DP_{1.7.1.}$	Vision system.
		$FR_{1.7.2.}$	Check functionality.		$DP_{1.7.2.}$	EOL test bench.
		$FR_{1.7.3.}$	Check part for scratches and damage.		$DP_{1.7.3.}$	Visual inspection station.
	$FR_{1.8.}$	$FR_{1.8.1.}$	Call up attention to important subjects within written text.	$DP_{1.8.}$	$DP_{1.8.1.}$	Capital letter and/or bold/bigger for important words.
		$FR_{1.8.2.}$	Call up attention to the area of interest on the worktable.		$DP_{1.8.2.}$	Thick outline of text and image fields.
Level 3	$FR_{1.1.4.}$	$FR_{1.1.4.1}$	Make the digital instructions easy findable and accessible.	$DP_{1.1.4.}$	$DP_{1.1.4.1.}$	Intuitive start of the software application.
		$FR_{1.1.4.2.}$	Make it wheelchair accessible.		$DP_{1.1.4.2.}$	Wheelchair compliant workstation and instruction monitor position.

3.3. FR1-DP1 – Enabling Inclusion of People with Mental Health Issues in Daily Work Life

The following lower level FR-DP pairs were derived by the research team based on the top-level FRs and DPs in order to determine physical health issues of elderly people and people with disabilities in daily work life.

Table 2. Decomposition of FR_2-DP_2 pair.

	FR			DP		
Level 1	*FR2.*	*FR2.1.*	Overcome visual problems.	*DP2.*	*DP2.1.*	Solutions for visual problems.
		FR2.2.	Overcome hearing difficulties.		*DP2.2.*	System that has loudspeakers.
		FR2.3.	Overcome difficulties in accessing the workplace and reachability.		*DP2.3.*	Reachability and accessibility of the workplace.
		FR2.4.	Overcome physical obstacles caused by heavy parts and/or devices.		*DP2.4.*	Physical worker assistance systems.
Level 2	*FR2.1.*	*FR2.1.1.*	Provide instructions acoustically.	*DP2.1.*	*DP2.1.1.*	Aural guide with the most important steps.
		FR2.1.2.	Provide instructions readable for blind people.		*DP2.1.2.*	Digital braille reading device.
		FR2.1.3.	Interact with the system in a non-visual way.		*DP2.1.3.*	Natural human interaction technologies.
		FR2.1.4.	Provide an audio signal for condition monitoring.		*DP2.1.4.*	Audible loud alarm/help system.
	FR2.2.	*FR2.2.1.*	Provide visual instruction and guidance.	*DP2.2.*	*DP2.2.1.*	Digital devices for visualization of information.
		FR2.2.2.	Provide a visual signal for condition monitoring.		*DP2.2.2.*	Visual alarm/help system.
		FR2.2.3.	Provide information regarding most important emergencies.		*DP2.2.3.*	Info-field on monitor.
	FR2.3.	*FR2.3.1.*	Provide accessibility.	*DP2.3.*	*DP2.3.1.*	Barrier-free design of the workplace.
		FR2.3.2.	Ensure ergonomic reachability of human machine interaction and reachability of tools, components and equipment.		*DP2.3.2.*	Ergonomic design of the workplace.
	FR2.4.	*FR2.4.1.*	Overcome physical workload caused by parts.	*DP2.4.*	*DP2.4.1.*	Cobots for handling of heavy parts.
		FR2.4.2.	Overcome physical workload caused by devices.		*DP2.4.2.*	Torque reaction arm for heavy devices.
Level 3	*FR2.1.3.*	*FR2.1.3.1.*	Interact with voice.	*DP2.1.3.*	*DP2.1.3.1.*	Voice control.
		FR2.1.3.2.	Interact with gesture.		*DP2.1.3.2.*	Gesture control.
		FR2.1.3.3.	Interact with touch.		*DP2.1.3.3.*	Touch control.

4. Final Design of an Assistance Workplace for Elderly People and People with Disabilities

The framework discussed shows a procedure to plan a systematic design of an assistance workplace for elderly people and people with disabilities. The final design of this workplace is illustrated in Fig. 1. It consists of three different worktable parts that can be connected via plug-in. The main worktable is equipped with a projection-based assistance system to show digital work instructions (workflow) in written form together with pictures and videos, enable automatic step recognition and to do vision checks. The industrial signal light tower can be used to let the person in charge know about the presence of the worker and as an opportunity for the operator to express the need of help by pushing the help/stop button. The monitor shows information, such as pop ups, counter, timeline and motivational content. The worktable itself is easily accessible for wheelchairs and has large legroom. For people without a wheelchair, it is equipped with an adjustable footrest and an ergonomic chair, and, in case the operator cannot sit for a long time, an ergonomic mat. Additionally, holding devices are available e.g., for crutches. The electric screwdriver is connected with a torque reaction arm. The second part of the workplace is equipped with a collaborative robot that can be gesture controlled by a device for gesture interaction (e.g., smart robot) to help the operator in the sense of a third hand. The third table consists of adjustable workplace lamps and additional concise lamps for quality control in different angles. Above is a quality control system, which can be used for end of line inspection. The additional monitor can be used for information provision and a counting gauge for the physical counting of parts. The main worktable, as well as the other parts can be easily reached by the operator. Due to the plug-in connection of the tables, it is also possible to arrange the tables in a line, e.g., for the use of this workplace by two workers instead of one.

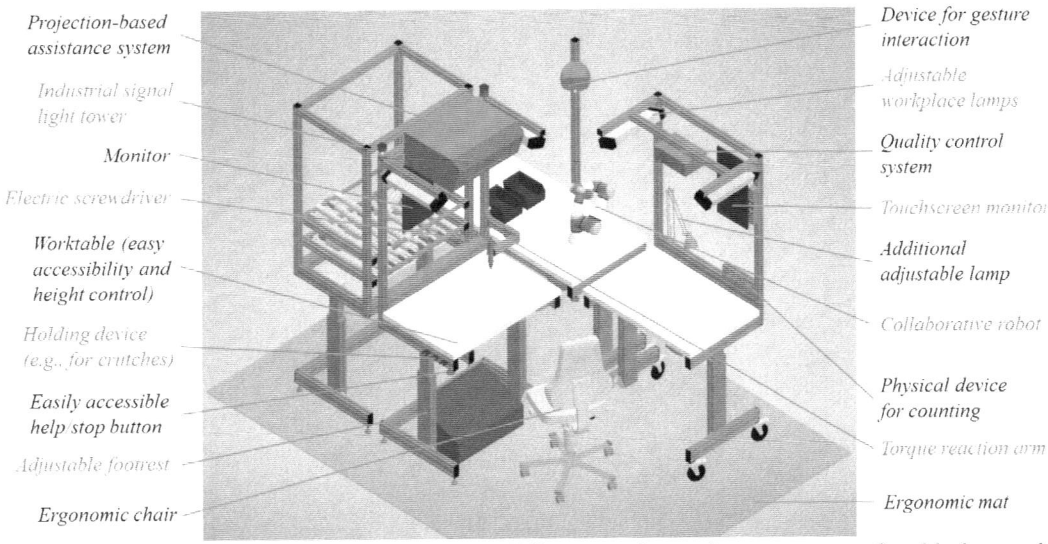

Figure 1. Design of a workplace equipped with different worker assistance systems for elderly people and people with disabilities.

5. Conclusion

The aim of this scientific contribution is to elaborate a design of a workplace equipped with worker assistance systems in production for elderly people and people with mental and/or physical disabilities. For this reason, the current situation had been analyzed by interviews with caregivers of elderly people and people with disabilities, elderly people and people with disabilities themselves, and companies. Based on the interviews and discussions with the different stakeholders, CNs were defined. In addition, these were transformed into a framework using AD mapping and decomposition in the form of FRs and DPs.

The input of this paper for practitioners is to split an initially complex problem, namely the design of a workplace equipped with worker assistance systems in production for elderly people and people with disabilities, into small packages, to discuss the possibilities and analyze their correlations. Based on this scientific contribution, a first step for the design of future oriented workplaces is to be laid, which gets more and more important due to the demographic change and the generally prevailing shortage of workers in manual production.

6. References

[1] Sendler U 2013 *Industrie 4.0-Beherrschung der Industriellen Komplexität mit SysLM* (Berlin/Heidelberg, Germany) pp 1-19

[2] Büttner S, Funk M, Sand O and Röcker C 2016 *Using head-mounted displays and in-situ projection for assistice systems – a comparison* (PETRA 2016) **44** pp 1-8

[3] Wolfartsberger J, Lindorfer R, Hallewell Haslwanter J D, Jungwirth M, Froschauer R and Wahlmüller D 2018 *Industrial perspectives on assistive systems for manual assembly tasks* (PETRA 2018) pp 289-291

[4] Habekuß F 2017 *Europas demografischer Wandel DOSSIER Demografischer Wandel* Available online: https://www.bpb.de/politik/ innenpolitik/ demografischer-wandel/ 196906/europas-

demografische-zukunft (accessed on 18 March 2020)
[5] Rütti N 2019 *Europa sucht händeringend Fachkräfte Neue Züricher Zeitung* Available online: https://www.nzz.ch/ wirtschaft/ europa-sucht-haenderingend-fachkraefte-ld.1507619 (accessed on 18 March 2020)
[6] Tomiyama T, Gu P, Jin Y, Lutters D, Kind C and Kimura F 2009 *Design methodologies: industrial and educational applications* (CIRP Annals) **58-22** pp 543-565
[7] Suh N P, Bell A, Gossard D, Suh N P, Bell A and Gossard D 1978 *On an axiomatic approach to manufacturing and manufacturing systems* (Journal of Manufacturing Science and Engineering) **100-2** pp 127-130
[8] Nam P. Suh 1990 *The Principles of Design* Oxford University Press
[9] Nam P. Suh 2001 *Axiomatic Design: Advances and Applications* Oxford University Press
[10] Brown C A 2005 *Teaching axiomatic design to engineers – Theory, applications, and software*, (Journal of Manufacturing Systems) **24-3** pp 186-195

Acknowledgements
This research project was funded by the Free University of Bozen-Bolzano; project title: Assist4Work: Social sustainability in production through age appropriate and disability friendly workplace design using assistance systems, grant number TN 200J.

IOP Conf. Series: Materials Science and Engineering 1174 (2021) 012014 doi:10.1088/1757-899X/1174/1/012014

Developing functional requirements for Temporary Housing by integrating Axiomatic Design with the 5 Gaps Model of Service Quality

C Pezzica[1,2*], C Bleil de Souza[2] and I V Dunichkin[3]

[1] Department of Energy, Systems Territory and Construction Engineering, University of Pisa, Largo Lucio Lazzarino, 56122, Italy
[2] Welsh School of Architecture, Cardiff University, Bute building, King Edward VII Avenue, CF10 3NB, Cardiff, UK
[3] Moscow State University of Civil Engineering, Yaroslavskoye Sh. 26, 129337 Moscow, Russia

[*] Corresponding author e-mail: pezzicac@cardiff.ac.uk

Abstract. Temporary Housing (TH) schemes are a controversial component of post-disaster recovery plans, and yet they offer a fundamental service to the homeless population. Their sustainability should be understood and addressed in terms of service quality for all clients, rather than as a matter of product engineering. Since the evaluation of service quality is different from that of goods, value in TH assistance should be measured according to how well it matches clients' expectations. This paper adopts the 5 gaps model of service quality as a framework for TH quality assurance and advances that closing the current performance gap requires tackling issues in the briefing, design, project execution and conformance phases, as well as in communication. Against this background, engineering methods such as Axiomatic Design (AD) can effectively be exploited to reduce the gap between what people want and what they get, considering the needs and objectives from humanitarian actors. Results indicate that AD can reveal conflicts and potential for cooperation between the many "clients" of TH, via the joint analysis of their different needs, and the associated Functional Requirements (FRs) and illustrates via a post-factum analysis what mechanisms need to be in place to ensure better preparedness for future disasters.

1. Introduction and background

This paper explores the integration of Axiomatic Design (AD) to the 5 gaps model of service quality, proposing a framework for Temporary Housing (TH) quality assurance using AD as a means to close performance gaps related to issues in the briefing, design, project execution, conformance phases and communication which happen throughout a TH project delivery. It illustrates the proposal using the 2016-2017 Central Italy earthquakes as a case study focusing on exploring Functional Requirements (FRs) from its two main 'clients': the affected communities (called 'Client 1'), the beneficiaries of the service, and the Public Administration (PA, called 'Client 2'), responsible to deliver the service.

Why are TH actually a service? In contrast with building-centered or product-centred approaches to housing emergency management and Disaster Risk Reduction, temporary housing is increasingly understood as a verb, more specifically as a continued effort to support people's health, security and

IOP Conf. Series: Materials Science and Engineering 1174 (2021) 012014 doi:10.1088/1757-899X/1174/1/012014

livelihood [1]. This means TH requires providing culturally and technically adequate housing and neighbourhood conditions to local communities, its main beneficiaries, but also promoting sustainable building construction practices and employing local resources, responding to governmental objectives related to efficiency and efficacy in project delivery considering its whole life-cycle [2]. These requirements qualify TH as an urban planning service (an intangible component having the distinctive function of supporting a full recovery post-disaster) to the affected communities, and a critical component of building sustainable cities and society to the PA.

In Italy, the TH supply and delivery process involves several actors and stakeholders, among which designers and engineers commissioned by the PA, including the National Department of Civil Protection (NDCP), which is responsible for delivering the service ('Client 2'), as well as TH beneficiaries, i.e., the affected communities displaced by the earthquake ('Client 1') [3]. For experts, reconciling the views and needs of these multiple clients in TH design and site planning is a complex task, which requires a careful management of operational decision-making components - especially those with potential for conflicts' generation [4] – and the consideration of organisational aspects, resources, plans and information from the part of the PA responsible for project delivery.

An important part of a service delivery process, in the built environment, is customer satisfaction, in which a considerable part of it is related to customer perception (ISO 55002), [5]. Customers' perception of the housing building service is important in housing construction projects, as ad-hoc adaptions to a set of non-standard situations are often needed [6]; what applies also to the construction of TH sites after disasters. *The quality of the TH assistance service depends on how it is perceived by all clients*, during the design and construction stages and beyond (e.g., if maintenance and/or buy-back options are included in the TH procurement document). Even when time an budget targets are met, several projects fail to realise the intended value –understood as a manyfold and multi-domain concept - due to approach inconsistencies, misalignments in decision-making, and a reductive focus on costs and project financing rather than on whole-life performance [7].

Thus, to improve, and possibly even measure, TH assistance value, we suggest linking AD [8] and the 5 gaps model of service quality by [9] within a novel TH delivery framework as outlined in Section 2. This will support a joint analysis of multiple clients' needs, revealing conflicts as well as potential for cooperation between top-down and bottom-up inputs. Specifically, in the analysis of a selected case study in Section 3, the 2 main clients of TH plans are separated to better administer the gathering of Customers' Needs (CNs) and discuss associated FRs. The conclusions situate the proposal within the need to support the reconciliation of TH schemes with the context-dependent value drivers of different clients and national disaster recovery policies. This is understood as the first step to address quality assurance in TH planning and control gaps in procuring, designing, and delivering sustainable TH solutions to resolve and/or prevent technical and community clashes associated to TH sites' location and layout choices, as well as TH units' design.

2. Hypothesis and methods: linking TH, AD and the 5 gaps model

[9] advance that clients' evaluation of services quality is different from that of goods, as their satisfaction depends on how well the perceived value of a services matches their initial expectations. Thus, in the 5 gaps model, they use this consideration to define the TH project performance gap (Gap 5), which is seen as the combined result of four different sub-gaps. Using the terminology of [10], who merged the 5 gaps construct with a "project as an information process" model, these four sub-gaps correspond to the briefing (Gap 1), design (Gap 2), execution and conformance (Gap 3) and communication (Gap 4) problems.

According to [11], in each gap service quality can be evaluated according to 5 key determinant dimensions - reliability, responsiveness, assurance, tangibles and empathy - which apply at different stages of service delivery and with different weightings. These, coincide with some of the issues identified as critical in [4], and respectively: (i) TH assistance consistency of performance; (ii) readiness of TH service; (iii) competence of humanitarian professionals to perform necessary tasks and/or deliver the service considering its whole lifecycle; (iv) physical outcomes for process tracking; and (v) attention to all clients' needs. Figure 1, shows how the 5 gaps model can be applied to the case of TH assistance delivery, for identifying where methodological innovations are needed to reduce current gaps and enhance TH service quality.

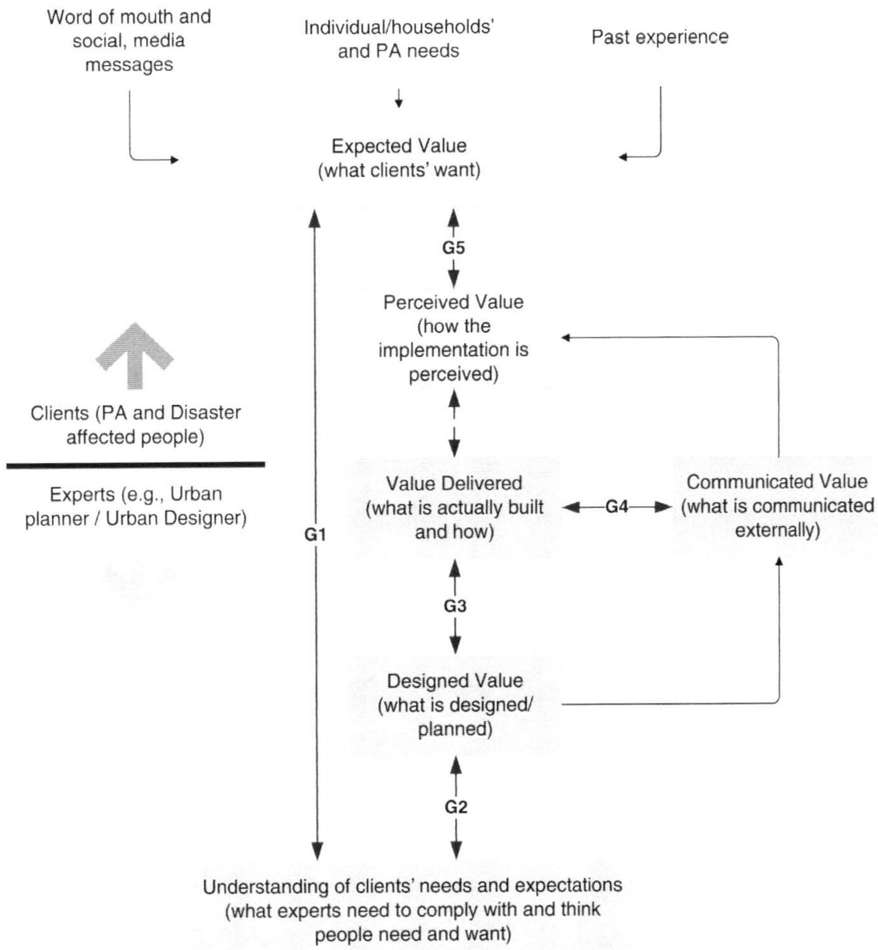

Figure 1. 5 gaps model of TH assistance service quality.

This study advances that AD can be linked to the 5 gaps model and support quality improvements in TH provision by closing Gaps 1, 2 and 3, ultimately reducing the TH performance gap. Figure 2 shows that some conceptual correspondences exist between AD and the 5 gaps models. In fact, both refer to multiple domains in which: people's needs are understood (costumer domain); functional requirements are specified (functional domain); design parameters are chosen to meet the brief (physical domain); and the implementation process is set up to adequately reflect the design intention (process domain). Accordingly, AD could effectively support the mapping of CNs for determining TH functional requirements and their systematic decomposition (Gap 1), the mapping of independent relations between these and candidate Design Parameters (DPs, Gap 2) and the TH supply and delivery process (Gap 3), considering input constraints, by specifying problems and solutions explicitly, in parallel, following a hierarchical order while maintaining data [8]. This proposal aligns with that of [12], who pose that AD could enhance the reliability of disaster response operations, as reliability is one of the 5 key service quality determinants identified by [11].

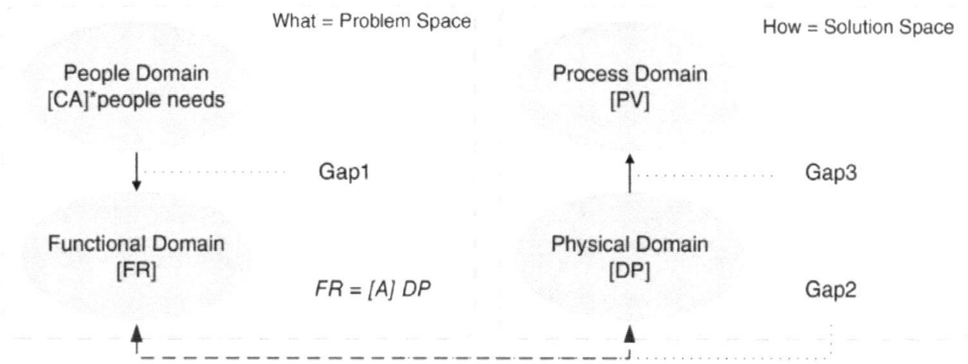

Figure 2. Axiomatic design model and associated Gaps.

Albeit originally intended for product design, AD has been successfully transferred to architecture to help problem-solving [13] and to service design [14]–[16]. For instance, in [14], the authors pose that the mirror decomposition through zigzagging between the functional, the physical and the process domains enables focusing designer' attention on CNs. In this case, AD is used to design web services for remote patient monitoring by systematically detecting the FRs of all the stakeholders considering also their interactions. [15] use AD to build a framework for optimising the flow of patients in hospitals to improve the efficiency of health services. In [16] AD is adopted to suggest improvements to the service offered to passengers with reduced mobility in airports by linking CNs with process components. These studies address specialised services which require complex process choreographies with several actors, while keeping a focus on costumers' needs. The authors of this paper pose that the TH assistance delivery process presents similar characteristics, and hence, could be effectively approached by adopting AD to close service quality gaps, while taking into accounts the CNs from all stakeholders.

This point is illustrated in what follows, where AD is exploited to address the TH briefing problem in a suitable test case, covering Gap 1 of the service quality framework. The case of the 2016-2017 Central Italy earthquakes is used to illustrate the approach and explore the setting and reconciliation of a selection of high level FRs from the service deliverer, the PA (Client 2), to the affected community (Client 1).

3. The two clients: conflicts and overlaps in the Italian case.

Identifying high-level FRs starts by independently mapping in detail the needs and aspirations of both clients, so they can be analysed to assess the presence of matches (common FRs), explore the overlap of tolerances when FRs converge (completely or partially) and flag potential conflicts between FRs from Client 1 and Client 2, so they are assessed with regards to their logical inconsistencies. The authors adopt a traffic light system to label these different types of FRs denoting:

- 'G' (green) to FRs which are either common or completely converge with regards to tolerances required from Clients 1 and 2.
- 'A' (amber) to FRs which partially converge with regards to tolerances required from Clients 1 and 2.
- 'R' (red) to FRs which are in conflict and are flagged for special scrutiny in a context-based analysis as: (i) they might be ignored by Client 2; (ii) their range of tolerances might have no overlap with the range of tolerances for Client 1 FRs; or (iii) their simple proposition might conflict with FRs from Client 1 altogether.
- 'N' (neutral) to FRs which are from Client 2 only and do not necessarily affect Client 1 expectations, needs or aspirations.

As pointed by [8] the writing of FRs is one of the most important parts of a design process and can take several iterations so conflicts are avoided and convergences are carefully crafted so the needs of multiple stakeholders are addressed. The authors therefore focus on assessing FRs from Clients 1 and 2 by inferring them based on CNs selected from different documentary sources. To detect the needs, and in some cases the FRs themselves, of the PA (Client 2), the TH procurement documents used in Central Italy, namely the "Capitolato Tecnico D'Appalto" [17] and [18] are used as core references together with the Sphere handbook [2], which is a key international standard used by most NGOs and governments in post-disaster situations. These documents also allow extracting systems constraints (Cs), i.e., bounds on acceptable solutions, which in this case are identified as regulatory restrictions related to the provision of TH in Italy.

In the Italian case, the definition of the TH brief involves multiple experts, who, in different phases, contribute to its refinement. The process begins with the loose drafting by the NDCP of the technical document of the strategic framework agreement for TH post-disaster procurement (in forth referred to as the "Capitolato"). This seeks to translate TH needs into a set of "solution neutral" FRs about TH supply and delivery. Tender's participants respond to the brief by presenting a technical and an economic offer (including the design of TH units), detailing the service they will provide. The actual number and type of TH units needed (including for instance special accessibility requirements) is decided by public authorities only after a disaster strikes, following a detailed assessment of damage to the housing stock and of TH needs. Next, candidate TH sites' locations are identified. When the plots become available for construction, the TH purchase order is raised, and the supplier is contracted to design the layout of the TH sites by the PA. The professionals working for the supplier, i.e., experts in the firms' technical office, need to translate the requests detailed in the supply order into a workable project brief, by refining the lower-level constraints and FRs, considering all the contractual, regulatory, site-specific, planning and design inputs. Within this framework, civil protection actors seek to include in the strategic brief of the "Capitolato", in later executive ordinances, and in the purchase orders their understanding of people needs, translated into Client 1 FRs. The same is presumably done by town planners and designers, according to their professional ethics, albeit normally within the framework set up by the PA, and possibly through a negotiation process with other stakeholders. Thus, the final list of FRs for TH units and sites is determined following what seems to be a mainly top-down and multi-staged process.

Although, in principle, the community should be directly involved in the process of defining the project brief to implement a truly people-centred TH programme, the way the TH assistance delivery process is currently organised makes this task particularly challenging, especially due to its tight time constraints. In addition, organisational and technical constraints may as well prevent civil protection actors from delivering what disaster-affected people expect. Thus, to detect the needs of TH occupants (Client 1) this work uses a mix of first [19] and secondary information [20] from official documents and semi-structured interviews.

3.1. Context independent common FRs

Some important common FRs, which open the door for setting converging tolerances in relation to DPs required to fulfil them, are illustrated in Table 1 (all assigned to group G). These mainly come from non-context specific documents such as [2] (used as guidance by humanitarian agencies) and – albeit in minor part - the "Capitolato", which attempt to respond to the needs and aspirations of both clients in coordination.

Table 1. Common FRs which are context-independent either at a national or international level.

FRs from Client 1 & Client 2
Site planning and design
G.1 - Optimise delivery timing.
G.2 - Provide water, sanitation facilities and drainage.

G.3 - Locate new settlements at a safe distance from actual or potential threats and/or remove hazard from sites and repair any serious environmental degradation.

G.4 - Provide acceptable distance and safe travel (or transport) to essential services and facilities.

G.5 - Provide essential services and livelihoods opportunities including child-friendly spaces, gathering areas, and spaces to respond to religious needs.

Housing Design

G.6 - Provide living space, toilet and bathing facilities, spaces for sleeping, food preparation, cooking and eating, socialising, and play areas[1].

G.7 - Provide optimal lighting conditions, ventilation, and thermal comfort.

G.8 - Consider local culture and lifestyles

3.2. FRs from Client 2 – Anticipating issues

FRs from Client 2, listed in Table 2, are found both in context-independent documents (in [2] Client 2 is not uniquely defined and could be the local PA as well as an international NGO) as well as in country-specific strategic documents related to the case study under examination [17], where they are stated as duties of the PA. These FRs mainly refer to speed of delivery and catering for social and environmental sustainability and can be understood as convergent with community FRs. However, since tolerances for these might vary between what is expected from the two different clients, they are labelled as 'A' rather than 'G'. These FRs should be carefully analysed with regards to the range of tolerances expected by each client, to ensure they have at least some alignments so they can be resolved throughout the design process.

Table 2. FRs from Client 2 assessed with regards to how they align with expectations from Client 1.

FRs from the Public administration (Client 2)
Site planning and design
N.1 - Optimise cost-efficiency.
N.2 - Ensure modularity in TH solutions and site layout.
G.9 - Restore TH site after occupation to allow the local population to use it.
A.1 - Consider the placement of essential services within the settlement, if possible.
A.2 - Consider the expected lifespan of the settlement to determine what essential services may need to be expanded or developed.
Housing Design
N.3 - Use environmentally sustainable construction techniques and materials.
G.10 - Ensure that price and quality meets market standards.
G.11 - Ensure any energy supply options meet user needs, and provide training and follow-up as needed.
A.3 - Salvage and reuse, recycle or re-purpose available materials, including debris.

In addition to these, and as part of the duties form Client 2, it is not uncommon to find constraints related to the urban context which imply working within a geographically defined area-based approach to better understand community dynamics and demands to follow standards for safety, protection and dignity in

[1] Here tolerances may need to be adjusted at lower levels as what is considered sufficient in terms of space by different clients is context dependent.

construction and planning, ensuring public health and safety as well as providing physical security, dignity, privacy, and protection from the local weather. For transparency purposes [21], it is also expected that Client 2 follows appropriate and traceable tender, bidding, procurement, contract and construction management processes and codes of conduct.

3.3. The reality brought by adding context – Mismatch of tolerances and potential conflicts.

When examining a real context and seeing FRs from the PA (Client 2) and the community (Client 1) involved in the Central Italy earthquake of 2016, the situation gets far more complex than what is assessed in sections 3.1 and 3.2. FRs in [18] and in the many disaster-specific NDCP ordinances issued from 2016 to 2018 get far more specific, and pragmatism needed from the PA to deliver TH plans using established resources can easily overwrite community FRs related to demographics and sense of place.

Table 3 uses the same traffic light systems adopted in sections 3.1 and 3.2 to assess commonalities, explore the overlap of tolerances and flag potential conflicts between FRs from Client 1 and Client 2. The reality of the Italian case study shows that when FRs are defined for a given context, special attention needs to be dedicated to their writing at lower levels, considering the range of tolerances expected by each client to ensure minimum alignment for DPs to fulfil them. For instance, some of the disaster specific FRs from the PA (e.g., A.7, A.8, A.13) present a rather limited range of tolerances which can well be in conflict with FRs from the community (e.g., R1, R4 and R2) once further decomposed into second level FRs. In fact, Table 2 shows that most of the conflicting FRs identified (R.2 - R.8) can arise from decomposing high-level FRs into low-level ones, exposing a set of community FRs not featured as part of TH project objectives.

Table 3. FRs from both clients extracted from [17], [18] and observations and interviews with the community.

FRs from the Community (Client 1)	FRs from the Public Administration (Client 2)
Site planning and design	
G.12 - Minimize land use change and keep people close to destroyed areas.	
R.1 - Provide design solutions which fit accessibility in mountainous context.	A.5 - Provide design solutions which respect site topography.
R.2 - Provide collective structures and plenty of green areas for urban gardening.	A.6 - Provide green infrastructure with local species or suitable to local climate inserted in the local landscape but with minimal plants and urban furniture.
R.3 - Provide paved walkways to access TH units, which don't get muddy when it rains.	A.7 - Provide hierarchical internal road and pedestrian path infrastructure with 2 parking spaces per TH unit.
	N.4 - Optimize construction speed.
Housing design	
G.13 - Connect TH units to the grid for water, energy electricity, sewage, phone, gas.	
R.4 - Give special consideration to accessibility for the elderly.	A.8 - Provide 3 TH units typologies (40, 60, 80 m^2 to host a max of 2,4,6 people respectively) with 20% of these ensuring disabled accessibility if necessary.
R.5 - Provide spaces to house caregivers to those without easy access to care services.	A.9 – TH units' structure needs to be prefabricated and units to be combined to reach two-stories with external individual access.
R.6 - Ensure energy systems provided are affordable.	A.10 - Use passive design strategies and minimise thermal energy demand.

R.7 - Provide climate responsive roofs & plumbing.	A.11 - Ensure TH units are flexible and adaptable to change of use in the same site and removeable.
R.8 - Provide complete design including finishes, and easy to maintain.	A.12 - Ensure TH with a minimum lifespan of 10 years fully furnished and ready to use.
	A.13 Provide external private spaces for THs with 25% of net floor area.

Table 3 shows that as soon as a context is brought into place and FRs need to be further decomposed towards the development of a tangible solution, the lack of community involvement on the process becomes not only evident but also a potential problem. Narrow tolerances for overlaps as well as conflicts are unfolded as potentially unsolvable whereas they could have been resolved through early consultations and iterative conversations among Clients 1 and 2 throughout the decision-making process. Since time is a tight constraint for emergency management in the aftermath of a disaster, a post-factum analysis of this sort is essential to extract lessons to be learned for informing future endeavours. An important one is that community interactions should be featured in preparedness plans in a pragmatic way to ensure a better post-disaster response.

4. Conclusions and suggestions for future work

The post-factum application of AD to the analysis of the case of the 2016-2017 Central Italy earthquakes enables lessons to be learnt from a strategic TH procurement, policy, and tactical planning perspective. Results indicate that the Italian PA should revisit its approach to problem definition. They also show local authorities should raise awareness of disaster risk and engage with communities likely to be affected by future hazards about TH response plans to discuss FRs with them early, as part of preparedness. In this way both clients would already know each other and be therefore in a better position to ensure FRs from Client 1 are extracted rapidly when needed so design objectives can be negotiated/reconciled promptly. The relationship between candidate DPs and the resulting FRs could then be systematically mapped on a case-by-case basis using the matrix representation $\{FR\} = [A]\{DP\}$, to check for factors' independence and enhance the completeness, consistency, and plausibility of processed information, enabling the delivery of more sustainable and context-based design proposals. As construction supply chains as well as climate and culture can change from region to region, this approach could support a shift of focus from the design of universal shelter solutions to the proposal of a sound design workflow without slowing down the process of TH assistance delivery.

The aforementioned case can well be used as a prompt to consultations with a focus on rehearsing mitigating mechanisms to refine and/or re-write conflicting FRs and poorly convergent tolerances, as proposed by [8], so that they can more flexibly accommodate a successful design response at a time of crisis, by for instance:

- Increasing the range of FRs' tolerances.
- Increasing the range of DPs' tolerances.
- Integrating DPs to respond to multiple FRs.
- Searching for independent FRs.
- Negotiating and establishing priorities between FRs so a strategic plan can be in place.

Notably, the analysis made a simplification in the number of clients and related needs to be considered in post-disaster situations which in reality is rather large and could extend, for instance, to comprehend NGOs, volunteers and other local residents in addition to TH occupants and the PA. The latter can also be further subdivided into national and regional governments, different local authorities, professional bodies, the army, different civic protection bodies etc. Nevertheless, AD does not differentiate according to their number. On the contrary, it enables to flexibly reconcile all clients' needs within a unique framework without considering the distinctive characteristics of different clients and their different power relations which can visibly play a role in the FRs' selection and refinement processes.

Future work should therefore focus on investigating and developing mechanisms to ensure all different stakeholders, especially Client 1, are involved from the beginning in the TH decision-making process so the 5 gaps framework in connection with AD can be effectively applied when a disaster strikes, preventing it to simply pay a lip service to the affected community. Better preparedness plans are indeed essential to enhance technical procurement documents for TH supply and delivery services. By embedding in those the possibility to include the value drivers of all the clients of TH schemes in different contexts, designers will be better instrumented to perform the hard task of rapidly delivering context-sensitive TH solutions which foster the sustainable recovery of settlements impacted by rapid onset disasters.

References

[1] I. Davis and D. Alexander, Recovery from disaster. Taylor and Francis, 2015.
[2] Sphere Association, The Sphere Handbook: Humanitarian Charter and Minimum Standards in Humanitarian Response, Fourth edi. Geneva, Switzerland: Practical Action Publishing, 2018.
[3] C. Pezzica, "Temporary Housing and Transitional Urbanism: planning for sustainable post-disaster recovery under uncertainty," University of Pisa, 2021.
[4] C. Pezzica, V. Cutini, and C. Bleil de Souza, "Mind the gap: state of the art on decision-making related to post-disaster housing assistance," Int. J. Disaster Risk Reduct., vol. 53, p. 101975, Nov. 2021.
[5] BSI, BS ISO 55002: 2018 Asset management - management systems - guidelines for the application of ISO 55001. British Standards Institution, 2018.
[6] P. Forsythe, "Modelling customer perceived service quality in housing construction," Eng. Constr. Archit. Manag., vol. 15, no. 5, pp. 485–496, 2008.
[7] Construction innovation hub, "An Introduction To The Value Toolkit." pp. 1–9, 2020.
[8] N. P. Suh, Axiomatic Design: Advances and Applications. New York: Oxford University Press, 2001.
[9] A. Parasuraman, V. A. Zeithaml, and L. L. Berry, "A Conceptual Model of Service Quality and Its Implications for Future Research," J. Mark., vol. 49, no. 4, p. 41, 1985.
[10] G. Winch, A. Usmani, and A. Edkins, "Towards total project quality: a gap analysis approach," Constr. Manag. Econ., vol. 16, pp. 193–207, 1998.
[11] A. Parasuraman, V. A. Zeithaml, and L. L. Berry, "Servqual: A Multiple-Item Scale For Measuring Consumer Perc - ProQuest," J. Retail., vol. 64, no. 1, 1988.
[12] A. Sutrisno, D. Handayani, W. Caesarendra, I. Gunawan, and R. Indan, "A framework of designing reliable disaster response operation using axiomatic design," in AIP Conference Proceedings, 2020, vol. 2227, no. 1, p. 040015.
[13] M. Marchesi and D. T. Matt, "Application of Axiomatic Design to the Design of the Built Environment: A Literature Review," in Axiomatic Design in Large Systems, Springer International Publishing, 2016, pp. 151–174.
[14] C. Parretti, F. Rolli, E. Pourabbas, and P. Citti, "Axiomatic Selection of Health and Social Care Web Services on the Basis of Use Cases," in Prooceding of the he 12th International Conference on Axiomatic Design (ICAD 2018), 2018, vol. 223, p. 01013.
[15] G. Arcidiacono, D. T. Matt, and E. Rauch, "Axiomatic Design of a Framework for the Comprehensive Optimization of Patient Flows in Hospitals," J. Healthc. Eng., vol. 2017, Jun. 2017.
[16] G. Arcidiacono, A. Giorgetti, and M. Pugliese, "Axiomatic Design to improve PRM airport assistance," Procedia CIRP, vol. 34, pp. 106–111, Jan. 2015.
[17] CONSIP & NDCP, "Allegato 5 - Capitolato Tecnico D'appalto AQ SAE," 2014. [Online]. Available: https://serviziosae.cnsofm.it/images/serviziosae.cnsofm.it/SAE2_Allegato_5_-_Capitolato_Tecnico_public.pdf. [Accessed: 14-Aug-2020].

[18] Presidenza del Consiglio dei Ministri, "Circolare della Presidenza del Consiglio dei Ministri del 16/01/2017." 2017.
[19] C. Chioni, "From Temporariness to Permanence.The case of 'Borgo di Arquata' after the 2016 central Italy earthquake", University of Pisa, 2019.
[20] Emidio di Treviri, Sul fronte del sisma. Un'inchiesta militante sul post-terremoto dell'Appennino centrale (2016-2017). DeriveApprodi, 2018.
[21] Italian Government, Legge n. 100 del 12 luglio 2012. Italy, 2012.

IOP Conf. Series: Materials Science and Engineering 1174 (2021) 012015 doi:10.1088/1757-899X/1174/1/012015

Robust Privacy Assessment in Transnational Healthcare Systems

C Parretti[1], E Pourabbas[2], F Rolli[1], F Pecoraro[3] and P Citti[1]

[1] Guglielmo Marconi University, Department of Engineering Science, Via Plinio 44 - 00193 Rome, Italy
[2] National Research Council, Institute for System Analysis and Computer Science "A. Ruberti", Via dei Taurini, 19 - 00185 Rome, Italy
[3] National Research Council, Institute for Research on Population and Social Policies, Via Palestro, 32 - 00185 Rome, Italy

E-mail: c.parretti@unimarconi.it

Abstract. Recent developments in Information and Communication Technology have paved the foundations for new forms of collaboration between health systems in different countries. These collaborations allow, on the one hand, to monitor recurrent health emergencies on territories, and on the other hand, they allow national health systems to share information about foreign citizens in transit on their territory. European legislation regarding the processing of personal data places strict constraints on the cross-border transfer of personal data. In this case, companies or organizations, operating on information interchange, must adopt robust mechanisms to verify the adequacy requirements, in order to allow monitoring of security levels, identification of possible intervention actions and preparation of updated security plans for inspection visits required by European standards. In this context, Axiomatic Design allows not only to design medical systems and equipment in compliance with current regulations, but also to provide representations of the design artifact already prepared to implant privacy risk assessment mechanisms. This makes it possible to identify the activities/components to be assessed up to a level of elementary granularity such as to allow risk assessment for the single module. At the same time, the axiomatic approach enables the overall recomposition of privacy violation risks on the basis of a modular representation of the whole system according to the well-known V model scheme. This recomposition allows to build the so-called risk privacy coverage matrix, in order to trace the risk level of the elementary modules, associating them with more and more complex components. In this way, the foundations to build a dynamic monitoring system of the privacy risk level of the system can be defined.

1 Introduction

The recent global health crisis induced by the coronavirus pandemic has brought to the attention of the international scientific community the importance of collaboration between countries in the field of public health. The World Health Organization (WHO) already provides specific communication protocols between the various national health systems and supranational organizations, in order to identify early outbreaks of infectious diseases, monitor endemic situations and define guidelines for intervention [1-3]. International collaborations between countries may concern both the sharing of data for exclusively scientific use, and the provision of information directly attributable to patients in order to allow their health treatment in another country, in case of mobility [3, 4]. This last specificity has always posed relevant questions regarding the respect of privacy and the personal data processing [4]. For this reason, the countries adhering to the European Union have adopted the so-called General Data Protection Regulation 679/2016 (GDPR) [5, 6]. The rules contained therein are binding for all countries in the EU area. This has laid the groundwork to enable the interchange of health information about citizens of member countries [7]. This regulation has impact not only at the level of personal data processing, but also it provides requirements for the design, configuration and evaluation of systems and equipment, which handle personal information [8, 9]. For this reason, European legislators have introduced the concepts of Privacy by design, Privacy by default and Privacy Impact Assessment.

Privacy by design means that the design of a healthcare system must be done in a way that "proactively" eliminates or mitigates issues that may lead to privacy violations [10]. Privacy by default, on the other hand, means that a medical procedure or device must be configured in such a way that its primary mode of operation does not allow a healthcare professional or other party to access information for which they are not authorized [8, 9]. Conversely, Privacy Impact Assessment (PIA) is a risk assessment tool, which involves all organizational and technological aspects that impact the processing of personal data [5]. Therefore, this type of assessment must be contextualized to a specific scenario, in which the system must operate.

In this paper, we will focus our attention on how Axiomatic Design (AD) can guide the design of systems operating interchanges of health information between EU member countries. In particular, our goal is to propose a methodological approach that proactively allows reducing the risk of privacy breaches according to GDPR requirements. Specifically, we will give particular relevance to how AD allows to provide representations of the output design, compatible with the adoption of an evaluation tool such as PIA. Based on these representations, we propose an application scheme of PIA based on the axiomatic decomposition of the functional requirements of the system. The goal is the construction of a risk privacy coverage matrix, which describes the privacy risk map for each component of the system. This matrix becomes a tool for monitoring the level of risk associated with the system. In particular, we illustrate how this methodology can be applied to the case of privacy risk assessment for a process of health data interchange for an Italian patient hospitalized in another EU country.

2 Methodological Approach

End-user desires can be formalized in terms of user requirements (Customers Attributes). They consist of functional requirements (FRs), which represent "what the system must do," and design constraints (CSs), which are limitations on operation. In the specific case of GDPR, these constraints are the rules governing the processing of personal data. Instead, Design Parameters (DPs) constitute "how" the functional requirements are to be executed. The possible relations between FRs and DPs can be also multiple. They can give rise to different combinations, but only some of them are acceptable, as they preserve the independence of the functional requirements. This means that only one-to-one relationships can be candidates to represent an acceptable solution of the system to be designed. In matrix terms, this means that the relevant design matrices must necessarily be square diagonal matrices (uncoupled relation) or triangular matrices (decoupled relation). Only these types of FR-DP relations guarantee the functional independence of FRs requirements. On these evidences, the axiom of independence has been formulated. It allows only a finite number of combinations between FR-DP to be defined as admissible solutions [11]. The information axiom, on the other hand, allows restricting the selection to the design solution with minimal information content, i.e., the solution with the lowest complexity [11]. The application of these two axioms takes on specific characterizations, depending on the particular operational context. In particular, the information axiom lends itself to multiple re-interpretations. In our case, the axiom of information can be interpreted in terms of Privacy Impact Assessment. This means that, design solutions that satisfy the axiom of information, can be assessed against the degree of implementation of GDPR requirements. In other words, this methodological assumption allows implementing the concept of Privacy by Design, i.e., designing a system by proactively considering privacy constraints. However, AD is a top-down design methodology, which is based on the decomposition of the functional requirements of the system. This process is carried out by the designers, down to levels of detail that allow the system to be designed. This allows us, also, to build a robust privacy risk monitoring mechanism. This mechanism follows a V-model implementation process, as shown in Figure 1. According to this scheme, we can distinguish two phases of this model. The first phase of the process consists of the axiomatic design of the system, starting from the decomposition of the user requirements (Customers Attributes). This part has been extensively developed in the literature [11-13]. The second phase starts, instead, with the privacy impact assessment of the single elementary modules. At this point, starting with the interactions between elementary parts, the level of risk is measured for increasingly complex components, until the overall risk is determined. All these steps can be tracked by building a risk privacy coverage matrix.

IOP Conf. Series: Materials Science and Engineering 1174 (2021) 012015 doi:10.1088/1757-899X/1174/1/012015

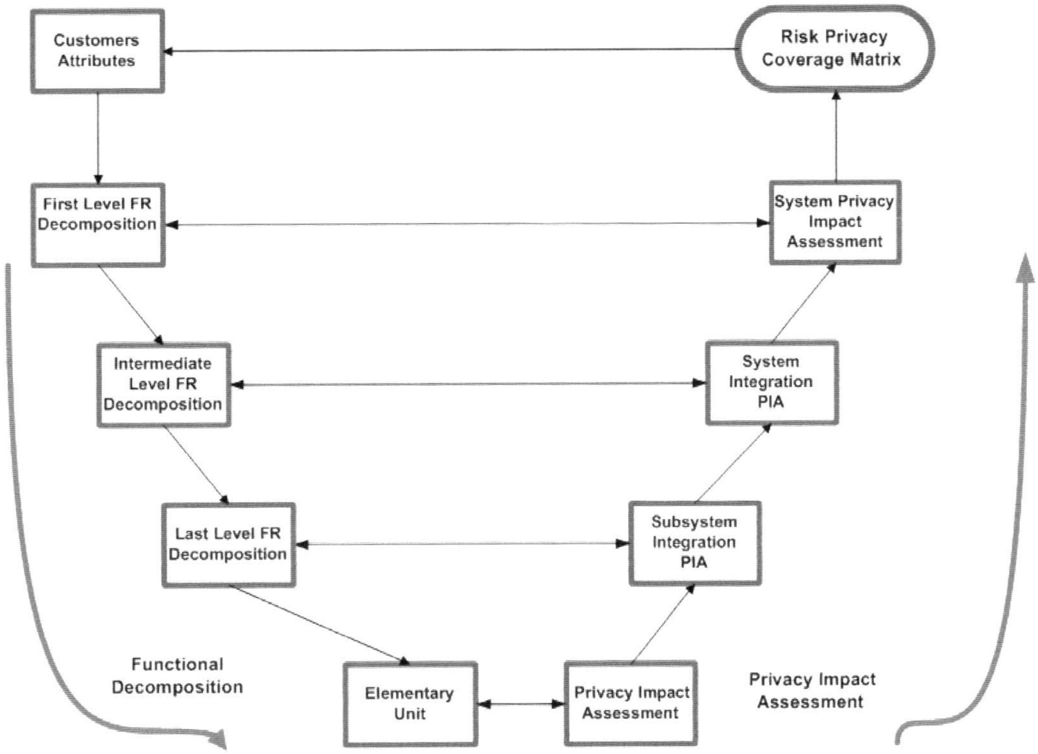

Figure 1 Axiomatic decomposition and Privacy impact assessment

3. Transnational exchange of health data for EU citizens

At this point we can introduce our case study in order to illustrate the application of the proposed approach.

The patient F. C., 65 years old, heart patient, works in Rome but as a dealer of a car company often travels to Greece. Due to his continuous travels, the patient always carries with him an App on his smartphone, where basic clinical data are stored (Patient Summary contained in the patient's medical record). The Patient Summary is generated by the general practitioner by accessing the information contained in his medical record (EHR).

During his most recent stay in Greece, F.C. has an illness and goes to the emergency department of a hospital in Athens. F.C. is conscious. He informs the physician he has the above App available. The doctor, through the data access service on the patient's mobile, views the information contained in the patient's medical record and sees that F.C. has a heart disease. Given the symptoms the patient has at that time, he requests access to F.C.'s additional information such as tests, specialist visits, etc. Thanks to a service offered by the App, after appropriate verification of privacy criteria, the information is directly uploaded from the medical record of the general practitioner. The doctor visits F.C., performs the diagnostic tests to verify the patient's situation and issues the report. The report is then stored on the phone. When F.C. returns to Rome, the general practitioner after examining him, views and stores the Greek doctor's report on his medical record.

In this scenario, therefore, we must take into account the following systems:

- General practitioner's medical record (EHR)
- Patient's personal medical record contained in their mobile (Patient Summary)
- App to access both the patient's personal medical record and the general practitioner's medical record.

3.1 Use-case scenario

The introduced case study can be schematized via UML representations [14, 15]. At this first level of detail, we define a representative form by the use case diagram shown in Figure 2. This representation allows us to define the conceptual design of the system [14]. It consists of the static representation of the system functions (use cases) and the actors operating on them [15]. For this purpose, we can introduce the following use cases:

1. Producing the Patient Summary (PPS). This activity is carried out by the general practitioner who treats the patient in his country of origin. It consists in defining the clinical picture and the therapeutic plan of the patient.
2. Store the Patient Summary (SPS). This use case consists of the GP recording essential health information to be made available in the patient's App for emergency situations.
3. Access the Patient Summary (APS). This use case consists of the Greek hospital center physician accessing summary health information in the cross-border patient's smartphone.
4. Access to Patient Record's (APR). This use case consists of the Greek physician accessing the health information on the cross-border patient's medical record.
5. Access to Italian Health Service (HIS). This use case consists in making available to the HIS the diagnostic and therapeutic information of the patient, treated in Greece.

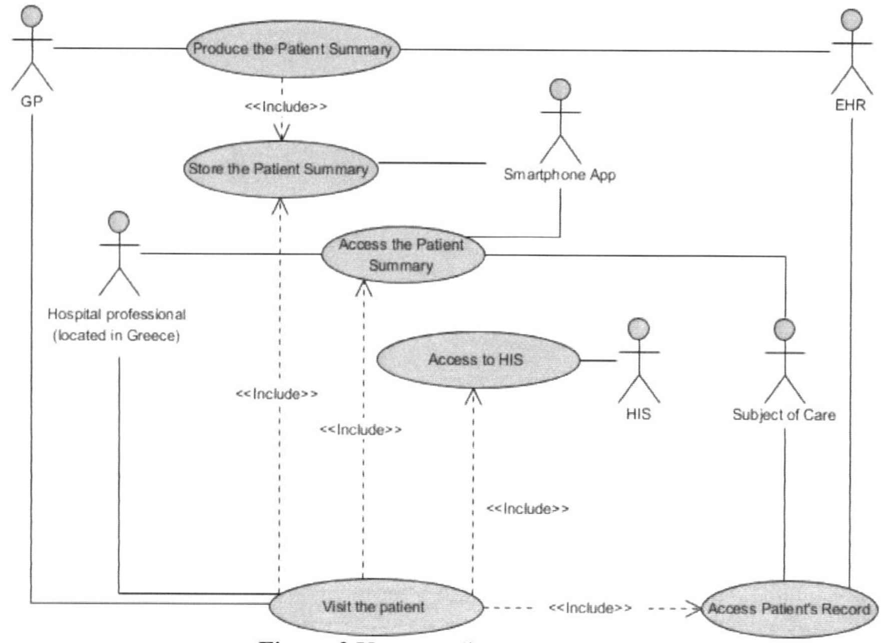

Figure 2 Use case diagram

3.2 Representative forms of the design artifact

For each level of functional decomposition AD provides a representation of the system in terms of design matrix [11]. From the high-level requirements of the system, it is possible to build design matrices representing increasingly detailed design details. This process is carried out to a level of detail sufficient to allow designers to implement elementary modules, which represent specific blocks of the overall design matrix. These modules can be software procedures, electronic components, mechanical parts or even human procedures [16]. AD allows the integrated design of modules of heterogeneous nature. The algebra of matrices, not only allows us to verify the logical coherence of the system, both in terms of functional independence of the individual component parts (Axiom of Independence), and of information complexity (Axiom of Information) [17, 18]. But it also allows us to translate the design matrix, into diagrams consisting of interconnected modules [11, 16]. These representational forms allow privacy impact assessment to be extended from the single elementary module to more complex structures based on the interactions of elementary components. In this study we will refer to only two levels of abstraction.

3.2.1 Conceptual design of the system

Figures 1 and 3 provide a summary description of the conceptual design of the system. But, in order to proceed to an initial application of AD we need to identify the FRs and DPs of our system. We can pose as FRs the use cases identified in §3.1. Instead, the DPs of the system represent the mutual interactions (collaborations) between the various use cases [12, 19]. Based on these two assumptions, we can provide an initial representation of the system in terms of a design matrix (Table 1). In this case, the relationships between use cases of the system are schematized using the matrix in Table 1. This matrix is constructed by reporting the use cases in Figure 3 along the rows: the functional requirements of the system [12, 19]. Along the column axis, the collaborations (interactions) between use cases are inserted: the high-level interactions between use cases. In terms of AD these are the design parameters of the system and define its behavior. The mapping between FRs and DPs is represented in Table 1 by the symbol X [11]. The cells of the matrix are valorized with symbol X if the use case indicated on the column axis is able to activate a process capable of modifying the state of the target use case [19]. By definition each use case can be modified by its target actor. For this reason, self-collaboration (VP collaboration with respect to VP) is represented in the matrix by the symbol X. As can be seen, the design matrix in Table 1 was constructed on an empirical basis, interpreting the description of end-user desires as given in the previous paragraphs. Our design process is iterative in nature. Therefore, as a starting point, this solution can be considered valid, also because it respects the axiom of independence, since the related design matrix is triangular [11]. With subsequent iterations, following the scheme summarized in Figure 1, we can proceed to improve the conceptual design of the system as well. However, the phenomenon related to sequential coupling will always turn out to remain. This is due to the fact that the activation of the use case FR_{i+1} always depends on the previous FR_i. This is a structural type situation [20]. On the other hand, this type of sequential coupling can be neglected because the relationship between FR-DP is of decoupled type. For simplicity, we do not apply the information axiom.

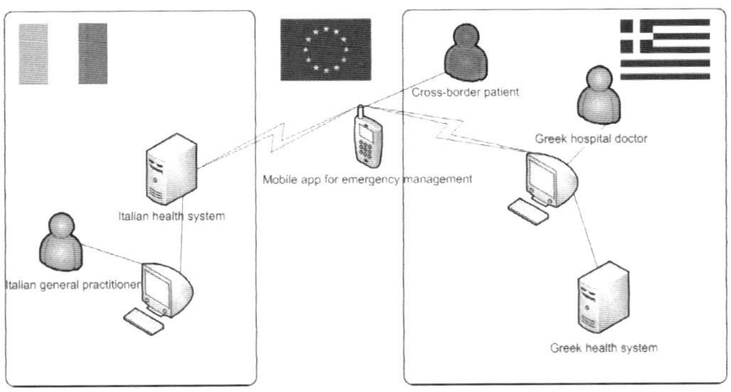

Figure 3 Summary representation of the mode of acquisition of medical information related to EU patient in emergency in Greece

Table 1. Conceptual system design matrix

	PPS collaboration	SPS collaboration	APS collaboration	APR collaboration	HIS collaboration
PPS	X				
SPS	X	X			
APS		X	X		
APR		X	X	X	
HIS			X		X

3.2.2. Modular representation of the conceptual design of the system

The design matrix in Table 1 can also be represented in equivalent form as a system of equations (eq.1).

IOP Conf. Series: Materials Science and Engineering 1174 (2021) 012015 doi:10.1088/1757-899X/1174/1/012015

$$\begin{Bmatrix} PPS \\ SPS \\ APS \\ APR \\ HIS \end{Bmatrix} = \begin{bmatrix} X & & & & \\ X & X & & & \\ & X & X & & \\ & X & X & X & \\ & & X & & X \end{bmatrix} \begin{Bmatrix} PPS\ col \\ SPS\ col \\ APS\ col \\ APR\ col \\ HIS\ col \end{Bmatrix} \tag{1}$$

This representation is possible because each row of the design matrix in Table 1 is a relationship between DPs and FRs. We can say that every functional requirement FRs is satisfied if there exists a function M (module) such that FR =M (DP). For simplicity of expression we can introduce the operator *. This operator allows us to represent in equivalent modular form the mapping relations (FR/DP), even in cases of non-linearizable systems and at high abstraction levels. In this way, we can use the following simplification: FR =M (DP) = M*DP. This assumption allows us to translate the design matrix of Table 1 into diagrammatic terms. In this context, we have resorted to flow chart diagrams to provide a modular representation of the system. However, before describing the flow chart associated to the system of Table 1, it is necessary to introduce three rules of composition of a modular system from a design matrix.

Rule #1:

If the elements of the design matrix are arranged along a diagonal, the system is said to be uncoupled [11]. These systems are characterized by the fact that the functional requirements FRs are all independent (eq.2).

$$\begin{Bmatrix} FR_1 \\ FR_2 \end{Bmatrix} = \begin{bmatrix} a & \\ & b \end{bmatrix} \begin{bmatrix} DP_1 \\ DP_2 \end{bmatrix} \rightarrow \begin{cases} FR_1 = M_1 * DP_1 \\ FR_2 = M_2 * DP_2 \end{cases} \tag{2}$$

In this case, the modules of the system can be represented in terms of flow charts as in figure 4. The symbol S stands for the sum relationship of the outputs of the two modules. This configuration corresponds to a parallel connection of the system modules.

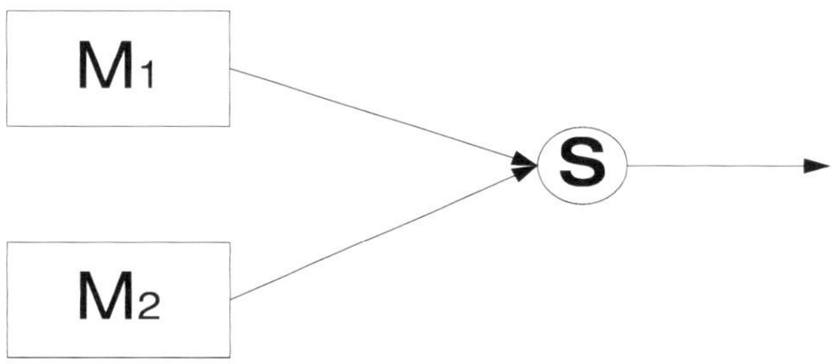

Figure 4
Flow chart diagram for an uncoupled system

Rule #2:

A design matrix of triangular type is defined as decoupled [11]. In this case, the next module always depends on the previous one. In equivalent form, we can say that the modules of the system are connected in series. The following system of equations illustrates this particular configuration (eq.3).

$$\begin{Bmatrix} FR_1 \\ FR_2 \end{Bmatrix} = \begin{bmatrix} a & \\ b & c \end{bmatrix} \begin{bmatrix} DP_1 \\ DP_2 \end{bmatrix} \rightarrow \begin{cases} FR_1 = aDP_1 = M_1 * DP_1 = f(DP_1) \\ FR_2 = bDP_1 + cDP_2 = M_2 * DP_2 = f(DP_1, DP_2) \end{cases} \tag{3}$$

where

$$M_2 = b\left(\frac{DP_1}{DP_2}\right) + c \tag{4}$$

In this case, the operation of the system is defined by an ordered sequence of actions. It begins with the M_1 module and progresses to the M_2 module. This means that the output of the module M_1 constitutes the input of the module M_2. Figure 5 shows the graphical representation in terms of flow chart diagram of the system of eq.2.

Figure 5 Flow chart diagram for an uncoupled system

IOP Conf. Series: Materials Science and Engineering 1174 (2021) 012015 doi:10.1088/1757-899X/1174/1/012015

Rule #3
A design matrix is called coupled when the functional requirements (FRs) are not independent [11]. In this case the behavior of the system can no longer be represented by simply connecting modules in series and/or in parallel. In order to define the state of the system it is necessary that some modules have an informative feedback (F-Feedback). In these situations, it is said that there is a feedback relationship between the various modules. In matrix terms connections of this type are represented by sparse matrices.

$$\begin{Bmatrix} FR_1 \\ FR_2 \end{Bmatrix} = \begin{bmatrix} a & d \\ b & c \end{bmatrix} \begin{bmatrix} DP_1 \\ DP_2 \end{bmatrix} \rightarrow \begin{cases} FR_1 = aDP_1 + dDP_2 = M_1 * DP_1 \\ FR_2 = bDP_1 + dDP_2 = M_2 * DP_2 \end{cases} \tag{5}$$

Where
$$\begin{cases} M_1 = a + d\left(\frac{DP_1}{DP_2}\right) = f(DP_1, DP_2) \\ M_2 = b\left(\frac{DP_1}{DP_2}\right) + c = f(DP_1, DP_2) \end{cases} \tag{6}$$

In graphical form the system of eq.5 is representable from the flow chart of figure 6.

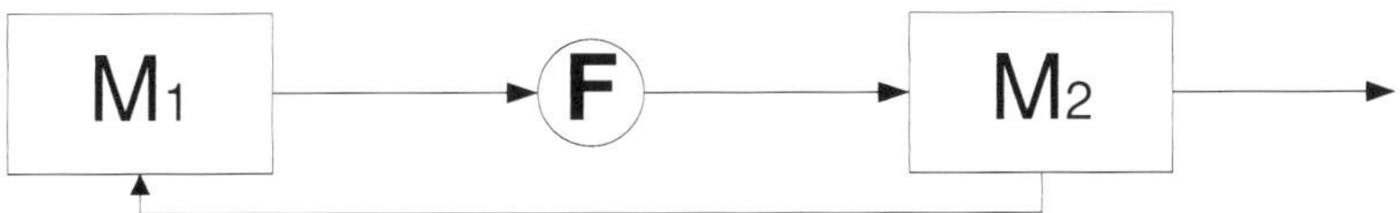

Figure 6 Flow chart diagram for a coupled system

From the previous three rules, the system of equations (eq.1) can be converted graphically in the flow chart diagram of figure 7. Each Mi-th block represents a row of the design matrix of Table 1 [16].

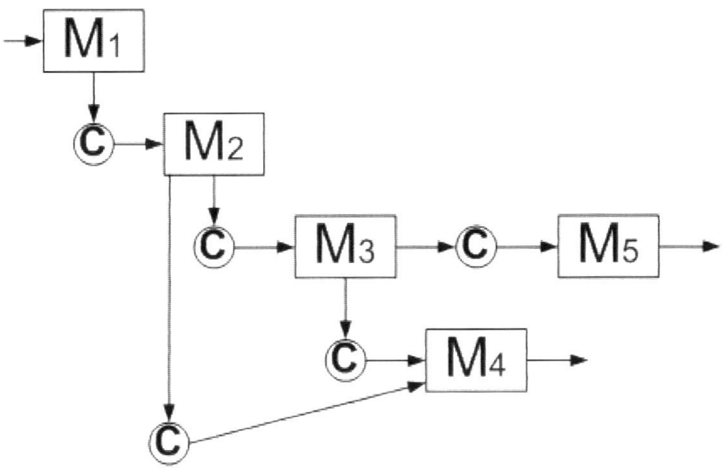

Figure 7 Flow chart of the conceptual design of the system

3.3 Representations of the logical design of the system
AD allows us to continue the functional decomposition of our case study. Starting from the matrix of Table 1 we can obtain, with two successive decompositions, the design matrix of Table 2. It represents in matrix form the logical design of the system and describes the behavior of the system itself in terms of sequence of activities. We can provide a twofold representation of the logical design of the system: dynamic or static. The dynamic representation comes described through translation of the design matrix of table 2 in terms of sequence diagram. In this case our attention is stopped on the sequence of necessary actions in order to define the main behavior of the system. The static representation of the system instead sets to the center of the analysis the activities that come executed in sequence. They come represented in modular shape through flow chart. At this level of detail, the phenomenon of sequential coupling is much more evident. As anticipated in section §3.2.1, it depends on the particular operational context

IOP Conf. Series: Materials Science and Engineering 1174 (2021) 012015 doi:10.1088/1757-899X/1174/1/012015

that involves the sequential activation of the actions constituting the main behavior of the system [20]. However, in this case, this phenomenon can still be neglected because the design matrix of Table 2 turns out to be decoupled. Therefore, the independence axiom is valid.

Table 2 System logic design matrix

			PPS collaboration			SPS collaboration				APS collaboration							APR collaboration						HIS collaboration					
			DPA		DPB	DPC				DPD		DPE	DPF	DPG		DPH	DPI	DPL	DPM				DPN		DPO			
			DP1	DP1.1	DP2	DP3	DP3.1	DP3.2	DP3.3	DP4	DP4.1	DP5	DP6	DP7	DP7.1	DP8	DP9	DP10	DP11	DP11.1	DP11.2	DP11.3	DP12	DP12.1	DP13	DP13.1	DP13.2	DP13.3
PPS	A	1	X																									
		1.1	X	X																								
	B	2		X	X																							
SPS	C	3			X	X																						
		3.1				X	X																					
		3.2					X	X																				
		3.3						X	X																			
APS	D	4								X																		
		4.1								X	X																	
	E	5									X	X																
	F	6										X	X															
	G	7					X						X	X														
		7.1												X	X													
	H	8													X	X												
APR	I	9														X	X											
	L	10															X	X										
	M	11					X											X	X									
		11.1																	X	X								
		11.2																		X	X							
		11.3																			X	X						
HIS	N	12														X							X					
		12.1																					X	X				
	O	13																						X	X			
		13.1																							X	X		
		13.2																								X	X	
		13.3																									X	X

3.3.1 Dynamic representation of the logical design of the system

Starting from the design matrix of Table 2 we can obtain a sequence diagram [19]. This particular diagram describes the main behavior of the system as a sequence of actions. They can be human procedures, such as *4: getPatientInfo*. In this case the doctor of the Greek hospital facility accesses, via App of the patient's cell phone, the patient's health information. In other cases, they are automated processes of the system triggered under specific conditions. For such reason this particular diagram previews the indication of the trigger of activation of the action: a stylized man, in the case of human intervention; a rectangle with the indication of the activating system, for the processes started from the system. In figure 8 we have brought back the sequence diagram that describes the dynamic behavior associated to the logical design of the system [15].

3.3.2 Static representation of the logical design of the system

From the design matrix in Table 2, it is possible to represent the top-level modules (M_1, M_2, M_3, M_4, M_5) as distinct systems of equations [11]. From this we have the following equations:

$$\begin{Bmatrix} 1 \\ 1.1 \\ 2 \end{Bmatrix} = \begin{bmatrix} X & & \\ X & X & \\ & X & X \end{bmatrix} \begin{Bmatrix} DP1 \\ DP1.1 \\ DP2 \end{Bmatrix} \tag{7}$$

$$\begin{Bmatrix} 3 \\ 3.1 \\ 3.2 \\ 3.3 \end{Bmatrix} = \begin{bmatrix} X & & & \\ X & X & & \\ & X & X & \\ & & X & X \end{bmatrix} \begin{Bmatrix} DP3 \\ DP3.1 \\ DP3.2 \\ DP3.3 \end{Bmatrix} \tag{8}$$

$$\begin{Bmatrix} 4 \\ 4.1 \\ 5 \\ 6 \\ 7 \\ 7.1 \\ 8 \end{Bmatrix} = \begin{bmatrix} X & & & & & & \\ X & X & & & & & \\ & X & X & & & & \\ & & X & X & & & \\ & & & X & X & & \\ & & & & X & X & \\ & & & & & X & X \end{bmatrix} \begin{Bmatrix} DP4 \\ DP4.1 \\ DP5 \\ DP6 \\ DP7 \\ DP7.1 \\ DP8 \end{Bmatrix} \tag{9}$$

$$\begin{Bmatrix} 9 \\ 10 \\ 11 \\ 11.1 \\ 11.2 \\ 11.3 \end{Bmatrix} = \begin{bmatrix} X & & & & & \\ X & X & & & & \\ & X & X & & & \\ & & X & X & & \\ & & & X & X & \\ & & & & X & X \end{bmatrix} \begin{Bmatrix} DP9 \\ DP10 \\ DP11 \\ DP11.1 \\ DP11.2 \\ DP11.3 \end{Bmatrix} \quad (10)$$

$$\begin{Bmatrix} 12 \\ 12.1 \\ 13 \\ 13.1 \\ 13.2 \\ 13.3 \end{Bmatrix} = \begin{bmatrix} X & & & & & \\ X & X & & & & \\ & X & X & & & \\ & & X & X & & \\ & & & X & X & \\ & & & & X & X \end{bmatrix} \begin{Bmatrix} DP12 \\ DP12.1 \\ DP13 \\ DP13.1 \\ DP13.2 \\ DP13.3 \end{Bmatrix} \quad (11)$$

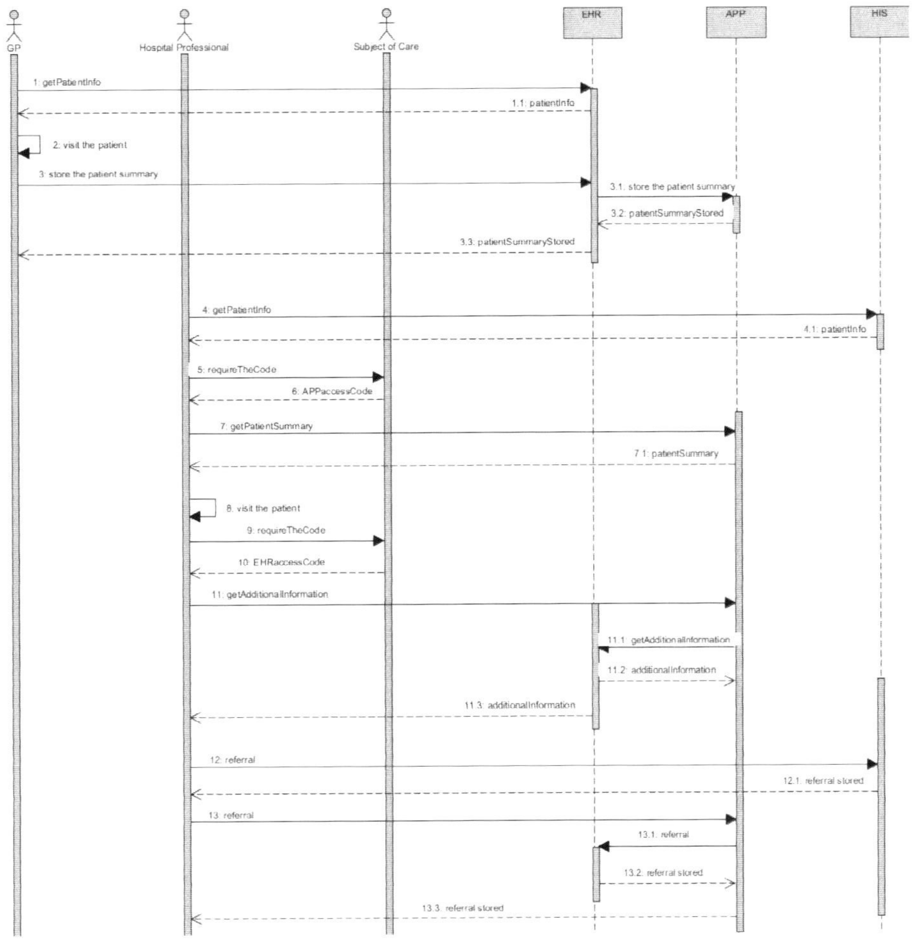

Figure 8 Sequence diagram of the logical design of the system

These equations (eq.7 through eq.11) allow us to treat the top-level blocks as autonomous systems. In fact, we can use the rules introduced in section §3.2.2 to decompose each block into lower level modules. In this way it is possible to construct a flow chart diagram representing the logical design of the system (Figure 9). The blocks M_{ij} are the rows of the matrices referred to the equations from eq.7 to eq.11. This process of modular decomposition can be carried out until identifying the activities or elementary components of the system, on which to carry out the first evaluation of the impact of the privacy constraints. The recomposition of the level of risk beginning from the elementary blocks follows the rules introduced in section §3.2.2. For this reason, the representation of the system in terms of flow chart allows to build the map of the overall risk level.

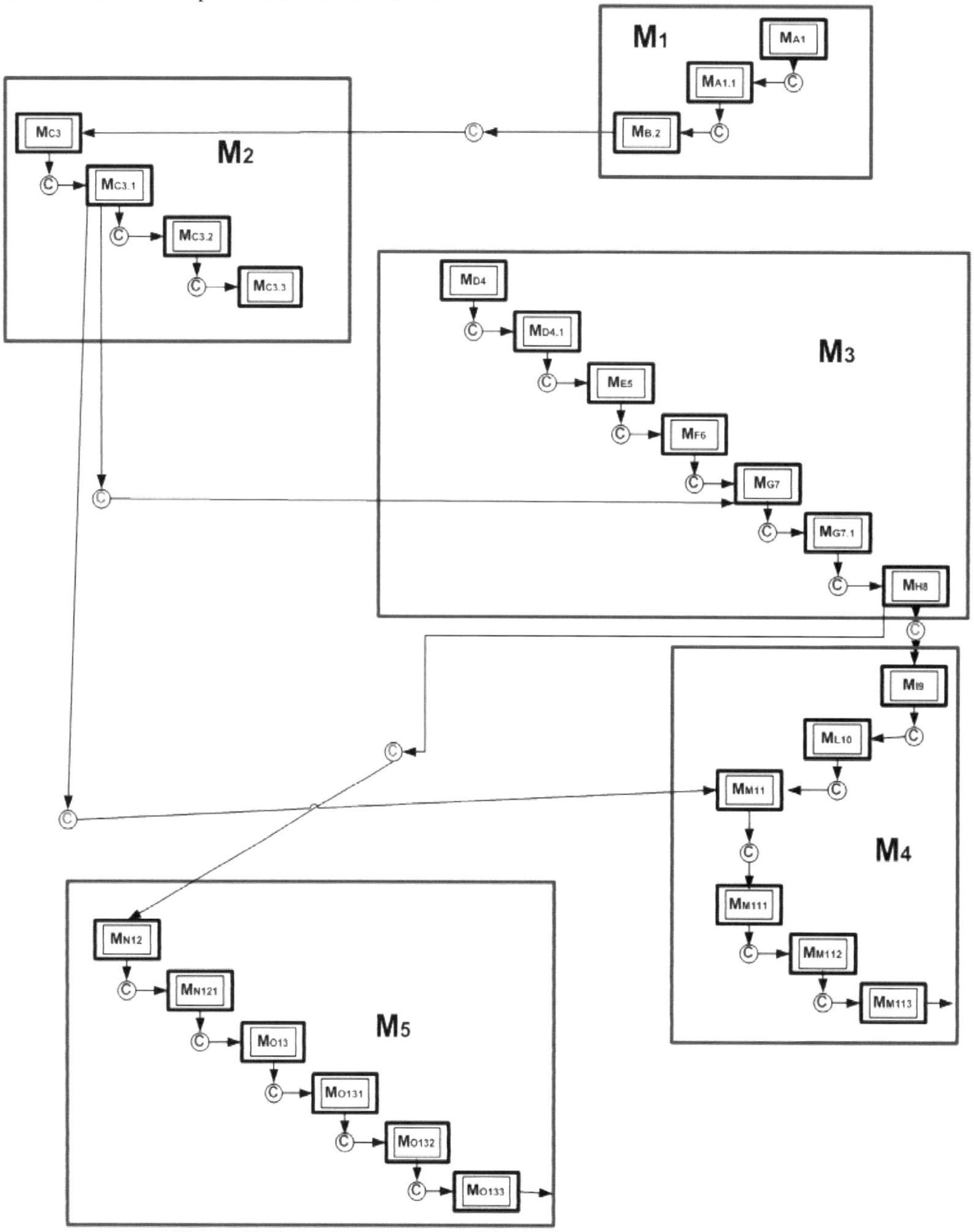

Figure 9 Flow chart diagram of the logical design of the system

4 Elements of the Privacy Impact Assessment plan

The concept of Privacy Impact Assessment, introduced in the European Data Protection Regulation [5], consists in the evaluation and systematic monitoring of the personal data processing that are most exposed to the risk of violation. European legislation has defined three macro-categories of privacy risk [8]:

- Illeg

- al access to data (CS_1);
- Unwanted data changes (CS_2);
- Loss of data (CS_3).

These three macro-categories may represent functional constraints for the AD methodology for the design and subsequent management of the system [11]. In this context, our goal is to build a flexible system monitoring mechanism that allows to estimate the impact of these three constraints in a preventive approach, as it supports software designers in the identification of any critical issues before problems arise. Furthermore, the evaluations criteria defined within the PIA should be adopted to define the overall plan of risks associated with the processing of personal data.

4.1 Modular assessment of the impact of privacy risk

The modular representation of our process allows us to estimate the impact of privacy constraints with respect to individual modules. In this case, to each elementary module M_{ij} we will assign an estimate aij which provides the measure of the impact of the privacy constraint CS_j [8, 9] compute as indicated in the following equation:

$$a_{ij} = S_{ij} + O_{ij} \qquad (12)$$

Where S_{ij} represents severity in case of the privacy constraint is violated with respect to the module M_{ij}, while O_{ij} describes the probability that the module M_{ij} does not comply with the constraint CS_j [9, 21].

4.1.1 Severity Assessment

The severity S_{ij} is assessed it is estimated in terms of potential negative impact on the patient deriving from non-compliance with the CS_{ij} design constraint. The values that S_{ij} can assume are integers between 1 and 4, as reported in Table 3 [9, 21]. The determination of these values can be made empirically on the basis of specific check lists [21]. In this regard, please refer to the information provided by the French Data Protection Authority (CNIL) [21].

Table 3. Matrix for assessing the severity S_{ij} [8, 20]

Severity		
Score	**Description**	**Assessment**
1	Negligible	The event did not cause any damage or caused minor inconveniences, such as the recompilation of some forms by the user.
2	Limited	The event caused temporary harm to the patient and necessitated additional interventions (additional costs, stress, temporary suspension of social services, ..)
3	Significant	The event caused temporary damage to the patient, which he should be able to overcome even if with serious difficulties (extension of hospitalization, withdrawal of social services ...).
4	Maximum	The event caused permanent harm to the patient (permanent disability, long-term psychological or physical ailments, death).

4.1.2 Likelihood Assessment

The probability Oij represents the feasibility that the constraint CSij can be violated. It is mainly estimated in terms of the level of vulnerability. The values that Oij can assume are integers between 1 and 4, as reported in Table 4 [9, 21]. The determination of these values can be made empirically on the basis of specific checklists [9]. Also in this case, please refer to the indications provided by the French Data Protection Authority (CNIL) [21].

Table 4. Matrix for assessing the likelihood S_{ij} [9, 21]

Occurence		
Score	**Description**	**Assessment**
1	Negligible	There are no known events.
2	Limited	Documented but not frequent
3	Significant	Documented and frequent
4	Maximum	Documented and very frequent

4.1.3 Risk assessment matrix

The assessment of the risk of a privacy breach is closely related to current legislation. It can only be done on a heuristic basis [21] [22]. Therefore, it is necessary to rely on professionals who are expert in both the legislation and the context in which the system must operate [22]. In this study, as anticipated in the previous sections (§4.1.1 and §4.1.2), we rely on the guidelines prescribed by the French Data Protection Authority (CNIL) [21]. This Authority prescribes a fairly rigorous evaluation process. It is based on personal expert judgments, but is in line with the requirements of the GDPR. As we have already seen, these judgments allow us to assess the level of severity of a privacy breach and the likelihood of it happening. In addition, the Authority itself provides guidance on how to aggregate the two types of ratings. In this case, it provides a two-dimensional representation of privacy risk based on a risk matrix as shown in Figure 9 [21]. Each coloured cell represents a specific risk level. This matrix allows to estimate the privacy risk level of the elementary blocks M_{ij} in Figure 8. By re-aggregating the elementary blocks (M_{ij}) into higher level blocks (M_i) we will be able to attribute the risk level to increasingly complex components. The process will end when the overall level of the process is estimated. This type of matrix is widely used in healthcare, as well as for other types of risk because it has the advantage of being easy to use [22]. However, it can be ambiguous and lead to misleading assessments [23, 24]. For this reason, approaches have been proposed based on quantitative measures of the type of risk to be assessed [24]. Unfortunately, such approaches are not applicable to privacy risk assessment, because the assignment of a value judgment cannot be correlated with quantitative measures. In this case, as already mentioned in sections §4.1.1 and §4.1.2, the various EU authorities propose a series of checklists, so as to allow the attribution of a balanced judgment [9, 21]. However, the combination of the PIA method with AD allows reformulating and refining the value judgments attributable in the first instance, due to the iterative nature of the whole process, as schematized in Figure 1.

Colored cells are the risk privacy categories	Low risk	Limited risk	Significant risk	Maximum risk

	Severity			
Likelihood	**Negligible (1)**	**Limited (2)**	**Significant (3)**	**Maximum (4)**
Negligible (1)	2	3	4	5
Limited (2)	3	4	5	6
Significant (3)	4	5	6	7
Maximum (4)	5	6	7	8

Figure 10. Risk assessment matrix

4.2 Risk privacy coverage matrix

The flow chart diagram reported in figure 9 does not only allow us to estimate the privacy risk of the single elementary module, but also to build the risk privacy coverage matrix. This matrix correlates the estimate of the single elementary module, with the higher level component of which it forms part. In this way, the privacy risk can be tracked throughout the process. As shown in Figure 1, the privacy impact assessment is carried out first on the elementary modules. Then, progressively proceeding along the right side of the V model of Figure 1. Thus, an estimate of the higher level module is provided. In this study, the attribution of values is carried out on an empirical basis as suggested by the CNIL [21]. Let's take as an example the module ME5 (RequiretheCode) shown in Figure 8 that corresponds to the request of the Greek doctor to the patient to have the access code to his smartphone. The patient is assumed to be conscious. This action is evaluated with respect to the three situations of privacy violations already introduced (section §4):

A. The CS_1 constraint consists in preventing unauthorized access to the patient's health information management App. In our case, the patient could lose the access code to the medical App of his Smartphone or the code could be used by an unauthorized person. Empirically, a maximum risk level (4) and a significant probability of occurrence (3) has been attributed to a possible violation of this type. Therefore, the level of risk associated with the constraint $CS_1 = 4 + 3 = 7$. If we consider the values of the risk assessment matrix, we have that the risk related to CS_1 has a maximum value. In the risk privacy coverage matrix, the cell corresponding to row ME5 and column CS_1 has the value 7.
B. The CS_2 constraint concerns the possibility that the Greek hospital doctor carelessly modifies the system data. For the M_{E5} module this risk is zero, because it involves requesting a code. No data writing or modification operations are allowed on the system. Empirically, a negligible level of risk (1) and a negligible probability of occurrence (1) have been attributed to a possible violation of this type. Therefore, the level of risk associated with the constraint $CS_2 = 1 + 1 = 2$. If we consider the values of the risk assessment matrix, we have that the risk related to CS_2 has a minimum value. In the risk coverage matrix, the cell corresponding to row M_{E5} and column CS_2 has the value 2.
C. The CS_3 constraint represents the possibility that operations performed by the Greek hospital doctor lead to data loss in the system. This risk is zero, because the ME5 module consists only of requesting the access code to the App of the cross-border patient's smartphone. No data writing or modification operations are allowed on the system. Empirically, a negligible level of risk (1) and a negligible probability of occurrence (1) have been attributed to a possible violation of this type. Therefore, the level of risk associated with the constraint $CS_3 = 1 + 1 = 2$. If we consider the values of the risk assessment matrix, we have that the risk related to CS_3 has a minimum value. In the risk coverage matrix, the cell corresponding to row ME5 and column CS_3 has the value 2.

For each module Mij, the total level of risk is the total risk level is equal to the maximum value of the risks related to the three privacy constraints introduced (CS_1, CS_2, CS_3).

Thus: $a_{ij} = max_{i=1}^{n=3}(S_i + O_i)$

Similarly, the privacy risk value of the higher-level M_i module is equal to the risk value of its higher-rated sub-module (Mij). If we consider figure 9, the M_3 module assumes the risk level of its M_{E5} sub-module, since it has the highest risk value compared to the other sub-modules. Following this reasoning, it is possible to construct the risk privacy coverage matrix of figure 11.

Risk coverage matrix		CS_1	CS_2	CS_3	a_{ij}	A_i
		S_1+O_1	S_2+O_2	S_3+O_3	$a_{ij}=max(S_i+O_i)$	$A_i=max(a_{ij})$
M_1	M_{A1}	2	2	2	2	3
	$M_{A1.1}$	2	2	2	2	
	M_{B2}	3	2	2	3	
M_2	M_{C3}	2	3	3	3	3
	$M_{C3.1}$	2	2	2	2	
	$M_{C3.2}$	2	2	2	2	
	$M_{C3.3}$	2	2	2	2	
M_3	M_{D4}	2	2	2	2	7
	$M_{D4.1}$	4	2	2	4	
	M_{E5}	7	2	2	7	
	M_{F6}	6	2	2	6	
	M_{G7}	5	2	2	5	
	$M_{G7.1}$	5	2	2	5	
	M_{H8}	3	2	2	3	
M_4	M_{I9}	7	2	2	7	7
	M_{L10}	6	2	2	6	
	M_{M11}	4	2	2	4	
	$M_{M11.1}$	4	2	2	4	
	$M_{M11.2}$	7	2	2	7	
	$M_{M11.3}$	7	2	2	7	
M_5	M_{N12}	3	7	6	7	8
	$M_{N12.1}$	2	2	2	2	
	M_{O13}	3	8	7	8	
	$M_{O13.1}$	3	8	7	8	
	$M_{O13.2}$	2	2	2	2	
	$M_{O13.3}$	2	2	2	2	

Figure 11. Risk privacy coverage matrix

5 Conclusions

The risk privacy coverage matrix makes it possible to trace the level of risk of the various modules of the system. This allows to lay the foundations for building a dynamic mechanism for monitoring the privacy risks associated with the process in question. A mechanism of this type has the great advantage of defining the critical areas of the system in advance, allowing designers and programmers to identify and apply the necessary adaptation interventions before the problems arise. In practice, this is the application of the concept of Privacy by Design provided for by the GDPR. This involves obtaining substantial economic savings, because it is always "better safe than sorry". But, as far as healthcare facilities are concerned, it guarantees the levels of adequacy of the processing of personal data required by EU legislation. This certificate of adequacy must be checked periodically. It constitutes the essential condition to allow the electronic exchange of health data relating to EU patients. Furthermore, this system is also dynamic, in the sense that the risk levels defined by the risk assessment matrix in Figure 11 can be reshaped on the basis of contingent situations. Moreover, the Covid-19 pandemic has brought to the attention of political decision makers the question of the temporary suspension of some privacy constraints, in favour of public health. Countries that promptly intervene in this sense have made it possible to contain the infections drastically. In this regard, the case of South Korea is very interesting. The South Korean regulatory system has allowed the use of big data and Apps on smartphones to track the movements of citizens who are positive for the virus. However, if this greater flexibility makes it possible to identify areas of potential infections, it makes it not difficult to identify people who are positive for the virus. This means subjecting some people to social stigma, seriously compromising the rights of citizens. An assessment of the impact of the privacy risk based on Axiomatic Design allows, instead, to proactively highlight all the possible effects of particular interventions. This can help public decision makers to consciously guide their choices.

References

[1] World Health Organization (1946), *The Constitution of World Health Organization, International Health Conference held in New York*;
[2] World Health Organization, (2018) *Communicating Risk in Public Health Emergencies*. A WHO Guideline for Emergency Risk Communication (ERC) policy and practice;
[3] World Health Organization, (2016) *Policy Statement on Data Sharing by the World Health Organization in the Context of Public Health Emergencies;*
[4] World Health Organization, (2017) *Policy on use and sharing of data collected in Member States* by the World Health Organization (WHO) outside the context of public health emergencies;
[5] *General Data Protection Regulation, Regulation* (EU) 2016/679 of the European Parliament and of the Council, Official Journal of the European Union;
[6] Enisa, (2016) *Privacy and Security in Personal Data Clouds*;
[7] European Commission, (2017) *Exchanging and Protecting Personal Data in a Globalized World*, Communication from the Commission to the European Parliament and the Council;
[8] Enisa, (2015) *Privacy and Data Protection by Design*;
[9] Enisa, (2017) *Handbook on Security of Personal Data Processing*;
[10] A. Cavoukian, S. Taylor & M. E. Abrams, (2010) *Privacy by Design: essential for organizational accountability and strong business practices*, Identity in the Information Society volume 3, pages405–413;
[11] N.P. Suh, *Axiomatic Design - Advances and Applications*, (Oxford University Press, 2001);
[12] C. Parretti, F. Rolli, E. Pourabbas, P. Citti, (2018) *Axiomatic Selection of Health and Social Care Web Services on the Basis of Use Cases*, the 12th International Conference on Axiomatic Design (ICAD 2018);
[13] P. Pimentel, C. Stadzisz, (2006) *A Use Case based Object-Oriented Software Design Approach using The Axiomatic Design Theory*. Proceedings of ICAD2006 Fourth International Conference on Axiomatic Design, 4:1-8;
[14] G. Booch, J. Rumbaugh, I. Jacobson, (1999) UML *User Guide*, Addison-Wesley;
[15] M. Fowler, (2010) *UML Distilled*, Addison-Wesley;
[16] S.H. Do, N.P. Suh, (2000) *Object-oriented software design with axiomatic design*, Proc. CIRP 49, 278-84;
[17] F. Rolli, A. Giorgetti, P. Citti, M. Rinaldi, (2016) *Information content evaluation to obtain robustness of the management in Italian fiscal process (Part 2): optimization of Italian income certification process of the year 2016*, Proc. CIRP 53, 63-69;
[18] F. Rolli, A. Giorgetti, P. Citti, M. Rinaldi, (2016) *Improvement of the compilation process of the Italian income certifications: a methodology based on the evaluation of the information content (Part 1),* Proc. CIRP 53, 56-62;
[19] C. Parretti, E. Pourabbas, F. Rolli, F. Pecoraro, P. Citti, A. Giorgetti, (2019*) Robust design of web services supporting the home administration of drug infusion in pediatric oncology*, MATEC Web of Conferences (Vol. 301, p. 00013). EDP Sciences;
[20] C.A. Brown, (2006) *Kinds of coupling and approaches to deal with them,* in Proceedings of 4th ICAD2006, The Fourth International Conference on Axiomatic Design, Firenze, June;
[21] CNIL, (2018) *Privacy Impact Assessment-Knowledge Bases;*
[22] NPSA/NHS, (2008) *A Risk Matrix for Risk Managers*, National Patient Safety Administration, National Health Service (NPSA/NHS): London, UK;
[23] L.A. Cox, (2008) *What's wrong with risk matrices?* Risk Anal. 2008, 28, 497–512;
[24] S. Vatanpour, S.E. Hrudey, I. Dinu, (2015) *Can Public Health Risk Assessment Using Risk Matrices Be Misleading?* Int. J. Environ. Res. Public Health, 12, 9575-9588.

IOP Conf. Series: Materials Science and Engineering 1174 (2021) 012016 doi:10.1088/1757-899X/1174/1/012016

Development of moulds for thermoforming using FFF additive manufacturing and axiomatic design

M A Boca[*1, 2], **L Slătineanu**[1] **and A Sover**[2]

[1]Machines Manufacturing Technology Department, "Gheorghe Asachi" Technical University of Iași, Romania
[2] Technology Departments, University of Applied Sciences, Ansbach, Germany

Corresponding author's e-mail address: marius-andrei.boca@student.tuiasi.ro

Abstract. Additive manufacturing processes are mainly used for prototype production. However, an innovative approach consists of producing moulds or tools, which can be used for other traditional manufacturing processes. The main problem with these plastic moulds is that they cannot achieve the traditional metallic moulds properties and quality. By using an axiomatic design, the study analyses different approaches available on the market and other unconventional alternatives, thus making it possible to meet the functional requirements. The obtained methodology presents advantages in production cost and time compared to traditional processes, allowing quick production with a high degree of flexibility in the design of the mould. Thus, the printed moulds obtained were tested, and the preliminary results for different process conditions were presented.

1. Introduction

The design and production of moulds, tools and accessories for classical technologies for plastics processing, using additive manufacturing, attracts the attention of both research communities and independent users. The realization of such moulds for thermoforming is a challenging technical problem due to the multitude of processes and materials that can be used. Another limitation is the combination of geometric and process constraints specific to the thermoforming and Fused Filament Fabrication process (FFF).

For this application, axiomatic design (AD) contributes to obtaining a constructive solution (in terms of design and technological process), by eliminating unacceptable solutions and by offering a good direction and systematization of the end-use of the product production.

Subsequently, the axiomatic design method, developed by Professor Suh Nam Pyo [1], has been a research topic for many researchers worldwide. For instance, Nguyen et al. [2] develop a simple and powerful tool that delivers precise optimal results without using complicated algorithms and software solutions for on-site engineers in medium or small scale production. The researchers conducted an experimental investigation based on AD in conjunction with the Taguchi method to develop a thermoforming apparatus to obtain products with uniform thicknesses. Ferreira [3] proposed a fully integrated framework in order to support tools design for injection moulding. The conceptual solutions obtained through axiomatic design guidelines are detailed and optimized to maximize customer satisfaction. Another group of researchers from Korea [4] evaluated and analyzed existing

IOP Conf. Series: Materials Science and Engineering 1174 (2021) 012016 doi:10.1088/1757-899X/1174/1/012016

manufacturing techniques and developed a new process by adapting the axiomatic design to designing and developing a thermoformed multi-layered plastic product.

Inspired by the positive results of the previous researches and motivated by the fact that none of those works integrates an additive technology in mould production, the current paper aims to satisfy all the functional requirements by transposing them into process parameters (from the idea up to a final product).

2. Axiomatic design methodology

According to Professor Suh Nam Pyo [5], in order to design or to obtain a concept, four domains must be taken into consideration: the customer domain (Costumer Needs- CNs), the functional domain (Functional Requirements- FRs), the physical domain (Design Parameters- DPs) and the process domain (Process Variables- PVs). Besides those, there are also Constraints (Cs), which interface with the domains and are strict requirements that must be avoided or taken into consideration. Also, the paper aims to meet Suh's design axioms [1] in order to minimize the information content (axiom two: information Axiom) and to maintain the independence of FRs through a suitable selection of DPs (axiom one: independence axiom).

For a proper application of the principles of axiomatic design, the processes involved must be fully understood as well as technological advances must be taken into account. Thermoforming is a generic term for a group of processes with identifiable processing steps [6]. The equipment is fed, the sheet is mechanically clamped and heated without any type of manipulation. After reaching the desired temperature, the heating source is closed or eliminated from the working area and the forming process can start. At this point, the sheet can be brought in contact with the mould, with or without pre-shaping. After a proper cooling of the sheet, the excess material is trimmed. On the other hand, like all additive manufacturing processes, FFF process starts from the idea of a 3D designed part and with the help of CAD software and a series of design rules, the part is generated. It is essential at this step to correctly identify the goals that the 3D printed part needs to achieve. The 3D shaped (.stl: Stereolithography file format) is sliced into layers with a CAM software. The obtained .gcode file provides information for the printer (where to move the nozzle and the build plate and also about printing parameters, such as temperatures, speeds, retractions, etc.). As the nozzle is precisely moved, at certain points, over the build plate/heated bed, it deposits a thin strip of extruded plastic. The deposed material solidifies quickly into the 3D printer chamber and come in contact with the build plate and/or with the previous layer. The process repeated until the whole piece is created. The components of the 3D printed part and the design rules will later be transformed into FRs and DPs.

2.1. Structure determination of the domains

For this application, the early imposed constraints and customers' needs can be seen in figure 1. They can be defined as:

Cs_1: The mould maximal gauge dimensions of 195x195x61 mm;
Cs_2: Use Fused Filament Fabrication technology to produces the mould;
Cs_3: Min. 0.8 mm foil thickness in order to ensure mechanical resistance of the foil to prevent the leaking of the pouring material;
CN_1: Adequate mould design (for both thermoforming and FFF technology);
CN_2: Mould production through FFF additive manufacturing process;
CN_3: Increase mould quality by post-processing;
CN_4: Preparing the foil for casting;
CN_5: Safety in production of the end-product.

The next step in the AD approach encompasses the translation of the previously identified CNs and Cs into appropriate first level FRs (figure 1), which are a minimum set of functional requirements that states in the functional domain.

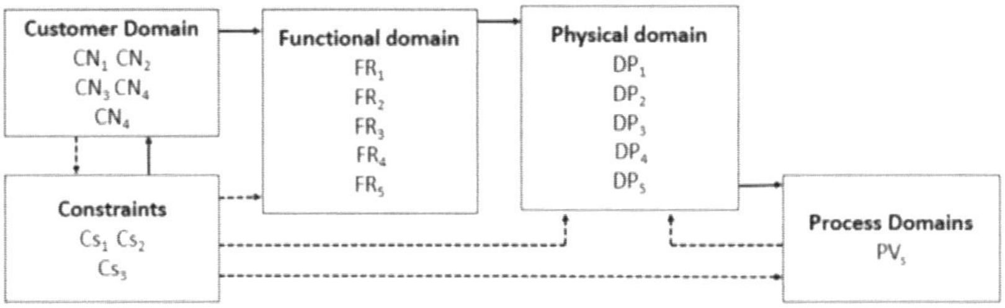

Figure 1. Axiomatic design domains with their components and relationships.

The customers' requirements were focused on mould design, production in a safe manner, and improved mould characteristics (aesthetic and functional). Thus, the following functional requirements presented in table 1, were established. Consequently, at the same step in the physical domain, the first DPs at the lowest level were described and can be seen in table 1 and figure 1. PVs will be further identified and explained.

Table 1. First level FRs and DPs.

Functional Domain: FRs	Physical Domain: DPs
FR_1: Create the mould design	DP_1: Design rules
FR_2: 3D print the mould	DP_2: 3D printing equipment and material
FR_3: Apply post-processing	DP_3: Additional equipment
FR_4: Obtained the thermoformed sheet	DP_4: Equipment and material for thermoforming
FR_5: Protect human operator	DP_5: Safety measures

From the previously stated FRs and DPs, the design equation can be written and as can be seen in equation (1). An uncoupled mapping for the first level AD elements was obtained. At this point each FR can be fulfilled independently by a correspondent DP.

$$\begin{Bmatrix} FR_1 \\ FR_2 \\ FR_3 \\ FR_4 \\ FR_5 \end{Bmatrix} = \begin{bmatrix} X & 0 & 0 & 0 & 0 \\ 0 & X & 0 & 0 & 0 \\ 0 & 0 & X & 0 & 0 \\ 0 & 0 & 0 & X & 0 \\ 0 & 0 & 0 & 0 & X \end{bmatrix} \begin{Bmatrix} DP_1 \\ DP_2 \\ DP_3 \\ DP_4 \\ DP_5 \end{Bmatrix} \tag{1}$$

From this working principle, the identified lower-level structures were splintered into second and third level FRs and DPs. The mapping, zigzagging, and decomposition processes presented in figure 2 continues until a solution can be implemented.

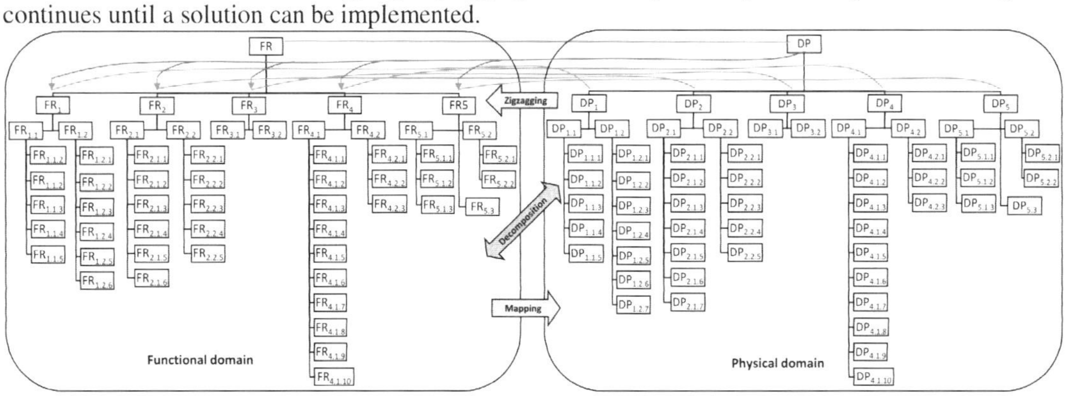

Figure 2. Mapping, zigzagging and decomposition of the FRs and DPs.

To obtain the detailed FRs and DPs, presented in table 2, the 3D printing equipment, the thermoforming machine, the printing and thermoforming parameters were decomposed into functional requirements and physical components.

IOP Conf. Series: Materials Science and Engineering 1174 (2021) 012016 doi:10.1088/1757-899X/1174/1/012016

Table 2. Detailed decomposition of the Functional and Physical Domain.

Functional Domain	Physical Domain
CN 1: Adequate mould design	
FR_1: Create the mould design	DP_1: Design rules
$FR_{1.1}$: Provides design requirements for FFF part	$DP_{1.1}$: Design rules for FFF
$FR_{1.1.1}$: Ensure the air extraction through a vacuum system	$DP_{1.1.1}$: Air extraction holes
$FR_{1.1.2}$: Improve air extraction	$DP_{1.1.2}$: Vacuum channels
$FR_{1.1.3}$: Increase the thermal resistance of the upper surface	$DP_{1.1.3}$: Min. 10 top layers
$FR_{1.1.4}$: Optimize the dimensions and the details of the mould	$DP_{1.1.4}$: CAD software
$FR_{1.1.5}$: Eliminate undercuts	$DP_{1.1.5}$: Solid walls without bridges
$FR_{1.2}$: Provides design requirements for thermoforming mould	$DP_{1.2}$: Design rules for thermoforming
$FR_{1.2.1}$: Eliminate sharp 3D corners	$DP_{1.2.1}$: Chamfer
$FR_{1.2.2}$: Eliminate sharp edges/2D corners	$DP_{1.2.2}$: Radius
$FR_{1.2.3}$: Avoid 90° angles to the base of the part walls	$DP_{1.2.3}$: Draft angles
$FR_{1.2.4}$: Compensate thermal expansion of the material	$DP_{1.2.4}$: Negative scaling
$FR_{1.2.5}$: Compensate material shrinkage	$DP_{1.2.5}$: Positive scaling
$FR_{1.2.6}$: Ensure mould cooling during cycles	$DP_{1.2.6}$: Cooling channels $DP_{1.2.7}$: Increase time between cycle
CN 2: Mould production through FFF additive manufacturing process	
FR_2: 3D print the mould	DP_2: 3D printing equipment, software and material
$FR_{2.1}$: Use adequate 3D printer	$DP_{2.1}$: BCN SigmaX R19 Desktop 3D Printer
$FR_{2.1.1}$: Consider build plate dimension	$DP_{2.1.1}$: Build plate XY dim.: 200 x200 mm
$FR_{2.1.2}$: Ensure a corresponding material supply	$DP_{2.1.2}$: 2,85 mm filament diameter
$FR_{2.1.3}$: Select a material with good thermal properties	$DP_{2.1.3}$: Material with heat resistance >70°C $DP_{2.1.4}$: Thermal conductive material
$FR_{2.1.4}$: Ensure a minimal printing temp. of 240 °C	$DP_{2.1.5}$: Heat block
$FR_{2.1.5}$: Ensure a minimal build plate temp. of 75 °C	$DP_{2.1.6}$: Heated glass build plate
$FR_{2.1.6}$: Ensure a constant flow of material	$DP_{2.1.7}$: Dual Drive Gear, Bowden Tube Extrusion System
$FR_{2.2}$: Appropriate selection of process parameters	$DP_{2.2}$: Printing parameters and software
$FR_{2.2.1}$: Use adequate slicing software	$DP_{2.2.1}$: BCN3D Cura
$FR_{2.2.2}$: Ensure an optimal ratio for printing speed and print quality	$DP_{2.2.2}$: Nozzle diameter
$FR_{2.2.3}$: Avoid or minimize surface imperfection	$DP_{2.2.3}$: Printing speed < 50 mm/s
$FR_{2.2.4}$: Provide a solid internal structure of the mould	$DP_{2.2.4}$: Infill structure
$FR_{2.2.5}$: Ensure a good adhesion between layers	$DP_{2.2.5}$: Printing temperature
CN 3: Increase mould quality by post-processing	
FR_3: Apply post-processing	DP_3: Additional equipment
$FR_{3.1}$: Increase plastic material thermal properties	$DP_{3.1}$: Convection oven for thermal treatment
$FR_{3.2}$: Increase mould top surface quality	$DP_{3.2}$: Sandpaper

IOP Conf. Series: Materials Science and Engineering 1174 (2021) 012016 doi:10.1088/1757-899X/1174/1/012016

CN 4: Preparing the foil for casting	
FR_4: Obtained the thermoformed sheet	DP_4: Equipment and material for thermoforming
$FR_{4.1}$: Use affordable adequate thermoforming equipment	$DP_{4.1}$: Desktop Vacuum Forming Machine
$FR_{4.1.1}$: Choose a proper type of thermoforming process	$DP_{4.1.1}$: Formech 450DT- One Step vacuum desktop forming machine
$FR_{4.1.2}$: Consider equipment working dimensions	$DP_{4.1.2}$: Min. forming aperture: 200 x 200 mm
$FR_{4.1.3}$: Ensure transport and easy manipulation	$DP_{4.1.3}$: Trolley
$FR_{4.1.4}$: Fixes the foil throughout the cycle	$DP_{4.1.4}$: Clamping frame
$FR_{4.1.5}$: Provide air extraction	$DP_{4.1.5}$: Vacuum pump
$FR_{4.1.6}$: Heat the foil until it gets soften	$DP_{4.1.6}$: 4x Quartz infrared heating elements
$FR_{4.1.7}$: Brought into contact the foil with the mould	$DP_{4.1.7}$: Sliding mechanism
$FR_{4.1.8}$: Provide air extraction over the entire work table	$DP_{4.1.8}$: Wire mesh
$FR_{4.1.9}$: Remove the thermoformed foil	$DP_{4.1.9}$: Pump exhaust
$FR_{4.1.10}$: Trim the thermoformed foil	$DP_{4.1.10}$: Cutters for manual trimming
$FR_{4.2}$: Appropriate selection of process parameters	$DP_{4.2}$: Thermoforming parameters
$FR_{4.2.1}$: Heat the foil until it softens	$DP_{4.2.1}$: Heating time > 40 sec
$FR_{4.2.2}$: Prepare the heaters between cycles	$DP_{4.2.2}$: Heating percentage between cycles >25%
$FR_{4.2.3}$: Varies the heating temperature	$DP_{4.2.3}$: Independent heating percentage for each zone
CN 5: Safety in the production of the end product	
FR_5: Protect human operator	DP_5: Safety measures
$FR_{5.1}$: Ensure a safety environment in the working space	$DP_{5.1}$: Particle (PM) retainer and cleaning
$FR_{5.1.1}$: Filter hazardous fumes and particles	$DP_{5.1.1}$: HEPA filter
$FR_{5.1.2}$: Reduce noise during 3D printing	$DP_{5.1.2}$: Enclosure
$FR_{5.1.3}$: Clean air after 3D printing and thermoforming	$DP_{5.1.3}$: Venting system
$FR_{5.2}$: Ensure fire prevention	$DP_{5.2}$: Fire protection features
$FR_{5.2.1}$: Eliminate the contact with the hot surfaces	$DP_{5.2.1}$: Metallic case for heating equipment
$FR_{5.2.2}$: Use smoke and fire detection system	$DP_{5.2.2}$: Smoke and fire sensors
$FR_{5.3}$: Supervision over the manufacturing process	$DP_{5.3}$: Monitoring system

From table 2, the basic design matrix for all level FRs and DPs, presented in figure 3 are yielding uncoupled design, except from the design equation for $FR_{1.2.s}$ and $FR_{2.1.s}$ and their correspondent DPs.

			Desig parameters DPs																																											
			DP1												DP2												DP3		DP4												DP5					
			DP1.1					DP1.2							DP2.1							DP2.2					DP3.	DP3.:	DP4.1										DP4.2			DP5.1			DP5.2		DP5.3
		Functional Requirments FRs	DP1.1.1	DP1.1.2	DP1.1.3	DP1.1.4	DP1.1.5	DP1.2.1	DP1.2.2	DP1.2.3	DP1.2.4	DP1.2.5	DP1.2.6	DP1.2.7	DP2.1.1	DP2.1.2	DP2.1.3	DP2.1.4	DP2.1.5	DP2.1.6	DP2.1.7	DP2.2.1	DP2.2.2	DP2.2.3	DP2.2.4	DP2.2.5			DP4.1.1	DP4.1.2	DP4.1.3	DP4.1.4	DP4.1.5	DP4.1.6	DP4.1.7	DP4.1.8	DP4.1.9	DP4.1.10	DP4.2.1	DP4.2.2	DP4.2.3	DP5.1.1	DP5.1.2	DP5.1.3	DP5.2.1	DP5.2.2	
		FR1.1.1	X																																													
		FR1.1.2		X																																												
	FR1.1	FR1.1.3			X																																									
		FR1.1.4				X																																								
		FR1.1.5					X																																							
FR1.1		FR1.2.1						X																																						
		FR1.2.2							X																																					
	FR1.2	FR1.2.3								X																																				
		FR1.2.4									X																																			
		FR1.2.5										X																																		
		FR1.2.6											X	X																																
		FR2.1.1													X																															
		FR2.1.2														X																														
	FR2.1	FR2.1.3															X	X																												
		FR2.1.4																	X																											
		FR2.1.5																		X																										
FR2		FR2.1.6																			X																									
		FR2.2.1																				X																								
		FR2.2.2																					X																							
	FR2.2	FR2.2.3																						X																						
		FR2.2.4																							X																					
		FR2.2.5																								X																				
FR3		FR3.1																									X																			
		FR3.2																										X																		
		FR4.1.1																											X																	
		FR4.1.2																												X																
		FR4.1.3																													X															
		FR4.1.4																														X														
	FR4.1	FR4.1.5																															X													
		FR4.1.6																																X												
FR4		FR4.1.7																																	X											
		FR4.1.8																																		X										
		FR4.1.9																																			X									
		FR4.1.10																																				X								
		FR4.2.1																																					X							
	FR4.2	FR4.2.2																																						X						
		FR4.2.3																																							X					
FR5		FR5.1.1																																								X				
	FR5.1	FR5.1.2																																									X			
		FR5.1.3																																										X		
	FR5.:	FR5.2.1																																											X	
		FR5.2.2																																												X
		FR5.3																																													X

Figure 3. Fundamental design matrix for all level FRs and DPs.

From equation (2), it can be concluded that the number of DPs is larger than the number of FRs and based on Theorem 3 [7], the design is called uncoupled redundant design.

$$\begin{Bmatrix} FR_{1.2.1} \\ FR_{1.2.2} \\ FR_{1.2.3} \\ FR_{1.2.4} \\ FR_{1.2.5} \\ FR_{1.2.6} \end{Bmatrix} = \begin{bmatrix} X & 0 & 0 & 0 & 0 & 0 & 0 \\ 0 & X & 0 & 0 & 0 & 0 & 0 \\ 0 & 0 & X & 0 & 0 & 0 & 0 \\ 0 & 0 & 0 & X & 0 & 0 & 0 \\ 0 & 0 & 0 & 0 & X & 0 & 0 \\ 0 & 0 & 0 & 0 & 0 & X & X \end{bmatrix} \begin{Bmatrix} DP_{1.2.1} \\ DP_{1.2.2} \\ DP_{1.2.3} \\ DP_{1.2.4} \\ DP_{1.2.5} \\ DP_{1.2.6} \\ DP_{1.2.7} \end{Bmatrix}$$

$$\begin{Bmatrix} FR_{2.1.1} \\ FR_{2.1.2} \\ FR_{2.1..3} \\ FR_{2.1..4} \\ FR_{2.1.5} \\ FR_{2.1.6} \end{Bmatrix} = \begin{bmatrix} X & 0 & 0 & 0 & 0 & 0 & 0 \\ 0 & X & 0 & 0 & 0 & 0 & 0 \\ 0 & 0 & X & X & 0 & 0 & 0 \\ 0 & 0 & 0 & 0 & X & 0 & 0 \\ 0 & 0 & 0 & 0 & 0 & X & 0 \\ 0 & 0 & 0 & 0 & 0 & 0 & X \end{bmatrix} \begin{Bmatrix} DP_{2.1.1} \\ DP_{2.1.2} \\ DP_{2.1.3} \\ DP_{2.1.4} \\ DP_{2.1.5} \\ DP_{2.1.6} \\ DP_{2.1.7} \end{Bmatrix} \quad (2)$$

2.2. Preliminary experiments and the influence of the process variable

In order to submit both axioms during the mapping of the AD elements, preliminary tests were performed to reduce the number of PVs and to choose the best DPs option. During this step, the process domain is introduced and PVs can be seen in figure 4, along with their correlations with DPs.

Physical Domain			Process Domain
DP1: Design rules	DP1.1	DP1.1.1	$PV_{1.1.1}$: Angle of inclination (from 0° up to 90°) of the axis of symmetry of the hole extraction to the bottom of the mould with hole diameter (from 0.4mm up to 1.8 mm)
		DP1.1.2	$PV_{1.1.2}$: Positioning of extraction holes, width and depth of extraction channels
		DP1.1.3	$PV_{1.1.3}$: Layer Heigh from 25% up to 80% of the nozzle diameter
		DP1.1.4	$PV_{1.1.4}$: Thermoforming foil thickness
		DP1.1.5	$PV_{1.1.5}$: -
	DP1.2	DP1.2.1	$PV_{1.2.1}$: Depth/Height of the mould cavity/wall and chamfer values
		DP1.2.2	$PV_{1.2.2}$: Draw ratio and radius values
		DP1.2.3	$PV_{1.2.3}$: Mould type (female/negative mould or positive/male mould)
		DP1.2.4	$PV_{1.2.4}$: Coefficients of thermal Expansion [10^-6 °C^-1] for Thermoformable polymers
		DP1.2.5	$PV_{1.2.5}$: Shrinkage Values [%] for Thermoformable polymers
		DP1.2.6	$PV_{1.2.6}$: Cooling channels dimension and type
		DP1.2.7	$PV_{1.2.7}$: Waiting time between cycles
DP2: 3D printing equipment, software and material	DP2.1	DP2.1.1	$PV_{2.1.1}$: Ultimaker/Prusa/BCN/Anycubic series
		DP2.1.2	$PV_{2.1.2}$: Printer configuration (1.75 mm and 2.85 mm)
		DP2.1.3	$PV_{2.1.3}$: Special designed heat resistent PET (up to 230°C), PESU (up tp 220°C), ABS (up to 85°C), PC, PMMA, etc
		DP2.1.4	$PV_{2.1.4}$: CF/GF PA Reinforced material (Ice 9 Rigid PA, PolyMide PA-CF20, PolyMide CoPa Nylon 6.6), Ice 9 Flex TPU, etc
		DP2.1.5	$PV_{2.1.5}$: BCN Sigmax R19 maximum Hotend temperature up to 290°C
		DP2.1.6	$PV_{2.1.6}$: BCN Sigmax R19 maximum printing surface temperature up to 100°C
		DP2.1.7	$PV_{2.1.7}$: -
	DP2.2	DP2.2.1	$PV_{2.2.1}$: Slic3R, Simplify 3D
		DP2.2.2	$PV_{2.2.2}$: Depending on the 3D printer (from 0.2 mm up to 1.2 mm)
		DP2.2.3	$PV_{2.2.3}$: Print speeds: wall speed, top/bottom layer speed, infill speed, retraction speed
		DP2.2.4	$PV_{2.2.4}$: Infill density, infill pattern, Infill line multiplier, Gradual Infill Steps and Height, etc.
		DP2.2.5	$PV_{2.2.5}$: Material type
DP3: Additional equipments	DP3.1		$PV_{3.1}$: Thermal threatment, Coating with thermo-resistant paste or resin
	DP3.2		$PV_{3.2}$: The number of repetitions and the fineness of the abrasive paper
DP4: Equipment and material for thermoforming	DP4.1	DP4.1.1	$PV_{4.1.1}$: Vaquform DT2
		DP4.1.2	$PV_{4.1.2}$: Maximal forming area: 430 x 280 mm or with Reducing window: 228 x 203 mm
		DP4.1.3	$PV_{4.1.3}$: -
		DP4.1.4	$PV_{4.1.4}$: -
		DP4.1.5	$PV_{4.1.5}$: Vaccum system that can provide a minimum presure of @22"Hg/ @0.745 bar
		DP4.1.6	$PV_{4.1.6}$: -
		DP4.1.7	$PV_{4.1.7}$: -
		DP4.1.8	$PV_{4.1.8}$: -
		DP4.1.9	$PV_{4.1.9}$: Presured air system
		DP4.1.10	$PV_{4.1.10}$: Trimming press, CNC trimming (Watter Jet Cutting, CNC Laser), In mould trimming etc
	DP4.2	DP4.2.1	$PV_{4.2.1}$: Material type of the thermoforming foil
		DP4.2.2	$PV_{4.2.2}$: Waiting time between cycles
		DP4.2.3	$PV_{4.2.3}$: Mould dimension and form
DP5: Safty measures	DP5.1	DP5.1.1	$PV_{5.1.1}$: >99.75% retention filter (H13 or H14)
		DP5.1.2	$PV_{5.1.2}$: Quality of the stepper mottor shield and thickness of the enclouser walls
		DP5.1.3	$PV_{5.1.3}$: -
	DP5.2	DP5.2.1	$PV_{5.2.1}$: The size of the perforated profile in the metal housing
		DP5.2.2	$PV_{5.2.1}$: -
	DP5.3		$PV_{5.3}$: Surveillance system (cameras, Webcam) or system created using development boards (Arduino, Raspberry Pi) and auxiliary accessories

Figure 4. Process variable with the correlation in the physical domain.

For the preliminary tests, the reliable and robust BCN3D SigmaX R19 ($DP_{2.1}$) professional Desktop 3D printer is used. This equipment has a dual drive gear Bowden extrusion system ($DP_{2.1.7}$) and works with a filament of 2.85 mm in diameter ($DP_{2.1.2}$). The used 0.4 mm brass nozzle size influences $FR_{1.1.3}$ and $FR_{2.2.2}$. Besides that, an enclosure that ensures noise reduction ($DP_{5.1.2}$) and a protected medium from external air currents keeps the warmth (up to 48ºC) from the heated bed inside the printer. The materials used ($PV_{2.1.3}$) to print the tested moulds and the extraction holes orientation test were the following: 2.85 mm brown PLA filament from ICE and a white heat-resistant PET filament from 3dk.berlin ($PV_{2.1.3}$). The thermoforming machine is the Model 450 DT ($DP_{4.1.1}$) from the manufacturer Formech. It has a maximum forming area of 430 mm x 280 mm ($DP_{4.1.2}$ and $PV_{4.1.2}$) and allows working with workpieces products with a maximum thickness of 6 mm (Cs_3), according to the technical specification [8]. The following materials and thickness sheets were used for the thermoforming process: 0.8 and 1 mm black PS and transparent PET (Cs_3).

The experimental study starts from a male type mould with a 35 mm deep female cavity, 61 mm vertical wall, sharp corners, round surface, which can be seen in figure 5 (left side). The mould presented in figure 5 (right side) is modified based on the thermoforming and 3D printing design rules (DP_1) for corners ($FR_{1.2.1}$), edges ($FR_{1.2.2}$), draft angle ($DP_{1.2.3}$), webbing and vacuum holes ($DP_{1.1.1}$) and channels ($DP_{1.1.2}$). In order to further improve the quality of the final thermoformed parts, new holes have been added to the vertical surface, and the bottom vacuum channels ($DP_{1.1.2}$) have been modified to improve the air flowing and air drawing.

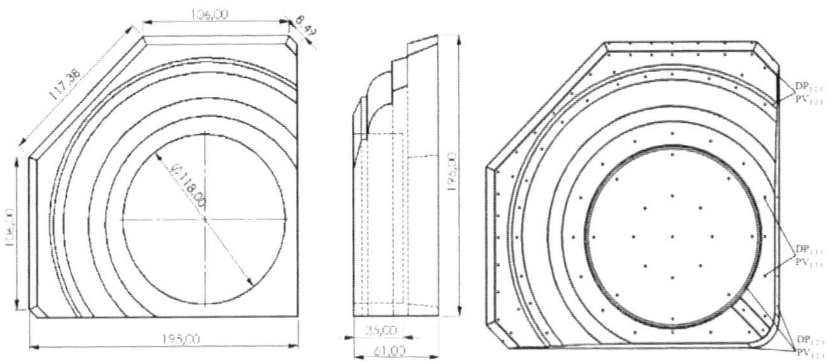

Figure 5. Initial mould design (left) and the modified mould design (right). (adapted from [8])

For proper 3D printing and to improve the lifetime of the mould, the following printing parameters have been chosen: the 80 % infill density was gradually decreased in 2 steps of 5 mm ($DP_{2.2.4}$ and $PV_{2.2.4}$), 3.2 mm top thickness equivalent of 16 Top Layers ($DP_{1.1.3}$ and $PV_{1.1.3}$) and 1.6 mm for wall thickness.

The printing of the mould, from figure 6 (left side), needed five days and 4 hours. In order to increase the crystallinity grade and make the mould temperature resistant up to 230 °C ($DP_{2.1.3}$), the printed object must be post-processed inside a convection oven ($DP_{3.1}$). To prevent the warping during tempering, the printed mould was placed at 60 °C inside the oven altogether with the printing table ($DP_{2.1.6}$). After 30 min. of preheating, the mould was further tempered at 110 °C for 2 hours. For cooling, the mould was let inside the oven to cool down slowly, without opening the door. The tempered mould was used to produce thermoformed foils (figure 6, in the middle) with thickness from 0.8 mm up to 1mm, without any mould deformation. The obtained foil was further used with success to pour and obtain the cement end-product (figure 6, right). All necessary safety measures (DP_5) were taken during all the processes described above.

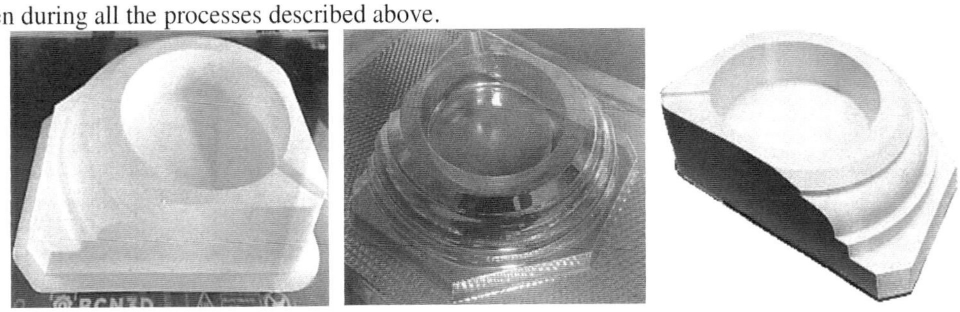

Figure 6. 3D printed mould, PET thermoformed sheet and the cement end product [8].

3. Conclusion

This paper proposes a framework for developing thermoformed mould obtained through FFF technology using axiomatic design. The method also helps to systematize and make decisions regarding the design of the mould, process parameters for both technologies involved, as well as to ensure the necessary measures to protect the human operator. The systemized accumulated database through AD, will reduce the required period of development, minimize failures and mistakes during the design steps and provide instant cause analysis to further develop a mould for industrial use.

By using this approach, it is possible to obtain a ready to use custom mould within hours or days, with a high degree of flexibility in the design (optimized air suction holes and channels, manufactured during the same process) and at a low cost for equipment compared to classical technologies. Furthermore, the proposed design technique and the manufactured mould were both successfully validated by testing the mould with 0.8 mm and 1 mm sheets. The results attained highlight the great potential of the proposed concept.

4. References

[1] Braun C A 2020 Axiomatic Design for Products, Processes and Systems *Industry 4.0 for SMEs Challenges, Opportunities and Requirements* ed Matt D et al. (Switzerland: Springer Nature Switzerland AG) chapter 13 pp. 383-401

[2] Nguyen T K, Pham A-D, Chau M Q, Nguyen X C, Pham H A D, Pham M H, Ngyen T P and Nguyen H S 2017 *Chinese Mechanical Engineering Society and Springer*

[3] Ferreira I, Cabral J A and Saraiva 2013 *7th International Conference on Axiomatic Design* **7** ed Thompson M K (Worcester: ICAD) 169-177

[4] Lee S K, Cha S W and Koo S M 2007 *Polymer-Plastics Technology and Engineering* **46** (Lomdon: Taylor & Francis) 781-788

[5] Suh N P 1990 *The Principles of Design* (New York: Oxford University Press Inc (Verlag))

[6] Throne J L 2008 *Understanding Thermoforming* vol 2 ed: Carl Hanser Verlag GmbH & Co. KG pp 1

[7] Coelho A M G, Neştian G, Cavique M and Mourã 2012 *International Journal of Precision Engineering and Manufacturing* **13** 1837-1843

[8] Boca M-A and Sover A 2020 *2nd International Conference of the University of Applied Sciences Ansbach*

Development of Fused Filament Fabrication desktop 3D printer enclosure using axiomatic design principles

V Ermolai[1,*] **and A I Irimia**[1]

[1]"Gheorghe Asachi" Technical University of Iasi, Department of Machine Manufacturing Technology, Blvd. D. Mangeron, 59A, 700050 Iasi, Romania

Corresponding author's e-mail address: ermolai.vasile@gmail.com

Abstract. Additive manufacturing through Fused Filament Fabrication is one of the most spread methods for obtaining functional prototypes and components, mainly for single parts demands or small manufacturing lots. This direction of use presumes the use of engineering polymers in the manufacturing process, materials that require precise control of the environment regarding printing temperature and moisture. A Do-It-Yourself enclosure mock-up for engineering polymers processing was obtained using the Axiomatic Design principles. The enclosure is compatible with low-cost Fused Filament Fabrication desktop 3D printers and satisfies the functional requirements regarding manufacturing and secure usage of engineering-grade polymers. In addition, the presented design solved the requirements regarding surveillance and remote control during the manufacturing process.

1. Introduction

Fused Filament Fabrication-FFF is an Additive Manufacturing technology with one of the most economic growth due to the low cost of equipment and the great range of thermoplastic material available on the market [1]. Using engineering polymers (e.g., ABS, PC) implies proper control of the environment regarding temperature and moisture [2] to reduce part curling and increase mechanical characteristics. Thermal sensitivity is a common characteristic of this group of materials, making parts shrink, warp, or detach from the printing surface. Problems related to the thermal stability of the environment occurs when hot thermoplastic material is deposited on the previously solidified layer, causing rapid cooling and warping. The temperature difference between solidified layers and molten thermoplastic increases as the part grows due to the distance from the heated deposition surface. This phenomenon creates internal stress in the part structure and can cause previous layer curl or led to layer delamination [3]. This drawback can be avoided using a heated chamber [4], but the total cost of FFF equipment increases with the printing capabilities [3].

Given these facts, the main goal of this article is to use the Axiomatic Design-AD principles in the conception activity of a Do-It-Yourself-DIY controlled environment that can accommodate different FFF equipment. The motivation for this project came from the need to extend the manufacturing capabilities of the 3D printing machines from our disposal for research activities.

2. Methods

As a method for the Computer-Aided Design–CAD modelling process, five key steps were established for the constructive design; after each stage, a validation process was applied. The key steps are:

- Step 1: Determinate the Bill of Materials-BOM for equipment and system using the AD;
- Step II: Volume allocation and preliminary positioning an orientation of the main components;
- Step III: Generate preliminary design and fastening method;
- Step IV: Complete the design and fastening method;
- Step V: Integrate all hardware components.

2.1. *Axiomatic design theory (principles)*

The AD is a creative design method that comprises four areas: the customer domain, the functional domain, the physical domain and the process domain. Followed in this order, firstly, the Customer Needs (CNs) must be established and reshape in Functional Requirements (FRs) and Contains (Cs). Then, based on specified FRs, the Design Parameters (DPs) are developed. Each physical solution is then characterized by Process Variables (PVs) to obtain the desired solution. Finally, the designer must outline two main questions: "What we want to achieve?" and "How do we choose to satisfy the need?" with respect to the two axioms of the AD method [5]:

Axiom 1: Maintain the independence of the functional attributes (FRs)–which provide adaptability, controllability and minimize iterations and unintended consequences;

Axiom 2: Minimize the information content–that maximizes the probability of success.

The AD method comprises three fundamental parts: axioms, structure and process. Regarding the structure, the design domains were, which must be decomposed vertically in design hierarchies and detailed as physical components. In other words, the design requirements are formulated, and design solutions are developed from ambiguous to detailed between the domains in a process called decomposition with respect to the axioms [6].

Thompson (2013a) provides a deeper understanding of the CNs by describing the need to identify the possible stakeholders (STCHs.) and their needs (SNs). Frequently multiple stakeholders benefit from the same design. Thus, it is imperative to identify them to search for all potential customer needs by asking "who?" and "what he wants?". Afterwards, requirements classification must occur to apply the AD method successfully. Therefore, it is essential to identify the non-functional requirements (nFRS) and consider them in parallel domains to the traditional domains [7, 8].

2.2. *Application of AD principles*

This design activity aims to enhance the manufacturing and control capabilities of current FFF machines types at our disposal, a Prusa i3 Mk3S+ and an Ultimaker 2+, to produce components from engineering-grade thermoplastic materials for research. The first group of stakeholders, Students and Researchers (STCH1), resulted by abstracting this statement.

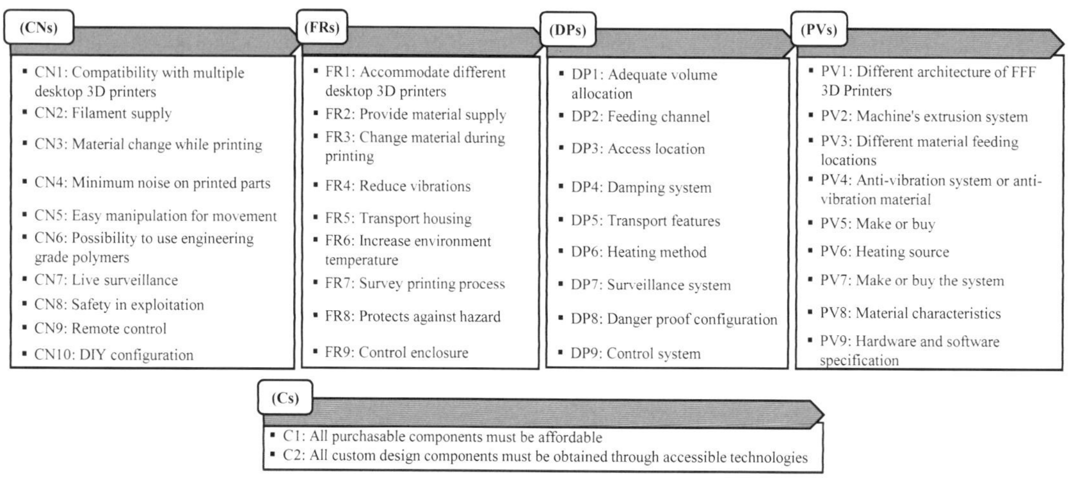

Figure 1. Customer, functional, physical and process domain of AD process for the heated chamber.

Other identified stakeholders are hobbyists (STCH2) and self-employed persons (STCH3). So, the resulting product must satisfy the needs of these users (with 3D printers from different vendors and various configuration); the final design must be more adaptive than specific.

In conformity with the above-identified stakeholder, the customer needs were identified. Then, those demands were used to formulate the principal FR0: Obtain a DIY controlled environment for

IOP Conf. Series: Materials Science and Engineering 1174 (2021) 012017 doi:10.1088/1757-899X/1174/1/012017

FFF desktop 3D printers and have as *DP0: DIY FFF desktop 3D printer heated chamber.* Afterwards, the first-grade attributes are determinate in the functional domain and Cs (figure 1). Finally, in concordance with the identified FRs, the first-grade DPs were developed with their PVs.

According to the DIY project specifics, regarding the need for customization [9, 10], some of the design parameters have generic names (e.g., webcam) to not to limit the freedom of choice, each potentially stakeholder having the possibility to use the specific device according to his needs.

As presented in figure 1, the first three PVs refer to the construction principles of the FFF machines and are important factors of the design process. This type of equipment combines an extrusion system with a Computer Numerical Control-CNC system. Starting from this assumption, all FFF technology vendors designed their products differently. Prusa and Ultimaker are two great examples of FFF cartesian 3D printers with distinct configuration. Firstly, Prusa produces i3 machines with direct extrusion systems (*PV2*) on 1.75 mm filament diameter and Ultimaker vends machines with indirect extrusion systems, Bowden (*PV2*) on 2.85 filament diameters, both of them having different feeding locations for the raw material (*PV3*). In the case of Ultimaker equipment, the extrusion system moves horizontally in the XY plane, and after each completed layer, the build platform moves down across Z-direction (*PV1*). For machines with i3 configuration, the extrusion system moves in the XZ plane, and the build platform translates in the Y-direction (PV1). After each completed layer, the extrusion system moves up in a vertical direction.

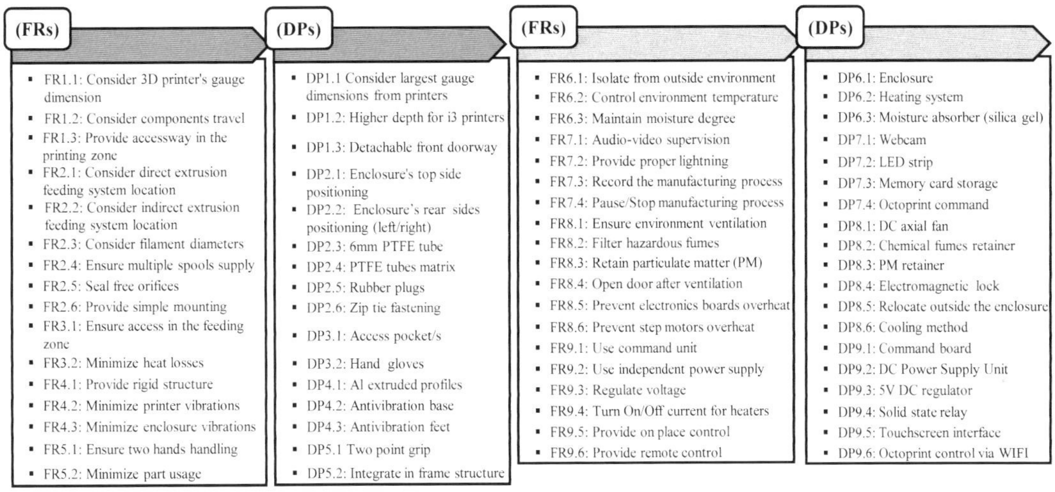

Figure 2. Functional, physical and process domain of second order for the heated chamber.

As observed in figure 1, the FRs and DPs are divided into two main groups. The first five refer to the product's constructive design of the FR0 and the remaining to the equipment integration and safety design, besides *FR/DP6.1.* This structural component was included in this field due to the requirement formulation (figure 2).

Regarding the product's physical design (figure 2, green label group), the CAD modelling started with the volume allocation (Step II) given by the gauge dimension of the considered 3D printing machines, *PV1.1:* Anycubic i3 Mega/Pro/S/X, Prusa i3 Mk2,5/2,5S/3/3S/3S+, Creality Ender3/3 Pro/5 and Ultimaker 2/2+/3, resulting a minimum required volume of 500x459x600 mm (WxLxH); thus 550x550x620 mm volume was adopted. Furthermore, another feature was identified by zigzagging between physical and functional domains, i3 type machines configuration, and moving build plate (*PV2.1: Anycubic* and *Prusa* printers). Therefore, the required volume was extended to 550x600x620 mm, and for design simplicity, the 620x620x620 mm volume was considered.

As for other DPs, several PVs were identified: *PV2.1*: Distinct location for the feeding mechanism, *PV2.2:* Raw material size, *PV4.2:* Vibration absorbent material, *PV4.3:* Damping system.

The second half of this AD process, yellow labelled charts, involved designing the equipment integration, system and safety features as presented in figure 2. The PVs for the DPs were highlighted as previously, and they are *PV6.1*: Manufacturing method, *PV6.2/PV7.1/PV7.3/ PV9.2/PV9.5:* DIY approach, *PV8.2:* Filtering method, *PV8.3:* Filtration degree, *PV8.6: Solution complexity*, *PV9.1:* Command board architecture.

Due to the insufficient information, for particular DPs, a third level decomposition was necessary. Thus, the identified requirements are presented as follows: *FR1.3.1:* Minimize heat losses, *FR1.3.2:* One-way fastening, *FR1.3.3:* Design locking feature, *FR1.3.4:* Provide clamping force, *FR1.3.5:* Ensure door mounting position, *FR1.3.6:* Facilitate door mounting/demounting, *FR1.3.7*: Minimize space usage, *FR3.1.1:* Provide access for direct extrusion systems, *FR3.1.2:* Provide access for indirect extrusion systems, *FR6.1.1:* Minimize heat exchange, *FR6.1.2:* Allow process observation, *FR6.2.1*: Heat actively, *FR6.2.2:* Measure environment status, *FR8.2.1:* Detect harmful fumes, *FR8.2.2:* Filter environment's air, and *FR8.2.3*: Minimize heat losses.

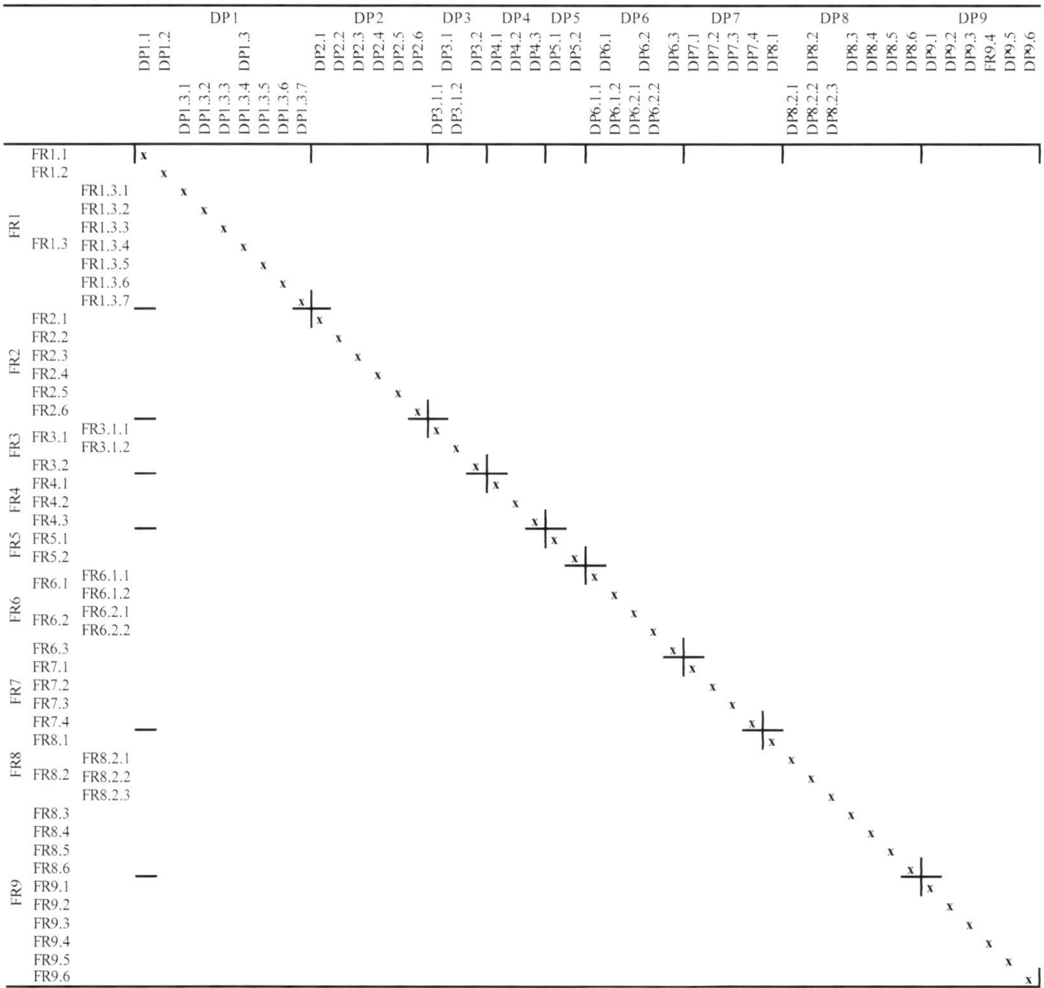

Figure 3. The resulted from the Axiomatic Design matrix.

The corresponding DPs are *DP1.3.1:* Sealing ring, *DP1.3.2:* Guiding tenons, *DP1.3.3:* Locking knob, *DP:1.3.4:* Compression mechanism, *DP1.3.5* Pre-mounting tenon, *DP1.3.6:* Ergonomic handle, *DP1.3.7:* Removable handle, *DP3.1.1:* Front side pocket, *DP3.1.2:* Rear side pocket, *DP6.1.1:*

IOP Conf. Series: Materials Science and Engineering 1174 (2021) 012017 doi:10.1088/1757-899X/1174/1/012017

Multilayer walls, *DP6.1.2:* Transparent wall, *DP6.2.1:* Heating device, *DP6.2.2:* Temperature-humidity sensor, *DP8.2.1:* Fume sensor, *DP8.2.2:* Active carbon filter and *DP8.2.3:* Automatic trap door, with their associated PVs: *PV6.1.1:* Material compatible with the manufacturing method, *PV6.2.1:* Heat source working principle, *PV6.2.2:* Sensor configuration.

3. Design results

According to the AD principles, it is necessary to determine which design matrix was obtained. Therefore, several questions must be asked: "which type of design the matrix represents?", "was the first axiom respected?" and if axiom one was not respected, "how the design should be changed to respect the independence axiom?".

The design matrix of *the DIY heated chamber for desktop FFF 3D printers* is presented above, in figure 3. As can be observed, the result is an uncoupled design matrix, each FR having a corresponding DP. Thus, the first axiom of the AD principles was respected. This design is the result of several reiterations of the first matrix. The main challenges in obtaining an uncoupled design matrix were given by the previous improper decomposition of the FRs and the trial of respecting the identified constraints.

Further on, the result obtained following Axiomatic Design principles will be presented. It consists of two parts, the constructive design and the equipment integration and safety system design.

3.1. Equipment integration and safety system design

The AD method was beneficial in defining which type of components are necessary to satisfy the identified FRs, and through PVs, possibilities for achieving the design results. Therefore, all electronic components highlighted in figure 4 are generic to respect the DIY project's characteristics (e.g., customization, standard component). This way, the stakeholder can adapt the presented schematic to his needs. For example, if the stakeholder chooses a Raspberry Pi as a command board (DP9.1), all the other DPs defined in the electronic scheme must be compatible.

It should be noted that parts of DPs can be changed or removed in particular cases. For instance, if the stakeholder decides that passive heating (heat generated through build plate), DP6.2.1 can be removed from the schematic, which influences DP9.2 (a smaller PSU can be used).

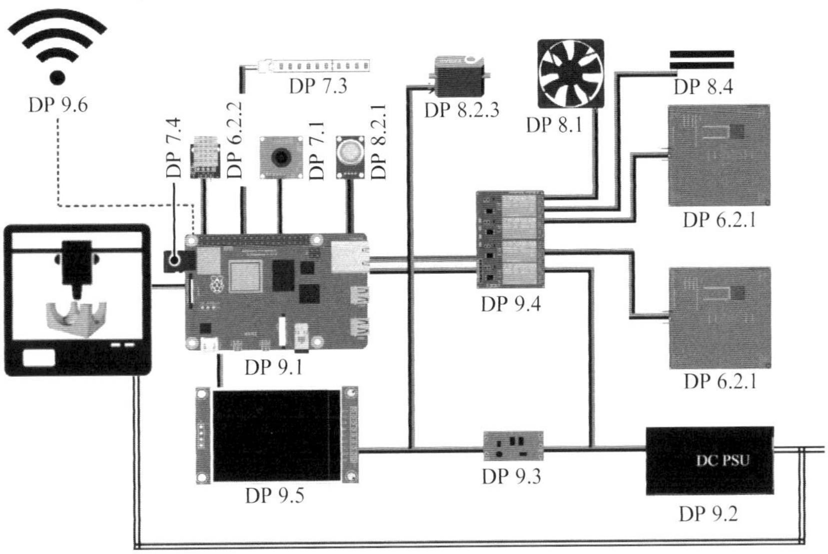

Figure 4. Heated chamber's electronic scheme.

3.2. Constructive design

Regarding the constructive design of DP0 (developed in Catia V5 environment), after the volume allocation step (Step II) of the inside environment (DP1.1/DP1.2), the next stage (Step III) consisted of the construction of the chamber frame (DP4.1) and walls (DP6.1.1/6.1.2) as shown in figure 5. Thus, a part of the DPs transposed in a CAD model suffered iterated changes (zigzagging between Step II and Step IV), part of them being at the third version (e.g., DP2, DP1.3, DP1.3.2).

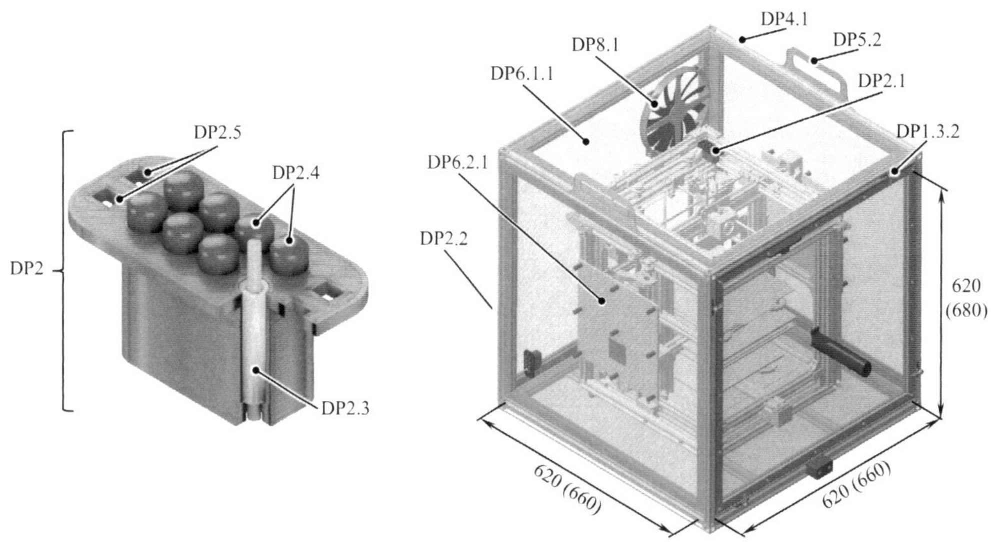

Figure 5. Partial representation of the DIY heated chamber (Step III to Step IV).

The insufficient decomposition of the FRs, because of the early stage of the CAD design process (Step II and III), determined several changes in the CAD model specification. This way, the AD matrix was updated several times.

All parts were developed according to C2, and components such as DP1.3, DP1.3.2, DP1.3.5 (which are in Step V of the constructive design) were designed by respecting Design for manufacturing (DMF) guidelines for laser cutting and FFF (figure 6). At this stage, the AD matrix presented in figure 3 is at the sixth version. However, it will require several add-ins as the CAD constructive design will be further developed in Steps VI & V (e.g., in the Electrical Harness Integration stage, where cable fixtures and passing points are necessary). Therefore, those extensions will be added as third or fourth level design features in the current AD matrix.

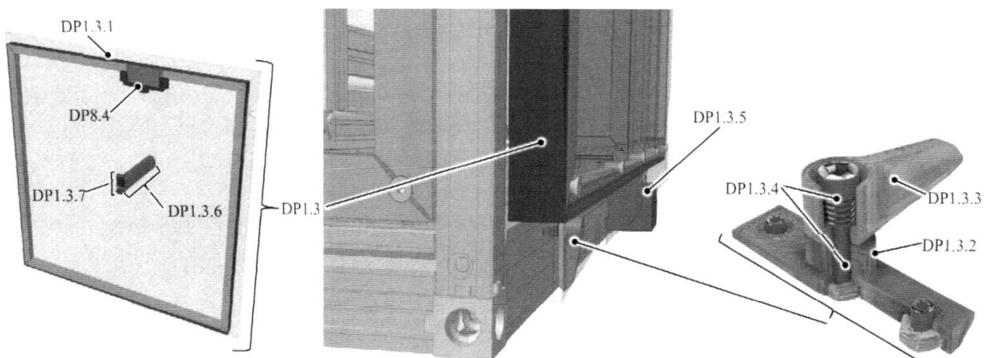

Figure 6. CAD mock-up of the detachable front door and the fastening mechanism (Step IV).

4. Conclusion

The Axiomatic Design method is a powerful instrument for enhancing the development of the constructive design process due to its systematic framework, which provides directions to be followed regarding what needs to be done and how to do it. Even if the AD matrix is comprehensive regarding functional requirements and the design solutions, the result must be filtered through a validation process to ensure the appropriate solution. An increased degree of attention should be paid to formulating the FRs. The decomposition process can be prematurely stopped if the requirements are not well formulated.

Regarding the association of the AD method with the CAD constructive design, it is very important to identify the best fit solution at the early stages of the modelling process to reduce the number of redesign iterations.

References

[1] Singh S, Singh G, Prakash C and Ramakrishna R 2020 Current status and future directions of fused filament fabrication *J. Manuf. Process.* **55** 288-306.

[2] Fang L, Yan Y, Agarwal O, Yao S, Seppala J E and Kang S H 2020 Effects of Environmental Temperature and Humidity facts of Environmental Temperature and Humidity on the Geometry and Strength of Polycarbonate Specimens Prepared by Fused Filament Fabrication *Materials* **13** 4414.

[3] Zawaski C and Williams C 2020 Design of a low-cost, high-temperature inverted build environment to enable desktop-scale additive manufacturing of performance polymers *J. Addit. Manuf.* **33** 101111.

[4] Harris M, Potgieter J, Archer R and Arif K M 2019 Effect of Material and Process Specific Factors on the Strength of Printed Parts in Fused Filament Fabrication: A Review of Recent Developments *Materials* **12** 1664.

[5] Suh N P 2001 *Axiomatic Design: Advances and Applications* (New York: Oxford Univ. Press).

[6] Brown C A 2020 *Industry 4.0 for SMEs Challenges, Opportunities and Requirements* ed Matt D T et al. (Cham: Springer Nature Switzerland AG) 383-401.

[7] Thompson M K 2013 Improving the requirements process in Axiomatic Design Theory *CIRP Annals* **62** 115-8.

[8] Thompson M K 2013 *Proc. of The 7th Int. Conf. on Axiomatic Design. ICAD* 107-112.

[9] Wolf M and Mcquitty S 2011 Understanding the do-it-yourself consumer: DIY motivations and outcomes *Academy of Marketing Science Rev.* 1 154-170.

[10] Bonvoisin J, Galla J K and Prendeville S 2017 Design Principles for Do-It-Yourself Production *Int. Conf. on Sustainable Design and Manuf.* **68** 77-86.

IOP Conf. Series: Materials Science and Engineering 1174 (2021) 012018 doi:10.1088/1757-899X/1174/1/012018

Using axiomatic design principles for the development of a device to measure the positioning error of an industrial robot

Răzvan Cosmin **Stavarache**[1,*], ***Vasile*** **Ermolai**[1], ***Alexandru*** **Irimia**[1], ***Mihaela*** **Etcu**[1]

[1] "Gheorghe Asachi" Technical University of Iasi, Machines Manufacturing Technology Department, Iasi, Blvd. D. Mangeron, 39 A,700050, România

stavaracherz@gmail.com

Abstract. Robots are part of the integrated production process, being widely used in manufacturing processes, including the positioning of parts. One of the factors that influences the accuracy of the products to be machined is related to the errors of positioning the workpiece at each stage of manufacture. To determine the amplitude and the medium value of the positioning error, we used axiomatic design to develop a device capable of measuring positioning errors. The result was a constructive solution for a device capable of taking the positioning data of the end of the robot arm along three axes of a spatial coordinate system. In this way, data are recorded using dial comparators positioned on existing channels in the three walls corresponding to the corner of a cube-shaped box. Statistical processing of measurement results should provide information on the positioning accuracy of a workpiece by an industrial robot.

1. Introduction

The use of industrial robots with arms for positioning and handling objects has already become a common practice for decades that helps increase productivity, by removing the human factor and multiple related risk factors and ensuring repeatability of operations that also involves continuity in the process. production. Over time, they have become increasingly accurate in providing smaller and smaller positioning deviations of manipulated objects.

To measure the positioning accuracy we know several devices, among which, we can use a tool known as a dial comparator. This is a mechanical device is used to measure or verify dimensions and deviations of the parts along a single axis, allowing the measurement of small and very small dimensions, with the help of a feeler rod that slides in the fixing arm in strokes up to about 10 mm, with a micron or micron fractions accuracy. This device has the disadvantage that positioning can be achieved by moving the feeler rod along a single axis and does not provide conditions for evaluating the positioning accuracy of an object by the robot's arm in a spatial Coordinate System-CS.

Another known coordinate measuring machine provides position measurements along three axes of a Cartesian CS. Measurement subsystems are provided along each axis. The shape of the part or its components is determined through point-by-point mapping in the reference CS. The coordinate values of points on the part are determined using a measuring head via a contact or non-contact method and their data are processed by an evaluation unit. For this machine type the measurement is usually made along a single axis which could be disadvantageous, and depending on the used tool type in measurement and its complexity, the purchase price can be very high. Another disadvantage is that the machine can't be used to evaluate the positioning accuracy of an object by a robot arm.

IOP Conf. Series: Materials Science and Engineering 1174 (2021) 012018 doi:10.1088/1757-899X/1174/1/012018

In view of these facts, the main goal of this paper is to create a device for evaluating the positioning accuracy of an object by the arm of an industrial robot by taking into account both the customer's requirements and the functional factors.

The main challenge of the desired device must solve is the positioning accuracy evaluation of an object by a robot arm and also the degree of repeatability of the positioning accuracy within a spatial CS.

2. Theoretical aspects concerning the measurement of the positioning accuracy of industrial robots

To further exemplify the use of the Axiomatic Design-AD method, we have briefly presented below the aspects that must be considered in the design process of a device that determines the positioning accuracy of industrial arm robots. For numerically controlled machine tools, after mounting the workpiece, the positioning accuracy can be measured, depending on the type of machine tool or its equipment, through numerically controlled axes with linear displacement (linear axes) or with rotational displacement (rotation) complete and multiple, or oscillation (axes of rotation) using Renishaw tools with palpatory head (figure 1) [2].

Figure 1. Renishaw with palpatory head [2].

The definition of the parameters, the conditions and the means to perform the measurements, the calculation/evaluation of the characteristic parameters are regulated by generally accepted standards in the world of machine tools, namely (SR) ISO 230-2 and VDI / DGQ 3441. These refer both to the verification linear and circular positioning accuracy [3].

In addition to the classic, well-known solutions of dial comparators (figure 2) a measuring system in polar coordinates can be used that can make measurements in horizontal and vertical directions using a reflector consisting of a system with mirrors that turns the laser beam back to the laser tracker. (figure 3).

Although these systems are accurate, they are complex, requiring a long time for set-up and adjustments, economically disadvantageous, and the measurement is performed on two axes simultaneously [1].

3. Principles specific to Axiomatic Design

Various design methods aim to choose the optimal constructive solution so that it meets all customer requirements. The axiomatic design method and its principles were developed by Korean scientist Nam Pyo Suh in during his time as a professor at the Massachusetts Institute of Technology in Boston, USA [5]. In using the axiomatic design method, the following sequence of activities corresponding to the design is considered:

- knowledge and understanding of customer requirements;
- conceptualizing solutions by using synthesis;

IOP Conf. Series: Materials Science and Engineering 1174 (2021) 012018 doi:10.1088/1757-899X/1174/1/012018

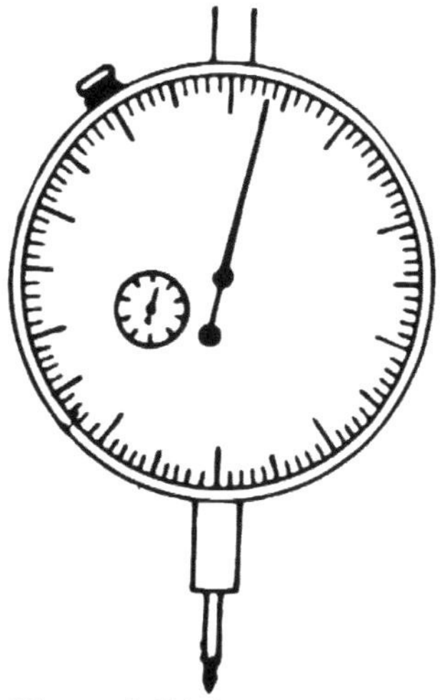

Figure 2. Dial comparators.

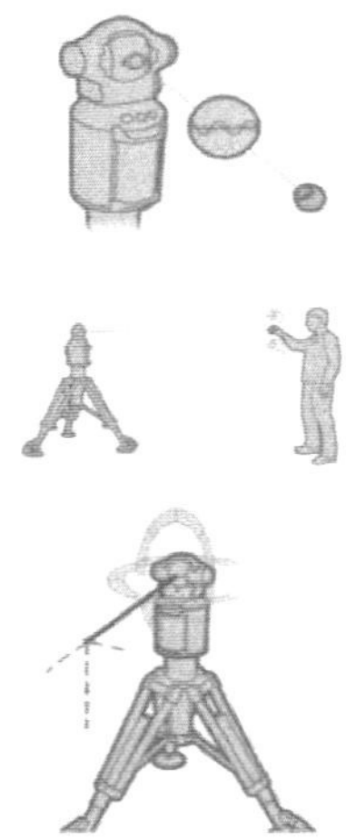

Figure 3. Laser tracker [1].

- development of analyzes for the optimization of previously identified solutions;
- verification of the extent to which the results obtained meet the initial requirements of the customer.

And the areas of use are the following:

- the client's domain, where the highlighted client's requirements are located;
- the functional area, where the functional requirements are defined;
- physical domain, where the established design parameters are;
- process domain, where are the determining process variables.

Also, there are some other customer needs to be considered in the production of the device needed; for example:

- the design technology must be adapted to the existing equipment;
- as many components as possible must be made at a single stage, in the realization of the device; if this condition is met, a higher machining accuracy will take place and the time required for the realization of the device will be shorter.
- a simplified shape is preferred to ensure and facilitate de construction of the device and to increase productivity.
- the size, shape, and mechanical characteristics of the device should be established in accordance with the material used.

The practice of design manufacturing technologies proved in time that the problem solved is not simple and first result usually adopt simplifications.

4. Functional requirements and design parameters of the device for measuring the positioning accuracy of an industrial robot

As it has been mentioned before, a device meant to measure the positioning accuracy of an industrial robot was designed taking into consideration the Axiomatic Design method to find the best-suited version of the device. To give an easier understanding of the way of using the principles of Axiomatic Design, have been applied some assumptions concerning the constructive solution. Further, will be mentioned some of the customer needs:

IOP Conf. Series: Materials Science and Engineering 1174 (2021) 012018 doi:10.1088/1757-899X/1174/1/012018

CN1: the device must be made relatively easy into a laboratory for mechanical machining, adaptable on machine tools and including by FDM printing;

CN2: for conclusive experimental results the device must have the possibility to measure objects of different dimensions and shapes;

Following the customer needs mentioned before, the main functional requirements are:

FR1: Ensure the movement of the measuring devices;

FR2: Secure the measuring instruments;

FR3: Ensure the position of the device;

Moving to the identification of functional requirements corresponding to the second level, consideration could be given to:

FR1.1: Ensure the movement of the measuring devices on several axes;

FR1.2: Ensure the movement of the measuring devices from 3 measuring directions;

FR2.1: Secure the measuring instrument on support;

FR2.2: Insert and prefix the support with the help of its sole that will enter the guide channel located on the wall of the device;

FR2.3: Fix the support on the device with a screw with nut;

FR2.4: Prefix the screw on a guide channel located on the wall of the device;

FR3.1: Ensure the position of the device by building it in one piece;

FR3.2: Ensure the position of the device by fixing it on a board;

FR3.3: Ensure the device on the board providing both the device and the plate with threaded holes.

Following the above-mentioned functional requirements, the following design parameters (*DPs*) were identified:

DP1: Procedures used to build the exterior case of the device;

DP1.1: lathe/drill / 3d printing machine used to process the wall of the device;

DP1.2: lathe/drill / 3d printing machine used to process the wall of the device;

DP2: Procedures used to fix the measurement device;

DP2.1: drill/3d printing machine;

DP2.2: prefix channel;

DP2.3: screw with nut for fixing;

DP2.4: prefix channel;

DP3: Procedure used to fix the fix the device;

DP3.1: 3d printing machine;

DP3.2: tightening screw;

DP3.3: threading machine.

The functional requirements and design parameters from above will be included in Table 1, to check the validity of the independence axiom. The correspondence between the functional requirements *FRs* and design parameters *DPs* can be written as a matrix relation:

$$\{FR\}=[A]\,\{DP\}, \tag{1}$$

where *A* is the design matrix by which the transfer functions *FR* and *DP* are taken into consideration.

In the case of the identified device, the relation (1) could be written by using Table 1.

In accordance with the Axiomatic Design theory, as it can be seen from the table 1, the design matrix is a diagonal matrix and it can be considered as an uncoupled matrix. This can be considered a result of a good design. The device resulted using the axiomatic design method can be seen in the figures below.

Table 1. Functional requirements and design parameters.

FRs \ DPs				Design parameters corresponding to the first level								
				DP1		DP2				DP3		
				Design parameters corresponding to the second level								
				DP1.1	DP1.2	DP2.1	DP2.2	DP2.3	DP2.4	DP3.1	DP3.2	DP3.3
First order FRs	FR1	Second order FRs	FR1.1	X								
			FR1.2		X							
	FR2		FR2.1			X						
			FR2.2				X					
			FR2.3					X				
			FR2.4						X			
	FR3		FR3.1							X		
			FR3.2								X	
			FR3.3									X

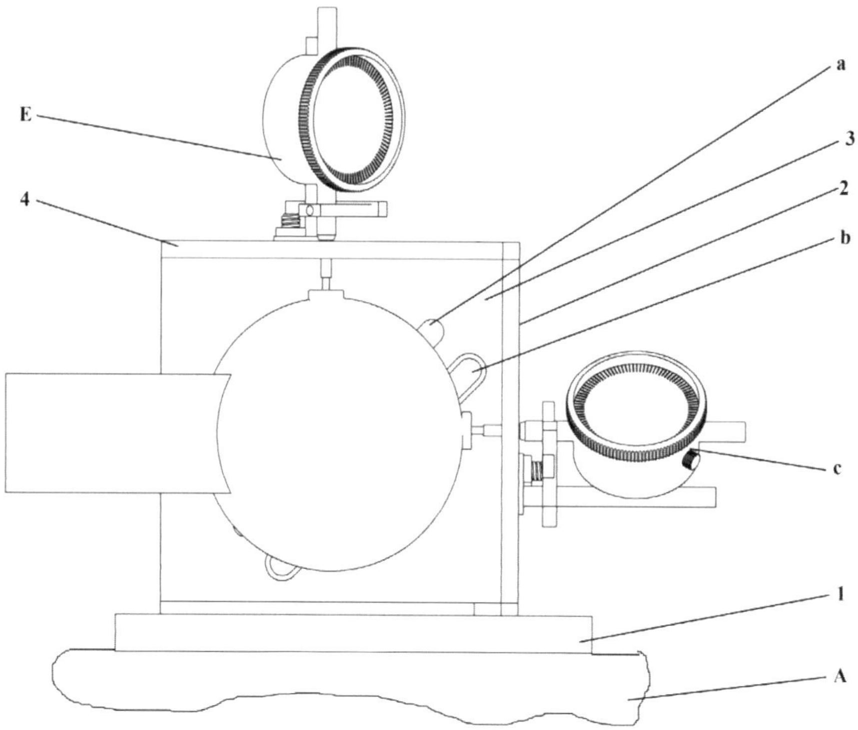

Figure 4. Principal view of the proposed device.

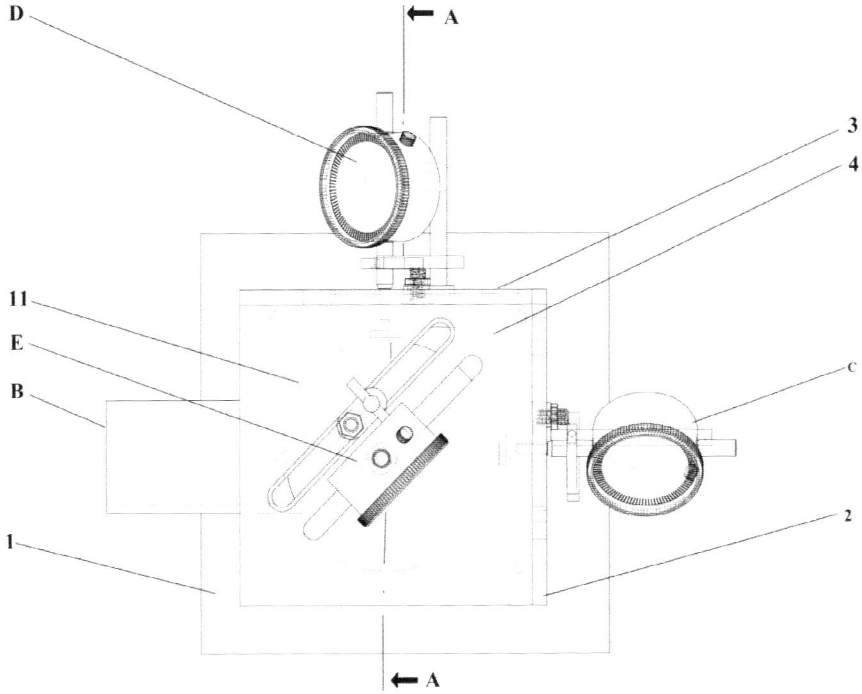

Figure 5. Lateral view of the proposed device.

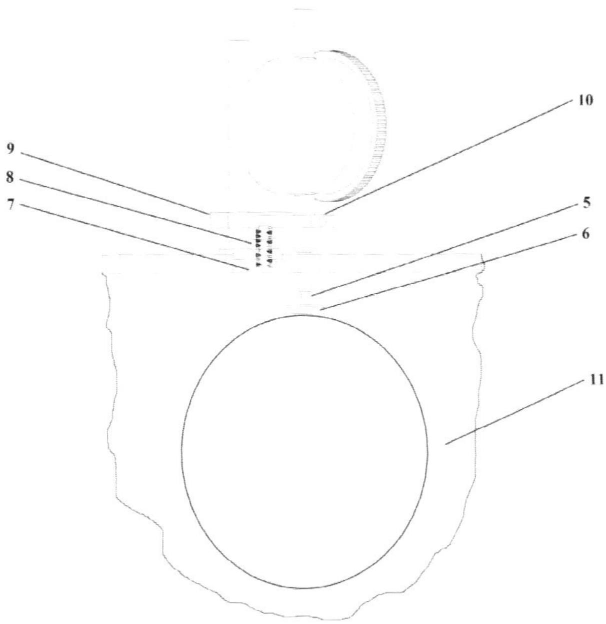

A-A

Figure 6. Section A-A of the proposed device.

According to figures 4-6, the device has as a basic part a base plate 1, located on an assembly A within the range of an arm B of an industrial robot. At the base plate 1 are joined, for example by welding, two walls 2 and 3, which, together with a wall 4, secured to walls 2 and 3, form a so-called corner. In each of the three walls 2, 3 and 4, of square shape, there are two perforated rectilinear channels a and b, the channel a being arranged along a diagonal of the square wall, and the channel b being parallel to this diagonal and at a certain distance from the respective diagonal. A disc 5 holding rod 6 in the form of a disc can move freely in the channel a. In the channel b a rod 8 supporting a part 9 can be moved and immobilized in a certain position using a nut 7 in which a guide bush 10 will be attached to guide the rod 5 supporting the probe 6 in the form of a disc to each of the three dial comparators, C, D and E.

5. Conclusion

Using the axiomatic design method, we obtained a device that measures the positioning accuracy of industrial arm robots in a spatial coordinate system such as the level of repeatability of spatial coordinates determined by several attempts to estimate the positioning accuracy of the object by the robot arm.

Another direction of research development can be the verification of the level of repeatability of the spatial coordinates of the positioned object by repeated measurements and the interpretation of the recorded data.

References

[1] Macovei M I, Laser traker procedures for experimental evaluation of the performace parameters for RI KAWASAKI FS10E (in Romanian), Student Scientific Session, 2016, Politehnica University of Bucharest. http://www.imst.pub.ro/Upload/Sesiune/Comunicari Stiintifice/Lucrari_2015/06.12/12_L07.pdf

[2] Renishaw - Elements of a CMM system, https://www.renishaw.com/en/elements-of-a-cmm-system--45561

[3] Gornic C, Checking machine tools. Measurements and interpreting the resuls, T&T, 2012, https://www.ttonline.ro/revista/masini-unelte/verificarea-masinilor-unelte-xvi-efectuarea-masurarilor-si-interpretarea-rezultatelor

[4] Grozav I, Improving quality through axiomatic design (in Romanian), Buletinul AGIR, 1-2, 2008, https://www.agir.ro/buletine/341.pdf

[5] Suh NP 1990 *The principles of design* (New York: Oxford University)

IOP Conf. Series: Materials Science and Engineering 1174 (2021) 012019 doi:10.1088/1757-899X/1174/1/012019

Visual representation aid in engineering design process: From axiomatic design towards knowledge intelligence

M Papachristopoulou, C Provatidis, AP Markopoulos* and V Spitas

National Technical University of Athens, Greece

*Corresponding author. E-mail address: mlpapazoglou@mail.ntua.gr

Abstract. The notion of function has been a key element in most of the design methodologies. Its definition has been controversial within the research community, with issues arising when methodologies rely solely on one universal strict definition. Axiomatic design relies on functional decomposition, which has as a perquisite the correct definition of the main function and subfunctions of a design problem. The inability of design methodologies to be incorporated into CAD and the absence of a systematic way to store precious knowledge occurring during the design process itself becomes a hurdle for true automation and intelligence in design. This paper proposes a visual representation process of the design problem, which verifies itself through relational graph representations and the physical laws governing the structure, where they are given valuable semantics. Knowledge can be inferred, stored and analysed with the scope of making engineering design more intelligent and potentially more automated.

1.Introduction

The conceptual phase of engineering design, is crucial for the quality of the final product. Computer Aided Design (CAD) is the most popular tool to assist designers with the visualization of their model but it is inefficient for true intelligence, because it conveys mostly mere geometrical information, an issue recognised since decades [1], [2]. CAD has developed tools like Knowledge Base Engineering (KBE) in order to become more intelligent, but only managed to automate routine design tasks. KBE Systems are not considered to be alternatives to CAD but rather a sophisticated blend of AI, CAD and object-oriented modelling, which has the potential to capture and efficiently reuse the knowledge [3]. Semantic knowledge in CAD is still poorly handled, and Product Life Management (PLM) systems still fail to capture significant semantic knowledge being generated during the design process.

At the same time academic research tries to fill the absence of semantics in CAD and systemizes design with the development of the field of the "Design Theory and Methodology [4]. The main objective of the design is the mapping of function to artifacts and products. Erden et al [5] offers an extensive review of Functional Modelling methodologies and their evolution through years. Among them, Axiomatic Design [6], has been particularly popular in the research community and has also been applied in industrial practice with success, compared to other methodologies.

Vermaas [7] and Dorst [8] believe that the notion of function itself cannot be objective as researchers would like to for a universal application, since it has been used differently among different theories. In AD, according to [9] very often mistakes in the notion of function happen, because there is a presence of physical information in the functional requirements, which prevents the creation of a solution-neutral design environment, it violates the first axiom of AD and trivializes the mapping process between the functional and physical domain. Similarly in other methodologies, ambiguities in the notion of behavior, function and structure appear [10].

The intuitive nature of the design process demands a flexibility of communication between different levels of abstraction in the ill-structured design problems. The standard approach to accumulate and represent abstract knowledge, demanded very hard work and the results were far from satisfactory. Systematic ways need to be found to induce, represent and manipulate large databases of structured, abstract knowledge, often in causal nature towards the way of robust intelligence [11]. The bottleneck is not the reasoning, but having the right information accessible in the context of real-world inference, which is very often the case of case- specific design methodologies which do not communicate between them through a dynamic flow of information.

In this paper, we propose the incorporation of visual graph elements during the design process, which by default denote functional semantics because of their visuospatial relationships. A strict definition of function is not enough to explore the design space alternatives, but the visual graph elements together with their relationships and the physical laws that govern them, may indicate knowledge and semantics essential to achieve any design intelligence.

2. Capturing the design process in methodologies

In the early design phase according to Kleban [12], designers tend to rely more on ad hoc representations (i.e., sketches) and communicate a lot to exchange information. Systematic methodologies may serve as a tool to capture knowledge in this phase and aid the designer to make decisions through a systematic way, however designers are reluctant to apply them to generate their initial decisions [13]. Most of the methodologies which developed to systematically aid and automate this phase, rely on the notion of function and in particular to functional decomposition, which has its roots in the 70's [14].

Function, according to Umeda and Tomiyama [15], is the bridge between human intention and physical behavior of artifacts and consider hierarchical decomposition the main task in design. In a conventional application of the methodology the procedure is top-down from abstraction to detail without the important interaction between levels of abstraction which actually occur in the design process [16]. The functions are decomposed into subfunctions until they can be associated with some physical features.

A lot of researchers tried to formalize functions into basic categories so as to automate functional modelling, by treating a device as a black box that transforms energy, information and signals flow [17], [18]. Kitamura et al [19] developed functional ontologies in the SOFAST software with main functions being described by verb and noun descriptions for flow-based modelling in order to provide a universal language between researchers to describe their devices, but difficulties on implementation arose due to the limited freedom these representations offer.

Gero [20] tackles the problem of static representation with dynamic change in the design context, introducing the Function-Behavior-Structure model (FBS) with function being the intermediate between the goal of the human and the behavior of a system and with structure being the means to achieve this function. The model supports functional decomposition by relating sub-functions to structural features and to their physical states in order to achieve the former. Dorst and Vermaas [10] make a critical review about the ambiguities introduced in the FBS model proving it to be unstable in consistent use.

Kannengiesser and Gero [21] recently revisited the design thinking methodology and its link to cognitive science. According to the dual-system theory on Kahneman's book that the paper revisits, the human cognition is governed by two systems, a fast, intuitive and effortless thinking and a slower analytic thinking which requires greater effort, with the first being the one which governs our daily activities. This finding, questions whether we are able to decide on problem solving in an objective manner. If dual-system theory plays indeed a role in design, then a design tool support for system 1 and system 2 design thinking is likely to differ. Most of the design methodologies and implementations rely heavily on system 2 without proposing anything for system 1.

Axiomatic Design (AD), offers a systematic way of analysing problems by considering high-level Functional Requirements (FRs) mapped with a zig-zag process to the Design Parameters (DPs) by fulfilling the independence and information axiom. The zig-zag process offers the designer a chance to modify the problem and explore the design space, having a significant representation flexibility benefit compared to other methodologies. AD has been proven to be particularly helpful in situations where designers are trained on how to apply the methodology and possess a great amount of knowledge for the design problem they want to solve. Successful industrial applications of AD have been mentioned, which concern mostly big enterprises [6].

Nakao and Iino [22] completed a survey with engineering students implementing AD. They found out that it is very hard for students to handle abstract concepts and are weak in dealing with FRs without physical substance. The students tend to list FRs chronologically and DPs spatially. FRs cannot be inferred from DPs and the students have to express invisible FRs with only verbal expressions. Some of occurring issues the methodology has, such as the FR being mixed with non-functional requirements have being mentioned in [9].

Little attention has been given to intuitive design and how to capture knowledge in this phase. By default the designer cannot have all the information it is required to solve the problem at the very beginning as proposed in most methodologies, but could benefit from various representation tools with less formal schema, which could provide a more convenient cognitive way for the human brain to extract potential FRs or DPs and explore the design space.

3. Visual Representation and ad-hoc modeling

Peirce with his semeiotic theory [23] who invented the most widely used notations of logic and believed that every mathematical (logic) reasoning is diagrammatic and especially found in graphs. Early Entity-Relationship (ER) graphs with the purpose of automating assembly design were able to represent basic hierarchical relationships, but didn't offer a way to escape explicit modelling of the graph prior to analysis [24].

Sowa and Majumdar [25] explore analogical reasoning by using conceptual graphs (CGs). Conceptual graphs are linked to relational databases, which represent a model with its entities, attributes and relationships. Sowa also believed that the rules in the CG are fundamentally graphical: they are easier to show than to describe, that's why he created graph grammar rules. The type of concepts and relations in a CG reflect the choices made by the original programmer, which in turn reflect the options available in the original programming tools. With class hierarchies and strict ontologies, the designer has some weak tools to represent the complex nature of the design process.

Yaner [26] extends the Structure-Behavior-Function model [27] and adds drawings and shapes for representation in the standard functional descriptions. Spatial relations are seen as an intermediate between shapes and behavior, so explicitly link functional knowledge to visuospatial knowledge [27]. Goel believes that the functional knowledge should be grounded to visuospatial knowledge.

Liang and Paredis [28] realize the importance of graph representation for a systematic and possibly automated exploration of design alternatives and classifies types of ports according to the FBS system. Ports are interfaces that connect elements of a system and through them energy, material flow and signals flow, a notation originating from the black box theories. In contrast to other methods in this work they don't consider only input and output flows but also internal flows which give semantics to the components themselves.

Kokotovich [29] explores the design thinking process during the design and the different way novice and expert designers operate. The network structure without class hierarchies, is the richest, assisting the designer in developing an understanding of the problem space and the dynamic interrelationships among the design issues. Hierarchical mind map in design fails to describe via text or graphic images the supporting rationale or complex symbiotic relationships between issues.

Towards this direction Mavrikas et al [30] applied their method of Synaptic Networks on a case study of a Helicopter gearbox by creating a topological graph model consisting of elements, interfaces and contacts in a non-hierarchical manner. By representing the system visually, they were able to trace the functional flow paths and realize solutions which were non obvious before the representation analysis and they access the idea of modularity by re-arranging the graph. Similarly on another implementation of the Idea Algebra and Synaptic Networks Amrin and Spitas [31] make a topological representation of an automotive drivetrain and trace again possible flow paths regarding force and torque equilibrium by extracting information of the relative position of the components. The model

IOP Conf. Series: Materials Science and Engineering 1174 (2021) 012019 doi:10.1088/1757-899X/1174/1/012019

however requires an explicit definition of its topology in order to perform automated reasoning about its function, which is often not easy to have in a design synthesis phase.

An intelligent model is the one which supplementary to our subjective thinking can include some objective elements that can be inferred not only by the visuospatial representation of components and its mapping to physical behavior (like in Idea Algebra) but also from an intelligent relational model expressing the possible different roles a component or its attributes can have in a system, without the restriction of a strict class hierarchy. The model should be able to support the gradual design thinking process by being dynamically expandable.

A potential tool for the application of this model in computers, could be one which utilizes knowledge graphs KG (intelligent graph databases). Grakn is developed by a startup company named Grakn Labs Ltd, offering an innovative KG tool which relates entities, relationships and attributes. Messina et al [32] give an early introduction to Grakn knowledge graph and its role in handling complex biological research data. It is not a traditional graph database, but a knowledge schema that may include nested relationships and infer non obvious ones. Grakn supports the gradual building of the model and provides automated reasoning while enabling the necessary flexibility to represent complex relationships, being found in a system/device. Grakn can offer an advantageous way to represent components in a system but additional schemas should be developed which support physical laws required for analysis as integral part of engineering design.

In figure 1 the fact of "Marry married Peter on 2/2/2020" can be represented in Grakn as depicted on the right, while compared to a standard entity representation on the left. The person in this Grakn representation plays the role of the spouse in a marriage relationship and attributes are linked to them. The representation is advantageous since now more marriages which happened on the same date can be attached to it and the graph can extend. An analogy can be found also for engineering relationships.

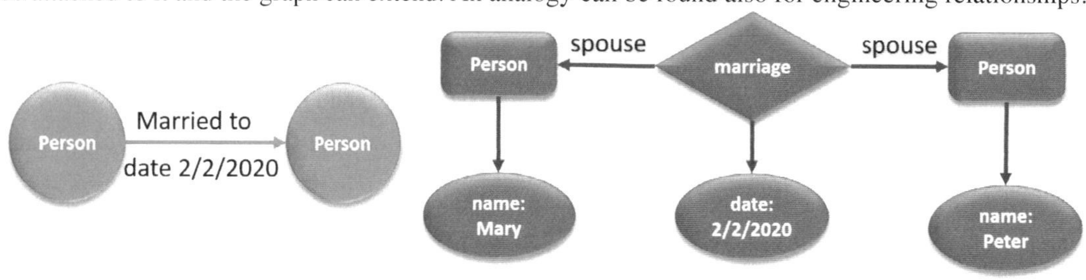

Figure 1. Grakn representational schema on the right compared to a standard entity one

4. A case study:-Comparisons with AD

The case study comes from an actual design project encountered by one of the authors. The client is an R&D firm in the field of optomechanics and needs assistance in designing some components in their optomechanics set. In particular the client asks for the design of a component which is going to attach a rotating mirror with a motor that provides motion to the mirror while the latter is located within one degree tilt from it. In engineering practice, such abstract descriptions about the design project and its goals are quite usual and designers are accustomed to these. It is widely accepted that the designers themselves with their expertise, have to interpret these into actual engineering requirements. In order to solve this problem with AD the designer would have to define the highest level FRs and then map them to DPs, probably by collecting more information to define this primary function. The scope of the case study is to present an implementation of a visual representation graph-like schema, which aids the design process by providing implicit functional information based on relationships between the components and indicating the physical laws of the problem.

This is achieved through the use of simple shapes which represent components, but spatially position them in 2d space, in a way to cognitively perceive the structure. The human brain is familiar with such a technique, which, by analogy, is similar to designing something in CAD for visual perception in the design synthesis phase, but without the tedious modelling process. By placing the motor next to the mirror (figure 2) an alignment relationship is implied, as well as rotation around the same axis (Axis 1), which was confirmed that it was indeed the case with the client, since motion can be transferred also when the axes of the components are not perfectly (or at all) aligned (alternative design solution). This is a type of knowledge that this form of visual representation is able to imply (semantics).

It is also required to represent the one-degree tilt requirement, so the client further explains that the mirror should be attached onto something that is in between the motor and the mirror and should be in one degree tilt relationship from the plane of this object, as depicted in figure 3.

Figure 2. Motor and mirror rotate around Axis 1.

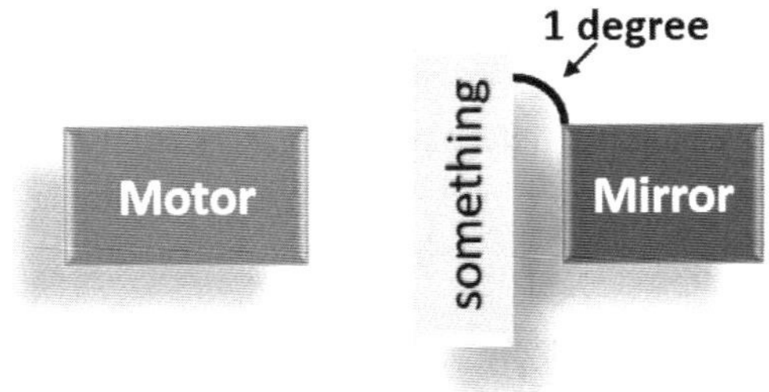

Figure 3. Mirror should be at one degree tilt from the plane vertical to the yet undefined 'something'.

The level of abstraction in the visual representation is a matter of design choice and the motor could still keep considered the motor as a black box or could be analyzed into basic components (figure 4).

The motor and the mirror are parts of a specific relationship A which indicates transfer of motion. The design becomes more detailed as more specific elements are gradually added according to the client's description; the whole set should be installed on an optomehanics base component in a mechanical fixation relationship, depicted by B in our representation (figure 5).

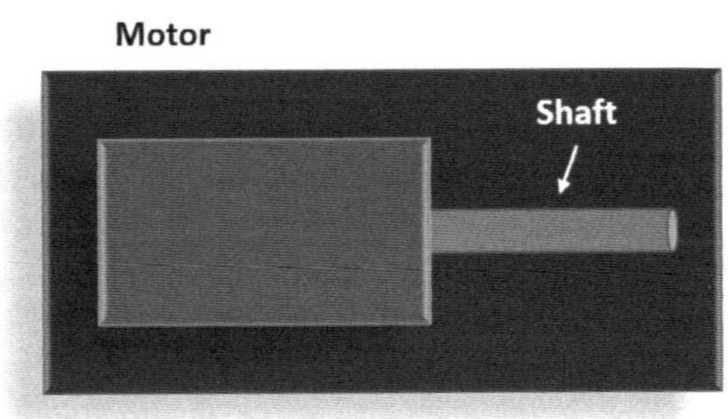

Figure 4. More detailed view of the motor.

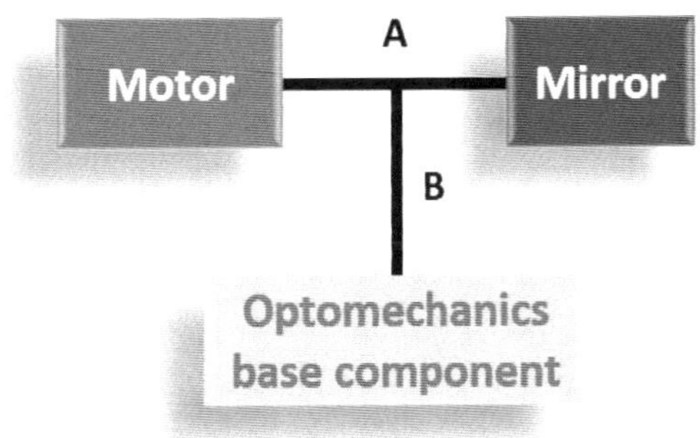

Figure 5. Relationships A and B defining the relational topology of motor, mirror and optomechanics base component.

The relationships in this graph like representation, do not comply with the functional basis taxonomies in their description. They also do not comply with the one-way attributions been given in class hierarchies. The system is perceived in its wholeness and the problem as underdetermined as it is, so the components are not limited in a unique relationship. For example, the motor can play the role of transferring motion in a transfer of motion relationship but could also play a different role in another type of relationship which could be later defined, following the Grakn relational schema, described in Section 2. This description enables a more realistic view on design problems.

The components are seen as members of functional modules, but not in the sense of the black box being mentioned earlier (input-output) but as graph modules being part of a relative topology which consists of a combination of relational and physical phenomena relationships. For example, a module that can potentially satisfy A, could be a standard shaft-coupler-shaft solution (figure 6) with one of the shafts being the component which shares relationships within two modules. The relationship A is a result of the relationships E, C, D, F. A suitability verification for introducing the shaft-coupler-shaft to take the place of A, would have been the crosschecking for satisfaction of physical laws (torque transfer, etc.) together with the satisfaction of relationships and roles. For example, the shaft should play the role of torque transfer medium in a transfer of motion relationship, instead of the shaft playing any role in a relationship which doesn't concern transfer of motion. If relationships and roles in the graph schema merge and if by laws of physics the solution is appropriate, modules could be inserted to substitute certain relationships of the graph like A.

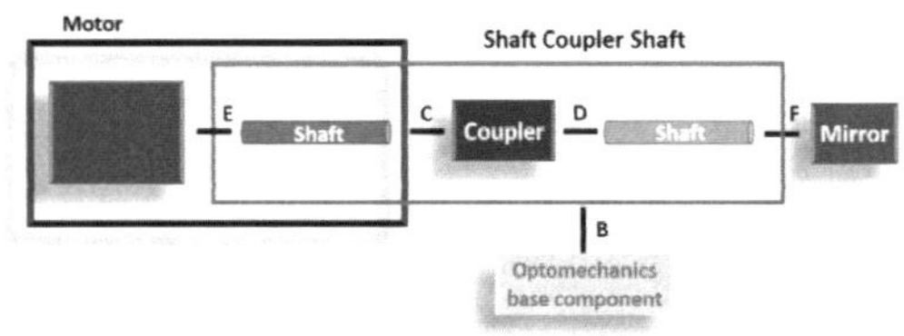

Figure 6. Nested relationships of all components.

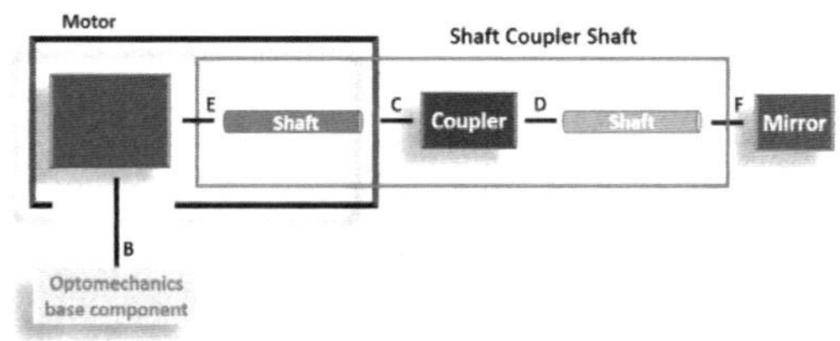

Figure 7. Moving Optomechanics base component to the static component.

In a similar manner, different location possibilities for the optomechanics base component can be explored and evaluated. The client has a standard post holder component as depicted in figure 7 and has to find a way to fix the whole structure on this base component. The optomechanics base component by default receives other components and secures them in place. If the optomechanics base component should be attached to the structure, a component that is geometrically suitable to fit in it should be found, which in this case could be only the shaft (cylinder). However, the optomechanics base has no moving parts so it cannot accept the moving components of the shaft-coupler-shaft module since they all represent a transfer of motion relationship while the optomechanics base component only served fixation relationships. In creative design though, different possibilities can be explored to include moving components on the optomechanics base component or static elements on the shaft-coupler-shaft module so as to satisfy the physics for the continuity of the model.

Since B cannot be attached to moving shaft-coupler-shaft module, the only static component of the structure is the main body of the motor. The relationship B should have an additional component which can satisfy the fixation relationship and the geometric relationship of attachment to the motor, since there is no matching geometry to assemble, already there. The motor has threads on the left side so the optomechanics base component together with the extra component of the B will have to share same positioning requirements since they are in a fixation relationship. This way of analysis is part geometric part spatial, part functional, which provides a more holistic view of the design problem. A fixation relationship can translate into physical spatial characteristics, which means that if the optomechanics base component is fixed with the motor in a way which will cause its rotation like in figure 9, the motor follows its spatial position and vice versa. This reverse relationship is something that could have easily not been taken into consideration as a fact as well in AD. By exploring the different spatial positioning possibilities of the optomechanics base component it intuintively appears that these components are placed on the base plate in a fixation relationship G as well. The base plate in this problem appears to be the absolute reference for spatial positioning.

Figure 8. Standard optomechanics base component

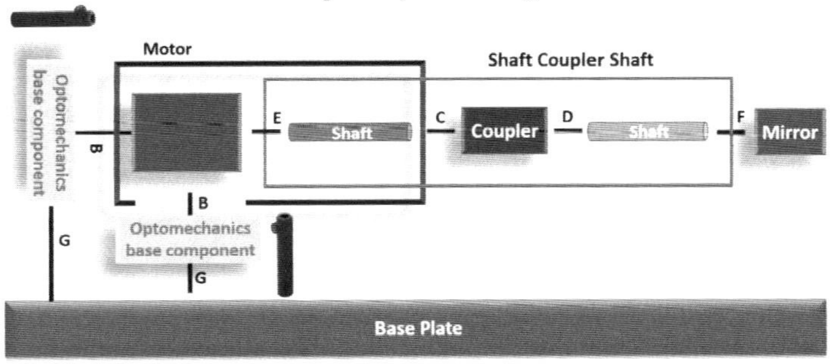

Figure 9. Moving optomechanics base component to the static component and exploring position possibilities according to geometric requirements

It is the first time that the relative position of the base plate to the whole assembly is introduced as a concept. The client has taken this for granted and didn't initially explain this, since what they tried to reach was a solution to fulfill relationship A. The optomechanics base component is in a positioning relationship (G) with the base plate. In the standard notion of hierarchic relationships, the base plate would give a position to the optomechanics base component and this would give the position to the rest of the assembly since the base plate is the absolute reference frame. Semantic-wise the E, C, D and F relationships, are all required to result in aligned and connected together components, since that is what the laws of physics impose for this type of coupling transfer of motion and if the mirror has to move, the whole block of Motor and Shaft-Coupler-Shaft would have to move together with the optomechanics base component as well, since it is in a fixed relationship with the motor.

After talking with the client, it was found that what appears to be the primary functional requirement which would have been required to be known for AD decomposition, is a lot more complicated and non obvious than what at first the client had explained. As a matter of fact, the first requirement that the client mentioned, is a combination of two potential subfunctions which would follow the independence axiom.

The Base Plate plays a specific role in this model, but can also play other roles in other models (following the Grakn schema). For every optomechanics component that is placed on the base plate in a fixation relationship (which could have a physical mapping to x,y,z coordinates), the base plate will always play that certain role of a "giving position" component in a positioning relationship, additionally to a fixation one. The motor and the mirror now are parts of the whole assembly of the optomechanics set, since the representation can expand, having the base plate as a common sharing semantics component.

The degrees of freedom of the optomechanics base component in relation to the base plate is its rotation around its axis and around a specific center of rotation on the base plate. If the whole current structure rotates around this center, it is possible that the structure will interfere with other components spatially. More information is required from the client whether it is the motor that has to rotate around that point or the mirror and whether they have predicted the spatial interferences.

The client answered that it is the mirror that should have the flexibility to rotate in the set up since it plays the role of collecting a light beam which comes from a laser and the rest of the components of the general assembly are placed spatially in such a way as to capture that. The specification of the one-degree tilt is for converting that beam into a light cone which is the main information for the design problem. Therefore, the client could have defined a more accurate primary functional requirement; the mirror should receive a light beam and convert it into a cone. This description would have been essential for performing AD functional decomposition, but it was not until this point that it was feasible to address the bigger picture of the problem by the aid of the visual representation of the model. The initial description was already decomposed into two FRs and unable to indicate the main physics problem behind the design.

So as to accurately control the position of the mirror with reference to the Base Plate, the optomechanics base component should be transferred towards the mirror. Since the mirror is rotating, it would be required to be connected with an intermediate component which is stationary, like a frame/base, which was already considered before. In this phase the representation can indicate two solutions:

- There could be a stationary component which contains the rotating mirror and provides a plane with respect to which the mirror can lie at one degree tilt.
- There could be the aforementioned, which also has that type of geometry that the motor can be fixed on (it has been left without fixation since the optomechanics base component moved)

Therefore, this procedure would lead to two (at least) different solutions, which implicitly denote different physical laws and semantics. Accepting that relationship-wise the components are in compatible connections from the relational graphs the second verification comes from the physics. For example, the solution of figure 7 was good enough before it was realized that the optomechanics base component had to move towards the mirror. However, the solution had formed a cantilever. Rotating components which belong in cantilever structures, cause vibration which can lead to the excitation of eigenfrequencies and lead to system failure. The client only later in the design process mentioned that the mirror has to rotate at 2200 rpm, which can lead to significant vibration phenomena which can affect the whole assembly and have a negative effect on the light cone. In that case, the graph model acts in a preventive way, as it alerts the designer for failures.

Eventually more than one acceptable solution can be found which should:

- Comply with matching relationships and roles of the components
- Comply with the laws of physics

The design problem as initially described, proved to be almost misleading for the main design goal that was to be resolved. This is not the fault of the client, neither the designer's, but it is the nature of ill- structured design problems. Universal vocabulary or strict class hierarchy descriptions may lead to partially right descriptions, however excluding the possibility to dynamically change the model while exploring the design space. One could consider that the information of this design problem did exist since the beginning, but it was not until the design synthesis between the different layers of abstraction, that managed to retrieve the main objective of the problem and on which parts to focus for proposing a design solution.

In this research paper the focus was on a general description of the effect the method can have on the gradual thinking process, but in future work, a relational schema will be developed utilizing Grakn which will be coupled with physical laws. In that way, knowledge about the model and the process of the sequential design decisions will be able to be stored in order to be analyzed to extract valuable insights with the scope of improving design automation and advance in the use of AI in engineering design [33,34].

Conclusion

A visual process for representing design problems and design thinking is proposed, which implies functional semantics. There is no need to explicitly define the primary function as demanded by functional decomposition methodologies, since this is a piece of information which is complementary for the design solution. The nested relationships between the components, the flexibility of the representation schema graph and the observation of the physical laws implicitly denoted by the representation, are all equally important to contribute to a design solution. The research community admits the need of hybrid representations to capture the semantics of models. A need to translate academic research to computer language while benefiting from state-of-the art software tools is necessary for the creation of next generation design software.

References

[1] Akman V, ten Hagen P J and Tomiyama T 1990 A fundamental and theoretical framework for an intelligent CAD system *Comput Aided Des*. **22**(6):352–67.
[2] Szykman S, Fenves S J, Keirouz W and Shooter SB 2001 A foundation for interoperability in next-generation product development systems. *Comput Aided Des*. **33**(7):545–59.
[3] Milton N 2008 *Knowledge technologies* (Monza, Italy: Polimetrica)
[4] Tomiyama T, Gu P, Jin Y, Lutters D, Kind C and Kimura F 2009 Design methodologies: Industrial and educational applications *CIRP Ann Manuf Technol*. **58**(2):543–65.
[5] Erden M S, Komoto H, van Beek T J, D'Amelio V , Echavarria E and Tomiyama T 2008. A review of function modeling: Approaches and applications. *Artificial Intelligence for Engineering Design, Analysis and Manufacturing*. **22**. 147 – 169
[6] Suh N P 2001 *Axiomatic design: advances and applications* (USA : Oxford University Press)
[7] Vermaas P 2013 The coexistence of engineering meanings of function: Four responses and theirmethodological implications. *Artif Intell Eng Des Anal Manuf*. **27**(3):191–202
[8] Dorst K 2004 On the problem of design problems-problem solving and design expertise Journal of design research **4** (2) 185-196
[9] Thompson M K A classification of procedural errors in the definition of functional requirements in Axiomatic Design Theory. 2013 7th Int. Conf. on Axiomatic Design Boston
[10] Vermaas P E and Dorst K 2007 On the conceptual framework of John Gero's FBS-model and theprescriptive aims of design methodology *Des Stud*. **28**(2):133–57.
[11] Marcus G 2020 The Next Decade in AI: Four Steps Towards Robust Artificial Intelligence ArXiv, abs/2002.06177.
[12] Kleban S D, Stubblefield W A, Mitchiner K W and Mitchiner J L and Arms M 2005 Collaborative evaluation of early design decisions and product manufacturability 34th Annual Hawaii Int. Conf. on System Sciences IEEE Comput. Soc
[13] Ivashkov M 2004 ACCEL : a tool for supporting concept generation in the early design phase, PhD Thesis, The Eindhoven University of Technology, Eindhoven
[14] Rodenacker W 1971 *Methodisches Konstruieren* (Berlin :Springer-Verlag)
[15] Umeda Y and Tomiyama T 1995 FBS modeling: modeling scheme of function for conceptual design Proc Working Papers of the 9th Int Workshop on Qualitative Reasoning About Physical Systems, pp. 271–278, Amsterdam
[16] Spitas C 2011 Analysis of systematic engineering design paradigms in industrial practice: A survey. *J Eng Des*. **22**(6):427–45
[17] Stone R and Wood K 2000 A Heuristic Method for Identifying Modules for Product Architectures *Des. Stud* **21**(1)
[18] Hirtz J, Stone R B and McAdams D A et al. 2002 A functional basis for engineering design: Reconciling and evolving previous efforts *Res Eng Design* 13 65–82
[19] Kitamura Y, Kashiwase M, Fuse M and Mizoguchi R 2004 Deployment of an ontological framework of functional design knowledge *Adv Eng Inform* **18**(2):115–27
[20] Gero J S 1990 Design Prototypes: A Knowledge Representation Schema for Design *AI Magazine* **11**(4) pp. 26-36
[21] Kannengiesser U and Gero J S 2019 Design thinking, fast and slow: A framework for Kahneman's dual-system theory in design *Design Science 5*.
[22] Nakao M and Iino K 2018 Students list FRs chronologically and DPs spatially, and need to integrate FRs functionally and DPs physically MATEC Web Conf. 223:01011.
[23] Peirce C S 1902 Logic Considered as Semeiotic MS L75 edited by Joseph Ransdell,
[24] Lieberman L I and Wesley M A 1977 AUTOPASS: An Automatic Programming System for Computer Controlled Mechanical Assembly. IBM Journal of Research and Development **21** (4) p 321-333

[25] Sowa J F and Majumdar A K 2003 Analogical Reasoning In: Lecture Notes in Computer Science. (Berlin, Heidelberg: Springer Berlin Heidelberg) p 16–36.2455 Teller Road, Thousand Oaks, California 91320: SAGE Publications, Inc
[26] Yaner P 2007 From shape to function: acquisition of teleological models from design drawings by compositional analogy. PhD Thesis. College of Computing, Georgia Institute of Technology
[27] Goel A K 2013 A 30-year case study and 15 principles: Implications of an artificial intelligence methodology for functional modeling *Artif Intell Eng Des Anal Manuf*. **27**(3):203–15.
[28] Liang VC, Paredis C J J 2004 A port ontology for conceptual design of systems *J Comput Inf Sci Eng*. **4**(3):206–17.
[29] Kokotovich V 2008 Problem analysis and thinking tools: an empirical study of non-hierarchical mind mapping *Des Stud*. **29**(1):49–69
[30] Mavrikas G, Spitas V and Spitas C 2018 Introduction to structural ageing-specific functions for computational models based on synaptic networks. In: *The Ageing of Materials and Structures*. Cham: Springer International Publishing p 197–215
[31] Amrin A and Spitas C Towards non-hierarchical system descriptions for automating functional analysis. In: DS 87-4 Proceedings of the 21st Int. Conf. on Engineering Design (ICED 17) Vol **4**: Design Methods and Tools Vancouver, Canada 2017 p 218–28.
[32] Messina A, Pribadi H, Stichbury J, Bucci M, Klarman S and Urso A 2017 BioGrakn: A Knowledge Graph-Based Semantic Database for Biomedical Sciences. In: *Advances in Intelligent Systems and Computing*. Cham: Springer International Publishing p 299-309
[33] Rahman M H, Schimpf C , Xie C, Sha Z 2019 A computer-aided design based research platform for design thinking studies *J Mech Des* N Y. **141**(12):1
[34] Rahman M H, Xie C, Sha Z 2019 A deep learning based approach to predict sequential design decisions. In: Volume **1**: 39th Computers and Information in Engineering Conf. American Society of Mechanical Engineers

IOP Conf. Series: Materials Science and Engineering 1174 (2021) 012020 doi:10.1088/1757-899X/1174/1/012020

A methodological approach for strategic evaluation - a response to water scarcity emergency

Miguel Cavique[1,3,5], António Mourão[2,3], Júlia Seixas[4], António Gonçalves-Coelho[3]

[1] Naval Academy, Base Naval de Lisboa, Alfeite, 2810-001 Almada, PORTUGAL
[2] Department of Mechanical and Industrial Engineering - DEMI, NOVA School of Science and Technology, NOVA University Lisbon, Campus de Caparica, 2829-516 Caparica, PORTUGAL
[3] R&D Unit for Mechanical and Industrial Engineering - UNIDEMI, NOVA School of Science and Technology, NOVA University Lisbon, Campus de Caparica, 2829-516 Caparica, PORTUGAL
[4] CENSE, NOVA School of Science and Technology, NOVA University Lisbon, Portugal
[5] Corresponding author: miguel.cavique@cest.pt; cavique.santos@marinha.pt

Abstract. The goal of the Paris Agreement to limit the increase of temperature may become an illusion. Within 20 to 30 years, Southern Europe may experience hot waves and broad dry periods. Water scarcity may turn into reality, and the Defense structures of each country must prepare for it. The Scenario has high uncertainty as well as what the response to the hazards might be. The Dempster-Shafer theory (DST) allows to handle uncertainty, and the Axiomatic Design (AD) is a tool to do good designs. This paper presents a methodological approach to handle with forthcoming events. It defines what is a scenario using the structured AD frame, defined by constraints and needs. In addition, it presents an approach to compute information for different developments of the design. Thus, it describes a Defense strategy to "provide the basic water needs to the population". The DST method is used to achieve belief and plausibility measures of the probability of success. Currently, the belief in the success of a response to hazards is almost zero. Plausibility can go up to 0.65 by defining a reaction structure, improving the technology, manufacturing new products.

1. Introduction

In December 2015, the United Nations Framework Convention on Climate Change (UN-FCCC) matched the so-called Paris Agreement. Paris Agreement aims to hold "the increase in the global average temperature to well below 2 °C" and ask all Member States to make the necessary efforts so that the increase of temperature at the end of the century will be 1.5 °C above the average values of the pre-industrial era (art.2) [1]. The UN's scientific committee, the International Panel on Climate Change (IPCC), defined four possible climate pathways for this century. Each pathway, a Representative Concentration Pathway (RCP), is a possible evolution CO_2 concentration in the atmosphere that increases the global temperature. The IPCC defined the RCP through many climate models of the earth depending on human behavior. The RCP index represents the balance between the average radiation received and emitted by the planet in W/m^2 in 2100.

The CO_2 concentration in the atmosphere might follow the RCP2.6. It leads to an average concentration of 400 ppm of CO_2 and a temperature increase in the range of 0.9 to 2.3 °C. The current

IOP Conf. Series: Materials Science and Engineering 1174 (2021) 012020 doi:10.1088/1757-899X/1174/1/012020

carbon concentration is already 410 ppm, higher than the acceptable by the end of this century. The RCP2.6 urges the World to be carbon neutral in 2070 and capture about 3 $GtCO_2$/y in 2100.

The climate uncertainty for the end of this century remains high. The range of temperature to achieve with RCP2.6 is wide, the uncertainty of climate models is high, and human behavior is unpredictable. Some authors check if climate models can predict recent climate data using past events. The predictions often run over the interquartile range [2].

If we maintain the current trend (RCP8.5), humankind will envisage a climate catastrophe in the following decades. Changing will depend on the behavior of the world leaders. In 2019 nuclear, hydroelectricity, and renewables account for less than 16% of the total world primary energy [3]. US, EU, and China give out more than 55% of the World's carbon emissions. US and EU are reducing emissions regarding the previous years, and China is still climbing [3]. EU aims to achieve carbon neutrality in 2050 and wants to reinforce the climate commitment by lowering carbon emissions in 2030 by 55% regarding the 1990 level [4].

The policy of China is to become carbon neutral by 2060. However, China relies on carbon to supply 58% of the primary energy [3], and carbon gives China a safe energy means regarding availability, stockage, and technological knowledge [5]. Moreover, China has a massive amount of thermal coal power plants with ten years average working time. If all infrastructures remain working during the following decades, the Paris agreement will not be accomplished [6].

The US returned to energy autonomy. Oil and gas fracking and sand oil extraction in Canada allow the US to look to their internal energy policy rather than the Middle East [7]. Betting on natural gas, US reduced emissions in recent years. But it needs extra effort. Energy and economy need decoupling in an energy-independent region. On the other hand, policy success regarding emissions depends on the ability to remain in the world leadership without extra energy needs.

If the World remains in the *laissez-faire* pathway (RCP8.5), all countries will be affected. South EU will require 20% extra water and will suffer from depletion of water resources [8]. Even stabilizing the CO_2 concentration, central and southern Europe will have hotter days in Summer by 4 °C and 20% rain reduction in South Europe [9]. The Mediterranean region will be one of the most affected areas by dryness in the globe. In a similar stabilization scenario, wetlands will reduce and may collapse by 2100 in the Western Mediterranean region [10].

In the RCP8.5, the international hydrological basins of the Iberian Peninsula will reduce water availability by 6% per decade [11], and dryness may feel a step increase by 2040-2050 [2]. Oddly, in the RCP2.5, rain may reduce in the Iberian Peninsula by 10% from the previous to the end of this century [12].

Dry periods and hot waves will be more common in the following decades. This work aims to define a defense strategy to provide water to the population. Countries in regions most affected by dryness variations need a force to act as the last warrant for the population survival. It is an adaptation measure that needs to be drawn despite infrastructure adaptation and reduction in carbon emissions.

The southern regions of Portugal, Spain, and Italy will be seriously damaged by dry periods. Similarly, Balkans counties, Greece, and Turkey will also await water problems.

In the next decade, these regions might undertake serious adaptation policies. The measures include water highways, river transfers, collect and recover grey waters, changes in agriculture water distribution, and in agriculture models. All adaptation actions that last for decades might apply.

Such uncertainty asks for the definition on how to create a scenario. Moreover, it is necessary to know how to handle uncertainty and how to define a strategy. Dempster-Shafer's theory (DST), described in section 2, is a recognized tool to handle uncertainty. Thus, section 3 presents the scenario creation calling for prospective thinking. Axiomatic Design (AD) can provide the means to define the high-level development of a design, called Strategy, in the context of this work. Using DST with AD allows computing the information content in the ill-defined design at a high decomposition level. Section 4 explains the theoretical framework to define a strategy using AD and DST for a specific scenario. Finally, section 5 shows how to define a force to fit the population's water needs. The last section presents the conclusions.

IOP Conf. Series: Materials Science and Engineering 1174 (2021) 012020 doi:10.1088/1757-899X/1174/1/012020

2. The Dempster-Shafer Evidence Theory

At the AD high levels of decomposition, the problem is ill-defined. There are uncertainty and ignorance about the solution. AD expresses the design using $[FR]=[A]\cdot[DP]$ that describes the mapping between the functional and physical domains [13].

The design equation is a peculiar representation of a mathematic equation, as the design unknowns are the DPs, not the FRs. Therefore, the DPs regarding "how to achieve" the FRs is in fact about "is a DP responsible for fulfilling an FR?". This type of question represents abduction reasoning.

DST is appropriate for handling uncertainty and defining the antecedent of an implication by knowing the consequent [14]. Using combined evidence, DST helps to reach the belief in a solution. DST is used in medical diagnosis, failure evaluation, and criminal investigation. Epistemologically, the type of problem and reasoning is covered by DST evidence equals the DP choice to fulfill an FR in AD.

DST introduces the concept of belief and plausibility. Belief is the measure of preconceived knowledge of a set of events; plausibility is a measure of what is the possibility of knowing about the events. The events are defined in a frame of discernment Θ with *n* mutually exclusive and exhaustive hypothesis, called singletons. The power set *P* of all subsets Θ has 2^n sets.
Suppose a frame of discernment with three singletons $\Theta=\{f_1,f_2,f_3\}$. Thus, the power set *P* has 2^3=8 sets, defined as $P=\{\varnothing,\{f_1\},\{f_2\},\{f_3\},\{f_1,f_2\},\{f_1,f_3\},\{f_2,f_3\},\{f_1,f_2,f_3\}\}$.
Now, let *A* be a proposition in evaluation such as $A\in P$. *A* is any set of sets *Ai* of *P*.

The basic probability assignment (*bpa*) $m(A_i)$, or mass probability, is the evidence. The A_i sets with non-zero mass are the focal elements in appreciation. The *bpa* is similar to the probability density function (*pdf*) in the sense that:

$$m(\varnothing)=0\ , \qquad \sum_{A_i\in P} m(A_i)=1 \tag{1}$$

However, the belief is defined over a power set and the probability over the crisp values of a universe. Moreover, the remaining probability of A is not the probability of non-A.
An expert can give evidence about *A* by defining the *bpa* to all focal elements A_i. The focal elements can express different causes or consequences of a proposition.
The belief in A_i, is the evidence in all subsets *B* of A_i, as $B\subseteq A_i$:

$$bel(A_i)=\sum_{B\subseteq A} m(B) \tag{2}$$

The plausibility $pl(A_i)$ is the evidence in A_i, plus the evidence in all subsets of A_i, plus the evidence in all sets *B* that intersect A_i.

$$pl(A_i)=\sum_{B\cap A\neq\varnothing} m(B) \tag{3}$$

It makes $pl(A)\geq bl(A)$ ensuring $bl(A)$ and $pl(A)$ are the lower and upper bounds of an evidential interval *Ev*=[$bl(A)$, $pl(A)$]. Plausibility is all we may know about A. Therefore, it is the complement of belief in what is out of A $pl(A)=1-bl(\bar{A})$. Ignorance is what is behind the belief of *A* and in the complement of *A* $Ig=1-\{bel(A)+bel(\bar{A})\}$, or $Ig=pl(A)-bl(A)$. If Ignorance is nil, plausibility is belief and both equal probability.

In the context of this work, belief and plausibility are the bounds for the probability of fulfilling an FR. The focal elements A_i create the universe of discourse of the FR.

Belief and plausibility are substitutes of the *pdf*. Therefore, they express a measure of probability and can help compute the information content of the design.

DST has been applied with AD to define the DP for a single FR-DP design [15]. It is necessary to stress the information content depends on the application of the AD first axiom. Therefore, the design equation influences the computation of the information content.

3. Methodological approach

Gaston Berger, in the 50s, was the father of "prospective thinking". Prospective thinking to "look at the future that disturbs the present". A look in the future needs to synthesize what we know about the future and create scenarios. Therefore, the idealization of a future scenario asks for innovative reasoning to synthesize ideas about the future [16]. Godet called scenarios the set of possible future events that asks for clarifying and define a decision point. The scenarios are "what could happen?". Scenarios and Strategy must not be confused. The Strategy is the way of solving a problem that asks for defining "what can I do?", the strategic options; "what will I do?", the strategic decisions; and "how will I do it?", the actions and operational plans [17].

This way, in what regards to AD, Strategy is the mapping between functional requirements (FR), design parameter (DP), and eventually process variable (PV) [19]. If Strategy asks for new PVs it will influence technology; the same way existing technologies help define the Strategy.

Scenarios are often seen as a cross of all future hypotheses. It raises multiple cases, most of them with scarce possibility. Consequently, it is better to choose the most plausible scenarios, or the risky ones, and work on them [18].

A strategy on climate change needs to consider adaptability to new conditions. In the future, it may not be necessary to maintain the same DPs, not even the same FRs. We must seek for needs and create the so-called "adaptive resilience", a creative transformation of functionalities to new environments, to create new opportunities [19].

The definition of a Strategy by Godet, as a means, has a counterpart in AD. AD maps between the Functional, Physical, and Process domains. Figure 1 shows the mapping between domains and Strategy thinking. Strategic thinking is the mapping at high levels of decomposition.

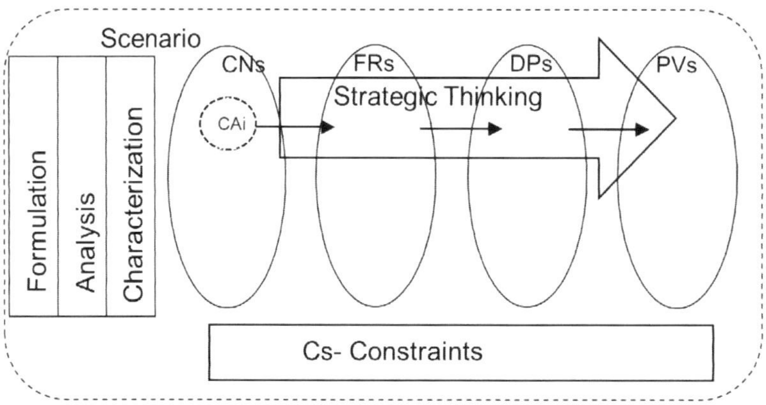

Figure 1. Scenario and Strategic thinking

The Scenario needs formulation regarding targets and time frame in evaluation. Moreover, it needs analysis of what concerns aggressors and hazards effects, evaluation points, and possible effects. Finally, the scenario characterization is the aggressors and receptor response. In what concerns to AD, the Scenario is the problem definition by the Customer attributes (CAs) and the input Constraints (Cs) of the design.

The description of a future scenario may contain available DPs that may help the design development. However, for AD a future Scenario needs to define:

- What are the customer attributes in the time frame in evaluation?
- The input Constrains of the design.

As an example, a CA can be "needs of drinkable water for the population". And the Cs defined by the climate conditions in the time frame in evaluation.

This paper intends to achieve a common purpose on how the *bpa* gives a measure of the fulfillment of the FR. Mapping between FRs and DPs regards how a DP fulfills an FR.

The DP may have various possibilities, say {d1}, {d2}, and {nd}. The basic solutions are called the belief nucleon. For this work, {d1} is a known physical issue, {d2} an improved one and {nd} is a non-existent or undetermined physical issue. Physical issues are decisions on the use of technologies or artifacts.

These technologies have images in the functional domain respectively {f1}, {f2}, and {nf} representing the modes of fulfilling the FR. The focal elements to choose are {f1}, {f2}, {f1, f2}, {nf} and {all}. The set {f1, f2} is a ill-defined solution that contains f1 and f2, the same way the {all} focal element is {f1, f2, nf}.

Regarding prospective thinking, the belief nucleon is our belief in fulfilling the FR at a certain level of decomposition in a time frame. The information of the design is the information content of the leaves FRs. Furthermore, DST helps to compute information for the different prospective solutions corresponding to each of the focal elements. It means that for the same mapping between functional and physical parameters, called the Design or the Strategy, there is more than one value for the information content. The information content - the probability of success of the design - varies according to the decisions on using physical artifacts, known or under development, for the same design equation.

The methodology proposed uses prospective thinking to define the future customers' needs and AD to stretch the Strategy. For the same Strategy, forthcoming decisions on the physical parameter decidedly influence the probability of success.

4. Application to water scarcity

In the future decades, a reduction in global pluviosity is expected, as well as a lower availability in underground water, the increase of higher temperatures than nowadays, and the occurrence of heatwaves.

The CA in the CN domain is "needs of drinkable water for the population". The design has the following Cs:

- Cs1: Unavailability of water in local rivers and in water wells;
- Cs2: Availability of water in nearby regions;
- Cs3: Sea bords a closer region;
- Cs2: Heatwave during the event;
- Cs3: Rural fires that ask for massive firefighters' intervention.

The Strategic Plan, or high-level design, has the following FRs and DP, according to table 1. FR2 and FR3 are decomposed until the second level of decomposition.

The Strategic plan needs a "System for water delivering" (DP0) to be created so that in extreme climate events population can adapt by receiving the basic water needs (FR0). The system gets (FR1), produces (FR2), transports (FR3), and delivers water (FR4). Get the water in dried dams, wells and rivers, pumps, and store the water (DP1). It filters and treats the water to remove contaminants and microorganisms (DP2). The system has transport means (DP3) and can distribute water in each village in small tanks and cisterns (DP4). The municipality services ensure a way to deliver the water flow to each family.

Drinkable water can be produced from available dams or the sea. This way, it can use "modular water treatment equipment" (DP21) or modular reverse osmosis (DP22). Moreover, water can be captured and produced in closer regions and transported by land in trucks or trains (DP31) or by sea (DP32).

Table 1. FRs and DPs for "needs of drinkable water for the population".

Functional Requirements		Design Parameters	
FR0	Deliver basic water needs to the population	DP0	System for water delivering
FR1	Get fresh water	DP1	Pumping and storing system
FR2	Produce drinkable water	DP2	Filtration and water treatment
FR3	Transport drinkable water	DP3	Transport means
FR4	Distribute water	DP4	Small tanks and cisterns
FR21	Turn drinkable water in nature	DP21	Modular water treatment equipment
FR22	Desalinate seawater	DP22	Modular Reverse Osmosis
FR31	Transport water by land	DP31	Tank trucks and tank trains
FR32	Transport water by sea	DP32	Water freighter

The strategic system described is the last line of defense of an adaptation policy that countries in southern Europe need to implement. Indeed, Member States need to execute large construction infrastructures, the same way they need to have a well-equipped and trained force to enter in action if necessary. Prevention pays well.

Equation 6 presents the design equation. It says it is compulsory to get the freshwater and then produce drinkable water. Transport and distribution can work independently. It is welcome to identify where to get the water and produce the water, as both induce the FRs at the second decomposition level. Therefore, DP at the second level of decomposition, modular system for treatment and osmosis, and transport means by land and sea depending on the quality of water captured and produced.

$$\begin{bmatrix} FR_1 \\ FR_2 \\ FR_3 \\ FR_4 \\ FR_{21} \\ FR_{22} \\ FR_{31} \\ FR_{32} \end{bmatrix} = \left[\begin{array}{cccc|cccc} X & & & & & & & \\ X & X & & & & & & \\ X & X & X & & & & & \\ X & X & & X & & & & \\ \hline X & & & & X & & & \\ X & & & & & X & & \\ X & X & & & & & X & \\ X & X & & & & & & X \end{array}\right] \cdot \begin{bmatrix} DP_1 \\ DP_2 \\ DP_3 \\ DP_4 \\ DP_{21} \\ DP_{22} \\ DP_{31} \\ DP_{32} \end{bmatrix} \quad (4)$$

Table 2 presents the *bpa* defined in the mass probabilities $m(A_i)$, the belief $bl(A_i)$, and plausibility $pl(A_i)$ for the high-level FR0 and the first level of decomposition, FR1, FR2, FR3, and FR4.

At the high level, suppose the water delivery has no legislation and no structure to handle it. The physical existence will give a mass probability for {f1} of 0,05. As the problem start to go on discussion it can make mass of {f2} equal 0,1 and a group {f1, f2} to be 0,1 as well. What is not yet known {nf} and {all} represents 0,75 of the mass probability.

Table 2 presents similar reasoning on the second to the fifth columns. FR1 and FR4 uses known, available and affordable technology. They depend on the willingness to put the system working. Therefore the mass probability of {f1} is high — the opposite occurs on FR2 and FR3. Regarding FR2, water treatment and reverse osmosis are known technologies that may undertake interesting developments.

Moreover, creating modular systems and putting them into action may take extra effort. Finally, FR3 asks for high investments and adaptations. It needs a huge quantity of tank trucks and personnel and the development of water freighters.

Table 2. Mass, belief, and plausibility for FR0 and the 1st level of decomposition

	FR0: Deliver basic water needs			FR1: Get fresh water			FR2: Produce drinkable water (subjected to got it)			FR3: Transport drinkable water (subjected to got and produce)			FR4: Distibute water (subjected to got and produce)		
	m(Ai)	bl(Ai)	pl(Ai)	m(Ai)	bl(Ai)	pl(Ai)	m(Ai)	bl(Ai)	pl(Ai)	m(Ai)	bl(Ai)	pl(Ai)	m(Ai)	bl(Ai)	pl(Ai)
{f1}	0.05	0.05	0.6	0.7	0.7	0.85	0.4	0.4	0.7	0.3	0.3	0.7	0.7	0.7	0.85
{f2}	0.1	0.1	0.65	0.1	0.1	0.25	0.2	0.2	0.5	0.1	0.1	0.5	0.1	0.1	0.25
{f1,f2}	0.1	0.25	0.7	0.05	0.85	0.95	0.1	0.7	0.9	0.2	0.6	0.8	0.05	0.85	0.95
{nf}	0.3	0.3	0.75	0.05	0.05	0.15	0.1	0.1	0.3	0.2	0.2	0.4	0.05	0.05	0.15
all	0.45	1	1	0.1	1	1	0.2	1	1	0.2	1	1	0.1	1	1
Σ	1			1			1			1			1		

The belief and plausibility of the solution are the product of all conditional beliefs and plausibilities of FR1, FR2, FR3, and FR4. Table 3 show the results regarding the known and improved situations, respectively {f1} and {f1,f2}. The belief in the current situation {f1} is near zero. Improving the technology development, creating and manufacturing new artifacts, can lead to the plausibility of success of 0.65.

Table 3. Intervals of Information

			Information	
1st level decomp	Belief	Plausibility	From	To
{f1}	0.06	0.35	1.5	4.1
{f1,f2}	0.30	0.65	0.6	1.7

5. Conclusions

The proposed methodological approach evaluates the future by prospectively envisage a disruption. The disruption event defines Constraints (Cs) and the population needs. The Strategy can determine how to accomplish the needs by mapping, at a high level of decomposition, between the Axiomatic Design (AD) domains. The mapping under the Cs is the design. For the same high-level design, known physical solutions or foreseen solutions give different probabilities of success. The Dempster-Shafer theory (DST) can define the bounds for the probability of success, between the belief and plausibility, for each solution.

In the prospective application, the World will see an increase in carbon in the atmosphere in the following decades. Despite EU, US, and China efforts, the climate will change in a couple of decades. The Mediterranean region will be one of the World's most affected regions. Hot waves, high temperatures, and low pluviosity will lead to water scarcity.

States need to focus on solutions to adapt to the new reality. It includes having a defense system to "deliver the basic water needs to the population". The system needs to get, produce, transport, and distribute water to the population. There is a need to define the force, implement it, use existing technologies and means, and develop new ones.

For the moment, belief in a suitable response in the event of scarce water is almost nil. Future development and implementation can lead to the plausibility of 0.65. Being aware of the problem is the first step to create the necessary conditions to establish a defense force.

Acknowledgments
The authors acknowledge Fundação para a Ciência e a Tecnologia (FCT, I.P.) for its financial support through the grant UIDB/00667/2020 (UNIDEMI).

References

[1] FCCC- Framework Convention on Climate Change 2015 *Report of the Conference of the Parties on its twenty-first session*, held in Paris from 30 November to 13 December 2015, United Nations, Paris

[2] Guerreiro S, Kilsby C and Fowler H 2017 Assessing the threat of future megadrought in Iberia, *International Journal of Climatology*, **37** pp 5024–5034

[3] BP 2020 *Statistical Review of World Energy*, 69th edition

[4] European Commission, 563 2020 *Regulation of the European Parliament and of the Council, establishing the framework for achieving climate neutrality and amending Regulation* (UE) 2018/1999 (Climate Law), Brussels

[5] Shaffer B 2017 A Guide to the Application of Energy Data for Intelligence Analysis, *Studies in Intelligence*, **61** 4 pp 43–52

[6] Tong D, Zhang Q, Zheng Y *et al.* 2019 Committed emissions from existing energy infrastructure jeopardize 1.5 °C climate target, *Nature*, **572** pp 373–377

[7] Cook S 2020 No exit, Why the Middle East Still Matters to America, *Foreign Affairs*, **99** 6 pp 133–142

[8] Konzmann M, Gerten D and Heinke J 2013 Climate impacts on global irrigation requirements under 19 GCMs, simulated with a vegetation and hydrology model, *Hydrological Sciences Journal*, **58** 1 pp 88–105

[9] Fronzek S, Carter T and Jylhä K 2012 Representing two centuries of past and future climate for assessing risks to biodiversity in Europe, *Global Ecology and Biogeography*, **21** pp 19–35

[10] Lefebvre G, Redmond L, Germain C, Palazzi E, Terzago S, Willm L and Poulin 2019 Predicting the vulnerability of easonally-flooded wetlands to climate change across the Mediterranean Basin, *Science of the Total Environment*, **692** pp 546–555

[11] Guerreiro S, Kilsby C and Fowler H 2016 Rainfall in Iberian transnational basins: a drier future for the Douro, Tagus and Guadiana?, *Climatic Change*, **135** pp 467–480

[12] Cunha L, Oliveira R, Nascimento J and Ribeiro L 2007 Impacts of climate change on water resources: a case-study for Portugal, Water in Celtic Countries: Quantity, Quality and Climate Variability, *Proceedings of the Fourth InterCeltic Colloquium on Hydrology and Management of Water Resources*, Guimarães, Portugal

[13] Suh NP 1990 *The principles of design* (New York: Oxford University Press)

[14] Ross T 2005 *Fuzzy Logic with engineering applications (Chapter 15 – Monotone measures: belive, plausability, probability and possibility)* (West Sussex: John Wiley &Sons Ltd.)

[15] Pappalardo M 2006 Fusion of belief in Axiomatic Design, *Proceedings of ICAD2006 4th International Conference on Axiomatic Design*, Firenze, Italy

[16] Goux-Baudiment F and Jones C 2013 Inventing the Future. In: Corsi P, Christofol H and Richir S (eds) *Innovation Engineering: The Power of Intangible Networks*, pp 3–24 (London: Wiley-ISTE)

[17] Godet M, Durance P, and Gerber A 2008 *Strategic Foresight – La Prospective – Use and misuse of Scenario Building*, Lipsor Research Working Paper (#10)

[18] Scoblic J and Tetlock J 2020 A Better Crystal Ball, the Right Way to Think About the Future, *Foreign Affairs*, **99** 6 pp 10–18

[19] Joakim E, Mortsch L and Oulahen G 2015 Using vulnerability and resilience concepts to advance climate change adaptation, *Environmental Hazards*, **14** 2 pp 137–155

IOP Conf. Series: Materials Science and Engineering 1174 (2021) 012021 doi:10.1088/1757-899X/1174/1/012021

Design Matrix as starting point to assess product sustainability

S Rizzuti[1] and L De Napoli[1]

[1] Università della Calabria, DIMEG - Dipartimento di Ingegneria Meccanica, Energetica e Gestionale, Ponte Pietro Bucci 46/C, 87036 Rende (CS), Italia

E-mail: luigi.denapoli@unical.it

Abstract. Product development is a time-consuming activity trying to answer, at best, the demanding customer needs. In general, the design of an industrial product needs several competencies that work conjointly, in the respect of many constraints that urge from the firm' departments. Last but not least sustainability is gaining a relevant role, since the recent past, and absolutely not deferred for the destiny of humankind on earth in the future. Design Matrix is the first array that designers should compose in order to verify the feasibility of product architecture since it put in relation functional requirements and design parameters. Clearly, designers are really far from the embodiment of valid proposals, at this moment. It is necessary to map this information to other data structures that allow them to follow the evolving architectural work, in which physical components or effective parts perform the actions that satisfy functional requirements. Design Structure Matrix (DSM) is a promising data structure that shows the relationship among all subsystems that compose the developing product. This structure can be arranged also to collect all the environmental parameters or indicators in order to have a rough sustainability assessment since the early design phases. The paper will discuss such kind of relation and apply it to a case study in order to clarify it.

1. Introduction

Over time, since the '80 years of last century, when the first systematic methods around product design and development appeared [1, 2], many approaches have been tried with the intent to support designers and stakeholders in this task. Many are the points of view that can enhance peculiar characteristics of the product in the phase of development and the competition between designers and project management may influence significantly the final structure of the product.

Many emphases may be put on designing, considering the more relevant aspects that a product must embed and that characterize it. The strategic essence of today and next future products must be sustainability [3, 4], considered as the basic characteristic that all engineering solutions to problems must have.

Axiomatic Design is a powerful method to recognize the consistency of a design solution, verifying the nature of the Design Matrix [5] after the identification of the relation between functional requirements and the associated design parameters. This should be strengthened in order to check and stress the sustainability characteristics of the product in the phase of development. Considering that Axiomatic Design was born with the intent to pose a more scientific basis to design it was not equipped to assess all possible scenarios in which a device should have worked. About 10 years ago two interesting applications of it to eco-design were published. Shin et al. [6] incorporated ecological issues into Axiomatic Design, including eco-factors as functional requirements and design parameters. Giorgetti et al. [7] proposed a smart eco-design platform able to support designers to develop products with a higher level of sustainability along all the product's life cycle.

The nature of the Design Matrix is extremely versatile, in that it can be integrated with other design methods in order to expand the concepts in many different directions. One step further designer must do is the choice of the components/subsystems that better will perform those functionalities. This step is strictly related to the building of architectural solutions, and several solutions may be invented. For this reason, new databases must be organized in order to collect information able to differentiate among design alternatives.

Design Structure Matrix (DSM) is one such database [8, 9]. DSM was born in a product management context and is employed for a variety of scopes. It collects the relationships among all components involved in a device. Knowing system interactions during the product development process is important for project management. Capturing system interactions between affiliated system elements is necessary for managing all phases of product development that must be performed efficiently and effectively. Although DSM is powerful in the analysis of design interactions, it is less effective in innovative design, where it is almost impossible to obtain a DSM for a new product that has never been designed before.

The integration of DM and DSM has been investigated recognizing a certain affinity between them, for the finality to support designers in product design and development [10-11]. The two theories have been integrated to augment the conceptual design process as a recursive interaction of AD's design matrix and the corresponding DSM [12, 13].

In any case, the original DSM does not contain all the information able to identify the sustainability level of the product under development. To this task, an augmented version of the original DSM (A-DSM) has been conceived [14], able to gather all such kinds of information and interrelation.

DM and A-DSM may be linked together and conjointly may guide designers to develop products able to solve problems for human beings today, avoiding that the next generations may not be able to solve theirs in future.

The paper discusses such kind of relation and briefly explains how to combine DM and DSM in a case of study.

2. Relation between Design Matrix and Design Structure Matrix

The form of both matrices DM and DSM coincides (see Figure 1). Their contents differ, even if the context in which they have been conceived remains almost the same: to describe reasoning throughout product design and development.

Figure 1a represents a diagonal Design Matrix in which the relation between seven FR_i Functional Requirements and the corresponding DP_i Design Parameters is defined. The necessary condition to continue to develop a valid engineering solution is made checking the squared form of the matrix, assuring since the first study that it is not invalidated by a lack or an over-dimension of parameters [5]. Also, a triangular form of the matrix may describe a valid solution, a structure that is frequent in product development. During this stage, the DP_is are identified as a set of quantifiable elements of which the values or the range must be accordingly chosen in order to satisfy the best behaviour of the device. The X present in the cells reports that a relation exists. The robust design method may be applied to search for the best performances [15]. In any case, every DP_i characterizes an actual component or a particular element of it.

Figure 1b represents the DSM squared matrix that puts the relation among them the components in which the DP_i have been embedded. The Design Structure Matrix has a squared form and it is formed as a non-diagonal matrix, in that it identifies the way in which the components interact with each other. The X present in the cells reports this interaction. The components are identified by a letter and moving along a line it may be discovered how/what a component gives to others, and moving along a column it may be discovered how/what a component receive from others. DSM is the first matrix that collects the components of a product and implicitly describes its architecture. During product development, a set of other matrices may be joined to this one, as DMM-Domain Mapping Matrix [17] and MDM-Multi Domain Mapping [18], in order to have a clear picture of all interactions among processes, people, organization, scheduling, which give for complex systems synthetic schemas with which operate to identify bottlenecks along the time.

There is a strict relation between both matrices, because DP_is and components are related and from one DM a set of DSM may be built, considering that several are the actual components that embed the parameter identified in the DM [16]. Several product architectures may be generated by a DM. This is obviously true if the decomposition of FR_i is stopped to a certain level, and a certain freedom of choice is allowed to designers.

In order to enrich the content of the DSM, each cell may be subdivided into four sub-cells where a set of information may be stored [19].

	DP1	DP2	DP3	DP4	DP5	DP6	DP7
FR1	X						
FR2		X					
FR3			X				
FR4				X			
FR5					X		
FR6						X	
FR7							X

(a)

	A	B	C	D	E	F	G
A	-	X					
B	X	-		X			
C			-				X
D		X		-	X		
E				X	-		
F						-	
G			X				-

(b)

Figure 1. Comparison between the matrices: a) Design Matrix; b) Design Structure Matrix.

3. Nature of the Augmented-DSM for sustainability assessment

The nature of the DSM has been further structured for its usage in eco-design [14, 18]. By reasoning around the product life cycle and considering "Raw Material", "Manufacturing", "Transportation", "Use" and "End of Life" as the main phases that characterize a product, the original DSM has been redesigned, and its name is now preceded by "Augmented".

The A-DSM is a matrix composed of two main parts (see Figure 2), where the designer has to manage different data types for the scope of product environmental assessment:

a) one DSM (the white square matrix), where n components are put in relation. In the diagonal cells, a set of Environmental Indicators (EIs) are stored that characterize each component in terms of eco-design. In the off-diagonal terms the flows (Material, Energy, and Signal) and the mutual interactions (Force) between components are recorded;

b) two extra columns (in yellow tonalities):

c) Columns Ω_1 (yellow), called "External Relations", where the designer collects interactions with the environment in input and output of each component that has interaction with the outside, with the same organization of the internal cells.

d) Column Ω_2 (amber), called "Sustainability Issues", contains in each cell the energy dissipation and the material wastes directed towards the outside.

This augmented matrix has an $n \times (n+2)$ dimension. Except that for the diagonal terms and the column Ω_2, each cell is divided into four parts. Moreover, the values inserted in each cell are numbers, where a value is positive if the link is outwards the component, whereas it is negative if the link is inwards; this property is not valid for Force connection, as it is considered a not directed link due to the mutual interaction of mechanical contacts. In column Ω_2, each element is divided into two sub-cells to collect only material and energy quantities (M, E).

IOP Conf. Series: Materials Science and Engineering 1174 (2021) 012021 doi:10.1088/1757-899X/1174/1/012021

The nature of each component may be characterized in terms of environmental indicators. The indicators give essential information that allows identification and quantifying of the environmental impact of each specific archetype related to the typical phases of the life cycle.

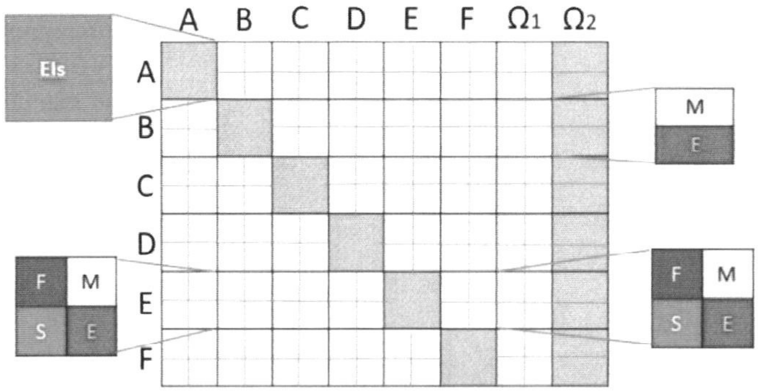

Figure 2. The structured form of the Augmented–DSM.

The indicators may be chosen in different ways. In a previous work [20] the authors have identified a set of 7 Environmental Indicators: Mass, Hazardous substances, Global Warming Potential, Impact in the manufacturing phase, Power in the use phase, Transport, Recyclability, even if many other parameters may be identified by designers and stakeholders in relation to the peculiarity of the product to be assessed.

4. Framework of a methodology for sustainability assessment

The strict relation between DM and A-DSM can be seen in Figure 3, where the main passages on product design are highlighted. The first step of the work is the functional analysis of the problem to be solved. The first data structures that must be organized are the graph of the interrelations of functions and the corresponding functional tree, in which the decomposition of functions is stored as dependence father-son (in up-left of the figure). A functional tree in the axiomatic design approach is associated with the zig-zagging process in order to characterize the associability of a functional requirement to its design parameter.

The functional study is the basis for the concept generation step, where a lot of product architectures may be conceived, each one characterized by a proper set of design parameters. Each product architecture is represented by a proper Design Matrix, because different may be the design parameters associable with a functional requirement. Or, on the other side, the identification of a set of design parameters requires a different decomposition of functions and the building of Design Matrices of different dimensions. In the figure, three solutions were conceived with different structures.

In order to be able to analyze a product by the axiomatic design approach, it is necessary to describe it by a Design Matrix. After this first step, a set of DMs is available for further analyses.

For each DM, the corresponding A-DSM may be built, considering the strict relation by these. This may be done only considering how each Design Parameter may be instantiated in a actual component. Each A-DSM is a collection of fluxes and values associated with the environmental elements that have been chosen to perform sustainability assessment.

The correspondence from the problem to the solutions is one-to-many, but the correspondence from Design Matrix to A-DSM is one to one because each A-DSM describes precisely what will happen in the associated device. The consistency of each Design Matrix is guaranteed after the application of the first axiom and the verification that its structure is at least uncoupled. The nature and structure of each A-DSM are also guaranteed because it derives from a solution validated by the Axiomatic Design method.

IOP Conf. Series: Materials Science and Engineering 1174 (2021) 012021 doi:10.1088/1757-899X/1174/1/012021

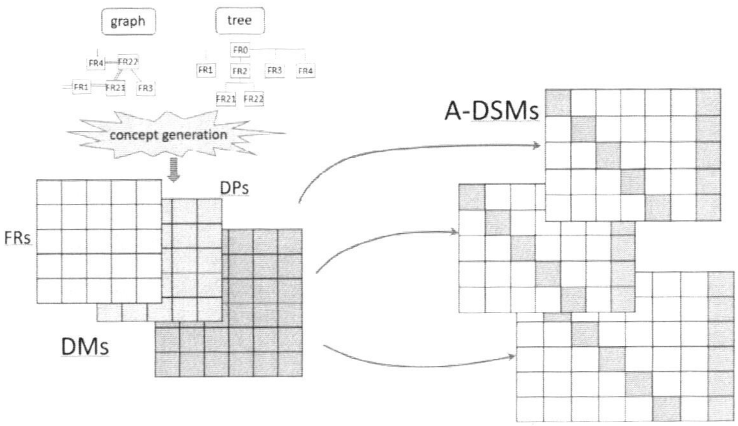

Figure 3. The framework for the sustainability assessment.

5. Case of study

In order to explain how the methodology can be employed during product development, a case study focused on the design of an orange juicer is presented. There may be several different types of such a device and this concept derives from a grinder; in fact, it can be used for all kinds of vegetables. It is based on another point of view about the juicer, which considers grinding all the orange parts, and mainly the white zone (the pith) where many vitamins are concentrated.

The user puts one whole orange inside the device and uses a pusher to push it towards the grinding. Figure 4 presents the functional model of such a solution. The schema is shown with a certain degree of detail, in fact, the functional model was created starting from the macro functionality "to squeeze an orange" expressed in the classical verb plus noun form. The macro function is then broken down into more sub-functions. The sub-functions form a graph (see Figure 4a) where connections are represented by functional links: Force (blue), Signal (green), Material (yellow), Energy (red). The associated functional tree is reported in Figure 4b.

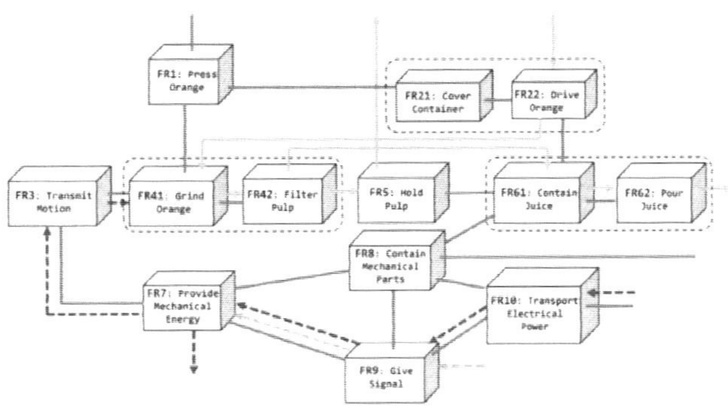

(a)

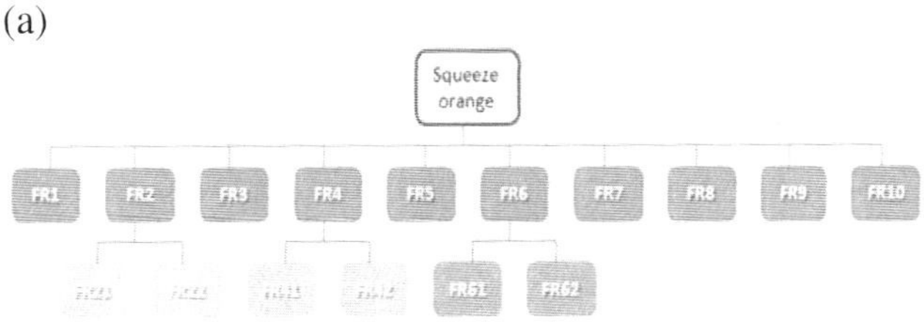

(b)

Figure 4. Functional description of the grinder: a) functional graph and b) functional tree.

During concept generation a set of Design Matrices may be identified, differentiating for the DPs chosen for each FR. The zig-zagging method may require further subdivision of FRs and consequently, the dimension of the matrix grows. Initially, the DM may have a diagonal form, which associates each Design Parameter to each Functional Requirement. The matrix must be further investigated in order to discover the possible conjoint influence of many design parameters on the same functional requirement. This is due to the way in which the component, where DP explains its task, may interact with the other components and the other functionalities.

Behind such a matrix, there is the instantiation of product architecture. As can be seen, Figure 5 shows a lower triangular matrix, derived from the choices made for the component/archetypes reported in Table 1. Some functionalities that were originally subdivided (FR2, FR4, and FR6) are now grouped into single components that explain them and became more complex items, respectively the cover (B), the perforated blade (D), and the bowl/spout (F).

Figure 6a represents the graph of the components, that corresponds to those of the functionality of Figure 4a, where such kind of aggregation has been made. In Figure 6b the sketch of the device is represented with the indication of each component.

Design Parameters

Functional Requirements

	DP1	DP21	DP22	DP3	DP41	DP42	DP5	DP61	DP62	DP7	DP8	DP9	DP10
FR1	X												
FR21		X											
FR22		X	X										
FR3				X									
FR41				X	X								
FR42						X							
FR5							X						
FR61							X	X					
FR62									X				
FR7				X						X			
FR8										X	X		
FR9												X	
FR10													X

Figure 5. Design Matrix built in relation the data of Figure 4.

Table 1. Relation between FRs and the DPs identified by a well specified component.

	Function	Archetype/Component	Design parameter	
FR1	Press Orange	A (Plunger)	diameter	DP1
FR21	Cover Container	B (Cover)	height	DP21
FR22	Drive Orange		diameter	DP22
FR3	Transmit Motion	C (Nut)	diameter	DP3
FR41	Grind Orange	D (Perforated Blade)	grid roughness	DP41
FR42	Filter Pulp		grid hole diameter	DP42
FR5	Hold Pulp	E (Container)	volume	DP5
FR61	Contain Juice	F (Bowl + Spout)	volume	DP61
FR62	Pour Juice		spout diameter	DP62
FR7	Provide Mechanical Energy	G (Electric Motor)	power	DP7
FR8	Contain Mechanical Parts	H (Housing)	volume	DP8
FR9	Give Signal	I (Switch)	position	DP9
FR10	Transport Power	L (Power Cable)	cable cross section	DP10

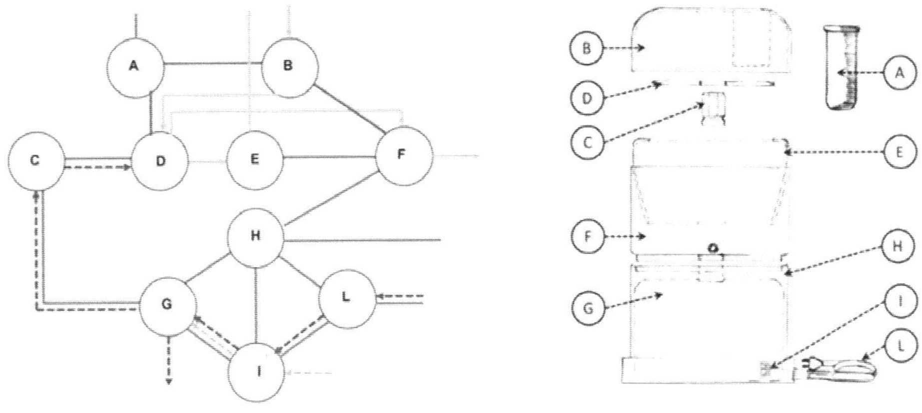

(a) (b)

Figure 6. The instantiation of the device built, in strict relation with the Design Matrix of Figure 5: a) graph of the concept in terms of components and links; b) sketch of the concept with component' label.

6. The associated A-DSM for sustainability assessment.

On the basis of the study performed so far, it is possible to build the A-DSM (see Figure 7). The content of such matrix is organized in strict relation to the graph of Figure 6a. It is in many parts symmetric and represents, for the interaction of Force, the adjacency matrix of the associated graph, in that it describes the relations that occur among the components. The diagonal terms, represented in the figure as grey-filled cells, contain the Environmental Indicators associated with each component, as designers and stakeholders decide to investigate the sustainability of their product. Each diagonal cell is structured as an array of seven positions. From these data, an environmental assessment may be done on each component or on the whole product.

In the off-diagonal terms and the column of the "External Relations" (Ω_1), each cell is divided into four sub-cells (as an array of four positions) where data are stored. Data have different nature and refer to the links of Force, Signal, Material, and Energy that the archetypes/components exchange among each other in the concept architecture. These data are derived from the links of the graph which represent the interaction among components (see Figure 6a). For example, node I interacts with the outside Ω_1 and with G, H ad L nodes. It gets from L energy that transmits to G. It is in contact with G, H, and L, and receives a signal from the outside that transfers to G.

Data related to functional links have to be stored following these rules:

- in the Force sub-cell, in blue, the value is Boolean, with the meaning: 1 if there is a physical connection, 0 (or "empty") else;
- in the Signal sub-cell, in green, the value is a number type: -1 if a signal or information is in input, 1 if it is in output, 0 (or "empty") else;
- in the Material sub-cell, in yellow, the value is a number type: it represents a quantity with a proper unit (e.g. g, kg, etc.); it is positive if the link is in output from the node/archetype, or negative if the link is in input; 0 (or "empty") if the link is not present;
- in the Energy sub-cell, in red, the value is a number type: it represents a quantity with a proper unit (e.g. kJ, J, etc.) and it is positive if the link is in output from the node/archetype, or negative in case of link in input; 0 (or "empty") if the link is not present.

The last column of "Sustainability Issues" (Ω_2) contains cells with only two slots in order to store links and quantities referred to material and energy and data follow the rules for these. It reports the quantity of energy that is lost and materials that are wasted.

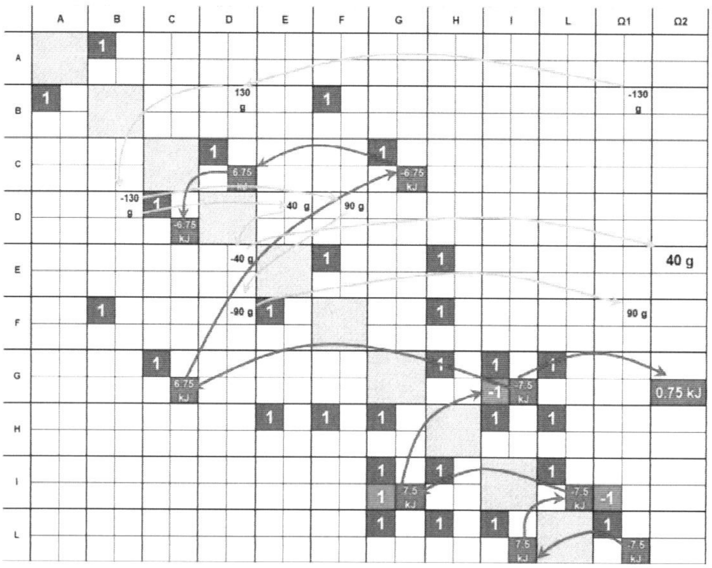

Figure 7. A-DSM built as collection of all the information associated to the device.

With the help of the arcs, it is possible to follow the flux of energy and material, how each of these comes from (or goes towards) the outside and is elaborated into the device. Further, it may be checked in correspondence of the elements where may be identified some problem, for the functionality in itself or for the influence on the environmental performance.

7. Extension of the Axiomatic approach.

A sustainability assessment may be done by Axiomatic Design characterizing the Functional Requirements with an environmental emphasis, codifying e-FRs [6]. As a matter of fact, it is important to start the design process by adopting a point of view oriented to sustainability, to be guided better along the process. Nevertheless, starting the design process with environmental Functional Requirements may not guarantee that products are built with the right degree of eco-design.

Fundamental is the necessity to assess, at the end of the process, the design solutions. For this, it is important to collect and use data in a structured way.

The association of DM and A-DSM allows designers to identify where problems are concentrated and where to intervene for improving the environment critical aspect hidden in a product. Working backward from A-DSM, identifying environmental problems, designers may find which DPs and components are responsible for them, modifying or changing them, and then the whole architecture of the product. Further, this may require a redefinition of the FRs from which the design process started.

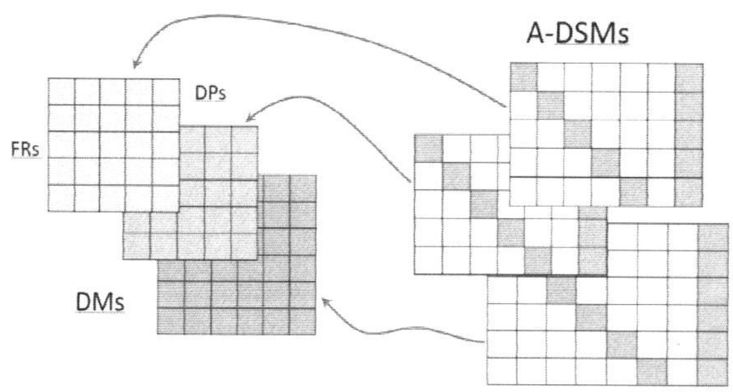

Figure 8. The backward process from A-DSM to DM.

Figure 8 reproduces such kind of situation, where changes may be guided in DM on the basis of the suggestions gathered by A-DSM. This process can be activated on a single product, or as present authors made in [20], on a set of A-DSMs that had been employed in order to compare several devices conceived for the same task. Attaining information from the enriched content of the A-DSM the comparison involved the Environment Indicators collected in the diagonal cells and the level of interaction with the outside, in terms of material wastes and energy losses.

8. Consideration and conclusion.

The A-DSM may be used in several ways. First of all, as a set of data that describe synthetically the nature of the interrelations present in the device.

Considering the possibility to implement a software check, immediately a set of contradiction could be highlighted on the device, i.e.: have a link of energy or material between two components, without a physical connection between them; at each node checks the balance of the global energy/material flux, and every time an unbalance happens in recognizing that something has been missed; jumping from a node to the other follows the direction of energy and material, also signal may indicate where to actuate some control.

Indubitably the link between DM and DSM during product design may open new ways that allow designers to take under control their work. Obviously, this can be done during the initial phases of designing, those where higher are the possibility to introduce mistakes or subtle inconclusiveness that may be unveiled with difficulty.

The authors would like to wish that other researchers may be encouraged to investigate such a kind of relation between DM and A-DSM and that this will open new possibilities to investigate product sustainability since the early phase of design.

9. References

[1] Pahl G and Beitz W 1996 *Engineering Design, A Systematic Approach, Second Edition* (London: Springer)

[2] Ulrich K T and Eppinger S D 2012 *Product Design and Development, Fifth Edition*, (New York: McGraw-Hill Education)

[3] Ramani K, Ramanujan D, Bernstein WZ, Zhao F, Sutherland J and Handwerker C 2010 Integrated sustainable life cycle design: a review. *J. Mech. Des.* **132**(9): 091004 (15 pages).

[4] Persson J G 2001 Eco-indicators in product development. *Proceedings of the Institution of Mechanical Engineers Part B. Journal of Engineering Manufacture* **215**(5): 627–635

[5] Suh N P 2001 *Axiomatic Design: Advances and Applications,* (New York: Oxford University Press).

[6] Shin M, Azhar M, Morrison J R, Lee T, and Suh H W 2011 On the Use of Axiomatic Design for EcoDesign. In *Proceedings of the 6th International Conference on Axiomatic Design*, Daejeon, March 30-31 2011, KAIST, pp. 79-87

[7] Giorgetti A, Citti P, Arcidiacono G and Delogu M 2011 Axiomatic Design for the Development of Eco-Sustainable Metaproducts. In *Proceedings of the 6th International Conference on Axiomatic Design*, Daejeon, March 30-31 2011, KAIST, pp. 88-94

[8] Steward D V 1981 The Design Structure System: A Method for Managing the Design of Complex Systems. *IEEE Transactions on Engineering Management* **28** pp 71-74

[9] Eppinger S D and Browning T R 2012 *Design Structure Matrix Methods and Applications,* (Cambridge, MA, USA: MIT Press)

[10] Cheng Q, Zhang G, Gu P and Shao X 2012 A product module identification approach based on axiomatic design and design structure matrix. *Concurr. Eng. Res. Appl.* **20**(3) pp 185–194

[11] Wang M, Beeh E, Höfer A and Horst F 2018 Function Integration for Lightweight Chassis Based on Axiomatic Design and Design Structure Matrix. *International Journal of Automotive Technology*, **19**(6), pp. 969–979

[12] Tang D, Zhang G and Dai S 2009 Design as integration of axiomatic design and design structure matrix. *Robotics and Computer-Integrated Manufacturing* **25** pp 610-619
[13] Tang D, Zhu R, Dai S and Zhang G 2009 Enhancing Axiomatic Design with Design Structure Matrix. *Concurr. Eng. Res. Appl* **17**(2) pp 129-137
[14] Rocco C, De Napoli L and Rizzuti S 2011 A Proposal for an Augmented DSM to Assess Product Sustainability. In: *Proceedings of the 13th International DSM*, MIT (Cambridge), USA, pp. 45-57
[15] Kuo T C and Wang C J 2019 Integrating Robust Design Criteria and Axiomatic Design Principles to Support Sustainable Product Development. *Int. J. of Precis. Eng. and Manuf. Green Tech* **6** pp 549–557
[16] Guenov M D, Barker S G, Hunter C, Horsfield I and Coleridge-Smith N 2006 An integrated approach to customer elicitation for the aerospace sector. *Syst. Engin.* **9** pp 62-72
[17] Danilovic M and Browning T R 2007 Managing complex product development projects with design structure matrices and domain mapping matrices. *International Journal of Project Management* **25** pp. 300–314
[18] Rizzuti S and De Napoli L 2014 The integration of DSM and Axiomatic Design in product design as part of a MDM process. In: *Proceedings of the 16th International DSM Conference: Risk and Change management in complex systems* pp 35-42
[19] Pimmler T U and Eppinger S D 1994 Integration analysis of product decompositions. *Proc. of the ASME International Design Engineering Technical Conferences* Minneapolis (MN) USA
[20] De Napoli L, Rizzuti S and Rocco C. 2017 An integrated model for the environmental assessment of industrial products during the design process. *Concurr. Eng. Res. Appl.* **25**(4) pp 360-380

IOP Conf. Series: Materials Science and Engineering 1174 (2021) 012022 doi:10.1088/1757-899X/1174/1/012022

Axiomatic design of matrix production systems

Petra Foith-Förster, Thomas Bauernhansl

Fraunhofer Institute for Manufacturing Engineering and Automation IPA,
Nobelstraße 12, 70569 Stuttgart, Germany
petra.foith-foerster@ipa.fraunhofer.de

Abstract. Manufacturing companies are operating in a turbulent business ecosystem that demands for product variety, product mix flexibility, volume scalability and high efficiency. Personalized production arises as new production paradigm to replace mass personalization and changeable matrix production systems are well suitable to cope with the challenges of this new production paradigm. In this paper, a method to design the value-add process modules of a matrix production systems is presented. The method is based on Axiomatic Design. The approach works process oriented, pursues a modular layout, and a flexible material flow in the production system. The paper concludes with a validation of the method performed in an industrial use case.

1. Personalized production in turbulent business ecosystems

Just like the 1990s European housing crisis and the 2008 global financial crisis, the Covid19 pandemic caused a sharp decline in the industrial production indices of economies worldwide, which had otherwise constantly been growing [1]. Studies show that intervals between crises became shorter in recent decades [2]. Independent of economic crises, production systems face day-to-day production volume fluctuations, caused by product lifecycles, as well as by seasonal or random ordering behavior of customers. Consequently, the capability for structural upscaling, the ability to follow a day-to-day variation of volumes, and the possibility to downscale abruptly in times of crises and upscale quickly after a crisis gives a competitive edge to production systems.

Furthermore, the number of variants to be produced has multiplied in recent decades. Surveys conducted in the automotive and machine building industries in Germany showed that around 75% of all surveyed original equipment manufacturers (OEMs) and first-tier suppliers had raised their offered variants in the last ten years before 2004 [3] and plan to raise or even double the number of their physically offered variants until 2023 [4]. On top of that, multiple product generations are produced in parallel [5]. The new production paradigm of personalized production is introduced to meet this booming demand for unique products, individualized by a customer and ordered in a lot size of one [6].

All in all, the utilization of production equipment with a limited number of certain dedicated product variants becomes unlikely. Production volumes are difficult to predict and designing a production system for a fixed range of output numbers bears a high risks of over- or underutilized equipment. Manufacturing companies need to prepare for change in order to stay competitive and prosper [7].

2. State of the art and problem statement

Prevalent flexible manufacturing systems (FMS) and lean, mixed-model production lines with coupled stations and a defined cycle time are tailored for the previously effective production paradigm of mass personalization [6]. Neither FMS nor lean mix-model production lines provide enough flexibility and reconfigurability to prepare manufacturing systems with the needed changeability to tackle the

IOP Conf. Series: Materials Science and Engineering 1174 (2021) 012022 doi:10.1088/1757-899X/1174/1/012022

challenges of a personalized production in a turbulent business ecosystem. Instead, modular reconfigurable production system are promoted as a suitable setup for personalized production [6].

Early works on reconfigurable manufacturing systems (RMS) concentrated on the technological concepts for modular multi technology machines [8] and (autonomous) production modules [9]. Koren et al. introduced RMS on the production system level [10]. They investigated configuration and conversion of manufacturing systems [11], and general modular layout concepts of computer numerically controled (CNC) machines [12]. The research group of ElMaraghy developed methods and algorithms for a periodic system configuration of capacity and functionality [13] and examined the topology of reconfigurable assembly systems (RAS) [14]. Albeit possibilities of using reconfigurable concept for assembly were examined [12], no methods to configure RAS exist so far.

Matrix production has been introduced as a concept of modular, reconfigurable production systems, built of flexibly linked process modules [15]. In a matrix production system, every variant processes through the system on its own, variant-specific path [15]. Process modules for the execution of the needed manufacturing and assembly processes are selected under consideration of process module functionalities, the possible sequence of operations and the current state of the system [15]. Existing methods on the design of matrix production have been developed for automotive assembly. Greschke [16] assigns assembly operations to process modules capacity-driven, aiming at an average system cycle time. Kern et al. [17] determine the functional scope of process modules with the help of an assembly precedence graph, considering variant-specific options. Neither of the two analyze production operations by their process characteristics, which could be applied to identify a possibility for functional integration of different operations, independent of the assembly sequence, to e.g. achieve a high utilization of investment-intensive or automated equipment.

Personalized production systems unite assembly and machining processes to realize the individualization of products as demanded by customers [18]. Assembly is very much affected by the increase of product variety and market turbulences, due to its proximity towards the customer in the value-creation chain [19]. Furthermore, the diversity of processes and tools is often higher in assembly than in machining, which restricts process integration and automation. Therefore, methods for the configuration and design of modular production systems for personalized production need to consider the specific requirements of both, machining, and assembly processes.

This paper presents a method to design matrix production systems with a process-driven approach for the functional segmentation of process modules and details the design of the value-add process modules. The method considering both manufacturing and assembly processes. To ensure a systematic and requirement-considerate design, Axiomatic Design (AD) is chosen as basic design method.

3. Introduction to Axiomatic Design

Axiomatic Design (AD) is a design methodology which guides the design process of complex technical systems with two fundamental axioms: The independence axiom (Axiom 1) seeks for functional separation of requirements and solution parameters. A design solution is considered ideal, if requirements are solved independently, i.e. each design solution only influences exactly one requirement. The information axiom (Axiom 2) delivers a decision-making principle between alternative design solutions by judging on their probability to successfully satisfy a respective requirement. The best design solution is considered the one with the least information content (I). [20]

AD proceeds through four design domains during the design process: The customer domain states the design goals, given as customer attributes (CA). The functional domain derives the functional requirements (FR) of the design from the CA. The physical domain maps design parameters (DP) to each FR. The process domain contains process variables (PV), which specify the needed parameters to realize the DPs. In addition to FRs, superordinate requirements, called constraints (C), set bound to possible solutions. Usually, a design needs to be completed before it can be checked regarding its Cs. During each mapping between two subsequent design domains, a design is detailed by decomposing each design state hierarchically into the next layer of requirements and design solutions, until an implementable design stage is reached. [20]

4. Axiomatic design method to design matrix production systems

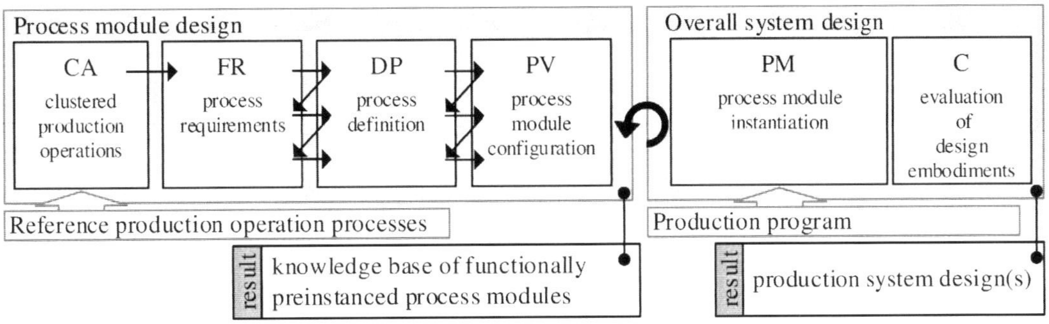

CA…Customer Attributes; FR…Functional Requirements; DP…Design Parameter; PV…Process Variables
PM…Process Module Instances; C…Constraints

Figure 1. Setup of the design method.

The presented method proposes a two-phases matrix production system design process (Figure 1).

The first phase follows AD along its four design domains for the design of process modules. In this phase, the system hierarchy and the system's functions are determined by decomposition of reference manufacturing, assembly, and material flow processes. Reference processes are derived from actual and conceivable future production operations. As a result of the process module design, functionally pre-instanced process module configurations are saved in a knowledge base of process modules. They are kept available for an initial system design as well as for later system reconfiguration designs.

In the second phase, process modules are selected from the knowledge base of process modules to fit the production program of a particular production system design project. The overall system design instantiates the process modules and relates them in a system layout.

The AD design parameters are assigned as follows:

- CAs represent the production operations.
- FRs characterize the functional process needs of production operations.
- DPs represent the actual processes of the production system, to execute the production operations.
- PVs represent the physical elements of a production system. On the topmost level, PV is a functionally configured process module.
- PM are process module instances of PVs.
- C represent the overall design constraints.

Both, the process module design as well as the project-specific production system design are checked for independence with Axiom 1 whenever a design solution is mapped. The method allows for a systematic design of alternative production system design embodiments. To support the decision-making of the production system designers, AD axioms and selected corollaries are provided in the method. For the final decision on one overall system design embodiment, defined cost and productivity key performance indicators (KPIs) are used as input constraints for the design. Finally, the method contains a comparative evaluation to judge on the best overall design. Discrete event material-flow simulation is suggested to estimate dynamic parameters of the production system for the KPI calculation.

The two-phases approach decouples the design of a system's functional concept from the design of output volumes and the consideration of variant-specific operation sequences. By doing so, the method helps the designer to achieve a flexible matrix production system, instead of limiting flexibility through a variant- and volume-driven segmentation of production operations.

This paper describe the method and the design process for the functional configuration of value-add manufacturing and assembly process modules, and their instantiation in an overall system design. The design of material flow process modules, the related selection of material supply strategies and buffer dimensioning, as well as the design of a production control process, are not part of this paper.

4.1. Process module design

4.1.1. Process module design goal clarification (CA-FR). Following the design process of AD, the clarification of the design goals is the very first step to achieve a successful design. Design goals are derived from CAs. The CAs of the process module design relate directly to the respective manufacturing and assembly operations, needed to produce certain reference products (Figure 2).

CAs cluster similar production operations. The subsequent design process designs a process module for each CA. Consequently, it must be conceivable that all operations of one cluster may be successfully executed through identical instances of production resources.

		CA1	CA2	...	CA n
Product A	operation 1	x			
	operation 2		x		
	operation 3	x			
	...			x	
Product B	operation 1	x			
	...				
	operation 20	x			
	...		x		
...	...	..	...	...	...
Product N	operation 45			x	
	...			x	
	operation n				x

Figure 2: Operations matrix for the clustering of production operations as CAs

To identify CA clusters, a change barrier analysis delivers the sorting criterion. A change barrier is defined as a category, by which the limits of use of a technical system can be quantified [21]. A change barrier may be any technological, technical or socio-organizational limitation of a process. A quantified change barrier is an inherent characteristic of the production system. It is related to specific product and process properties, given by the product design, or tied to a system property, given by the system design.

A change barrier prevents the change from one process variant to another, or the integration of additional processes into a process module if the given limits of the change barrier are exceeded. Hence, the existence of a change barrier makes universality of process modules impossible. If there are technical, technological or organizational solutions conceivable, so that it seems possible to functionally execute all customer-specific manufacturing operations on one manufacturing process module and all assembly operations on one assembly process module, there are no relevant change barriers. If operations need to be split for technical, technological or organizational reasons, the criteria that force the split are the relevant change barriers of the system design. Very often, automation is a change barrier. Accordingly designed-for-automation operations need to be split from operations that are only designed for a manual execution, in order to implement automated production cells.

Each CA is assigned with a FR in a one-to-one mapping. The CA-FR mapping delivers the design ranges for each CA. Each design range covers the individual values or ranges of all clustered operations. Design ranges are specified in the dimensions of change barriers. There is one design range per change barrier for each CA.

The specification of design ranges differentiates between ranges that describe a permissible tolerance and ranges that are associated to flexibility and reconfigurability requirements (Figure 3). Design ranges of flexibility and reconfigurability are specified by uniting the ranges of clustered operations in a so-called changeability design range. Design ranges of permissible tolerances are specified by intersection of the ranges of clustered operations in a so-called process capability design range.

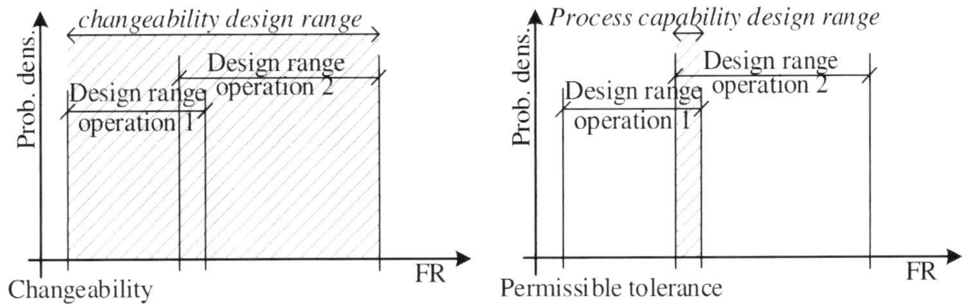

Figure 3: Specification of design ranges for CA-FR mapping from clustered operations

There are four AD corollaries that are relevant to the process module design goal clarification:

- Corollary 1 *Decoupling of coupled designs*: The CA-FR mapping must result in collectively exhaustive and mutually exclusive (CEME) [22] FRs, i.e. disjunct clusters of operations.
- Corollary 2 *Minimization of FRs*: An integration of a large number of multiple operations into one cluster of CEME CAs must be pursued by design.
- Corollary 6 *Largest design ranges*: An integration of multiple operations to achieve a broad range of (changeability) design ranges must be pursued by design.
- Corollary 7 *Uncouple design with less information*: If a conceivable solution exists that covers the design ranges of previously separated CAs, these CAs must be united.

4.1.2. Process Module hierarchy (FR-DP). The FR-DP mapping assignes and decomposes processes to each FR. The mapping effectively ensures, that all needed processes and activities are incorporated into a value-add process module to successfully execute all operations clustered in the associated CA.

A generic process module hierarchy was published as a reference model for process module design in a previous paper [16]. For the FR-DP mapping, an appropriate reference hierarchy must first be selected from the model. The system designers need to review the selected model's FR-DP design tree. The design tree needs to be checked for independence, relative to the requirements of clustered operations. Functional necessities may require for an adaption of the reference model, in order to tailor the process hierarchy to the needs of the previously determined design goals. It is the task of the designers to identify the inevitableness of an alteration and adapt the model accordingly. There are three principal reasons for an adaption (Figure 4):

1. Operation-specific process requirements: The reference model is built to generically fit any production technology. Depending on the project-specific technologies, it is possible that certain branches of the reference's model FR-DP design tree can be eliminated or must be duplicated.
   Example: A screwing process does not produce residue. The *eject residue* activity is deleted.
2. Coupled FRs: FR may be coupled due to technological constraints. In this case, a production operation associated to one FR cannot be successfully executed independent of a second operation associated to a different FR. The result is a lack of factory load case readiness. There is no feasibility of a material flow between coupled processes, and the respective process module instances cannot be separated spatially in the production system design.
   Example: Heating and shrinking operations might be clustered in different CAs, but must be designed as decoupled process for a successful execution of the shrinking operations.
3. Integrated CAs: As an alternative to couple FRs, it is possible to iterate into the cluster definition and split the specific production operations into separate CAs. This results in an ideal design, when an according FR integrates both the dependent operations. The FR-DP design tree needs to be functions-integrative, ultimately adapting the design tree with additional activity branches.
   Example: Instead of separate clusters for heating and shrinking, an integrative CA is built to cluster heating/shrinking operations separate of other heating and other pressing operations.

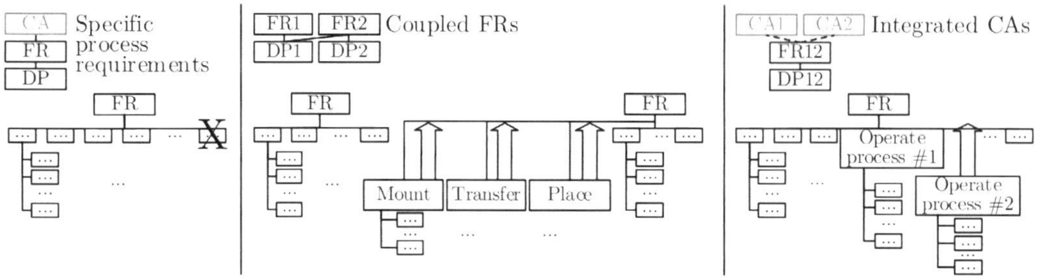

Figure 4: Example process module hierarchy adaption

4.1.3. Process module configuration (DP-PV). The DP-PV mapping finishes the functional configuration of value-add process modules by mapping production system elements to each process and activity (Figure 5). Each fully configured process module is saved in a knowledge base of functions to be kept available for a subsequent production system design. Production system elements are kept available for the configuration in a manufacturer-specific knowledge base of functions. The knowledge base of functions contains a selection of (market-)available production system elements for each reference activity, as defined by the process module reference model. Production system elements are specified in the knowledge base by their system ranges in the dimension of change barriers, according to the design ranges introduced in section 4.1.1. A production system element is a production resource, such as a specific tool, robot, machine, workstation, material bin or even a human operator, specified by its specific knowledge, if opeator knowledge was identified as a change barrier.

The PV selection is guided by AD's Axiom 2. Conceivable solutions, that led to the clustering of operations into CAs, may not necessarily be selected. There might be economical or technological reasons to select another solution, or simply different alternatives are designed.

$I_{overall}$ is calculated by summation of $I_{process\text{-}capability}$ and $I_{changeability}$ (Equation 1). $I_{process\text{-}capability}$ describes the fulfillment of the permissible tolerance requirements. It is calculated according to the calculation rules of AD's Axiom 2 (compare [20]). $I_{process\text{-}capability}$ thus evaluates, how much of a system range is covered by the common range, built by ssystem- and design range. $I_{changeability}$ describes the fulfillment of flexibility and reconfigurability requirements. $I_{changeability}$, on the other hand, evaluates, how much of a design range is overlapped by the common range. The calculation rules to compute $I_{changeability}$ have been published in a previous paper (compare [21]).

$$I_{Overall} = I_{process-capability} + I_{changeability}$$

$$= -log_2 \frac{|\, process\ capability\ common\ range\,|}{|\, process\ capability\ system\ range\,|} - log_2 \frac{|\, changeability\ common\ range\,|}{|\, changeability\ design\ range\,|} \quad (1)$$

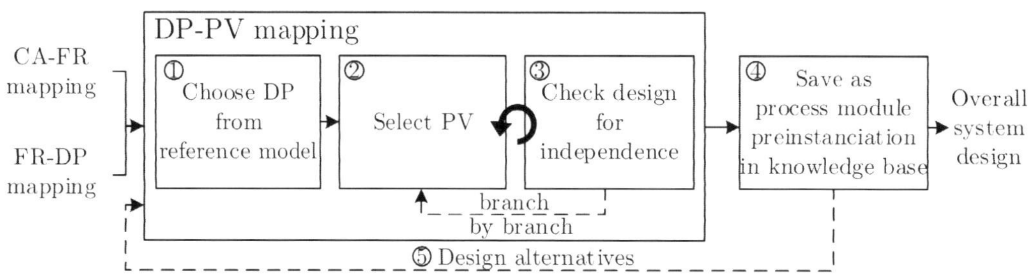

Figure 5: Process module configuration process

4.2. Production system design

Albeit the matrix production system must be designed reconfigurable to equip it for future developments in volatile business ecosystems, a functionally and capacity-wise fitting production system setup can only be designed under consideration of an actual or planned production program. As preliminary step to the production system design, this project-specific production program needs to be prepared and input C must be clarified.

To instantiate the value-add process modules of the system, the process-oriented design approach selects process modules from the knowledge base of process modules by function first, before considering capacity needs in the second step. The selection is a mapping of a process module to each individual production operation of the design project-relevant production program, considering its related process in the dimensions of change barriers.

For the actual selection, all value-add operations are compared to all PV from the knowledge base of process modules. A tabular display may be used for the comparison (Figure 6). The cells of the table contain the $I_{overall}$ of each pairing. The decision process is guided by AD's Axiom 2 and Corollary 3 *Integration of physical parts*. According to Axiom 2, any production operation should be assigned to a process module, for which the mapping results in the smallest I. At the same time, the validity of Corollary 3 reduces the total number of different process module configurations in the production system. For an integration of different operations into one process module, the system designer should select the most universal process module (e.g. PV1 and PV8 in the example of Figure 6). It is nevertheless possible, to deliberately select different PVs for different operations, even though the same PV would be possible. Such a choice is reasonable, if certain operations are automatable, but can also be executed manually, while other operations can only be executed manually. Automatable operations could, thus, be assigned with an automated PV, even though a higher-universal manual PV would also be capable of their execution (e.g. operation 8 in the example of Figure 6). It is furthermore useful to separate operations with non-configuration options from highly configurable operations on different PV. During operation, orders with low configurations are then able to omit certain process modules, reducing the throughput time.

process module preinstance / value-add production operation	PV1	PV2	PV3	PV4	PV5	PV6	PV7	PV8	PV9	...
	man.	auto.	man.	man.	man.	man.	auto.	auto.	auto.	...
operation 1	0	∞	∞	∞	∞	∞	∞	∞	∞	∞
operation 2	∞	0	0	∞	0	∞	∞	∞	∞	∞
operation 3	0	∞	∞	0	∞	∞	∞	∞	∞	∞
operation 4	∞	∞	0	∞	0	∞	∞	∞	∞	∞
operation 5	0	∞	∞	∞	∞	0	∞	∞	∞	∞
operation 6	∞	∞	∞	∞	∞	∞	0	0	∞	∞
operation 7	∞	∞	0.5	∞	∞	∞	∞	0	∞	∞
operation 8	0	∞	∞	∞	∞	∞	∞	0	0	0

◯ Lowest I of operation-PV mapping → pointing towards selected PV

Figure 6: PV selection table for function selection of process modules

The CAs and FRs of the production system design cam be concluded from the selection of process module configurations in the example of Figure 6 (Equation 2). From the mapping of the production operations, the FR-DP design matrices can be concluded (Equation 3).

$$CA1 = \{op1, op3, op5\}; CA2 = \{op2, op4\}; CA3 = \{op6, op7, op8\} \tag{2}$$

$$\begin{Bmatrix} CA1 \\ CA2 \\ CA3 \end{Bmatrix} = \begin{bmatrix} x & 0 & 0 \\ 0 & x & 0 \\ 0 & 0 & x \end{bmatrix} \begin{Bmatrix} FR1 \\ FR2 \\ FR3 \end{Bmatrix}; \begin{Bmatrix} FR1 \\ FR2 \\ FR3 \end{Bmatrix} = \begin{bmatrix} x & 0 & 0 \\ 0 & x & 0 \\ 0 & 0 & x \end{bmatrix} \begin{Bmatrix} DP1 = PV1 \\ DP3 = PV3 \\ DP8 = PV8 \end{Bmatrix} \tag{3}$$

Only change-barrier-relevant requirements matter for the process module design. Other process requirements, linked to the production operations of each cluster, need to be analyzed before the subsequent step of value-add capacity harmonization. If a selected PV is not capable to fulfill those process requirements, the respective process requirement is either a new change barrier, which has not yet been identified, or a different, process-capable PV must be selected for the cluster. Both cases cause an iteration to previous design steps.

To address the differing capacity requirements of the production operations, a capacity harmonization of selected PVs is done and leads to the process module instances (PM) of the system design. The harmonization considers the overall capacity needs of a production scenario as well as the distribution of capacity needs between different system functions. The tabular display of functional suitability for process module selection may be advanced to judge on the average utilization of each selected PV (Figure 7). Whenever a utilization reaches a target value (e.g. set to 85% to cover overall equipment effectiveness (OEE) losses), a new instance of this respective process module PV must be added to split the capacity load.

Volume: 210pc/workday		man.		auto.		man.		man.		man.		man.		auto		auto		auto		...	
		PV1		PV2		PV3		PV4		PV5		PV6		PV7		PV8		PV9		PV10	
value-add op	Vol %	PT	Capa need [min]	PT	Capa need [min]	PT	Capa need [min]	PT	Capa need [min]	PT	Capa need [min]	PT	Capa need [min]	PT	Capa need [min]	PT	Capa need [min]	PT	Capa need [min]	PT	Capa need [min]
op 1	85.7%	2	360	–	–	–	–	–	–	–	–	–	–	–	–	–	–	–	–	–	–
op 2	38.1%	–	–	1	80	1.5	120	–	–	1	80	–	–	–	–	–	–	–	–	–	–
op 3	85.7%	4	720	–	–	–	–	4	720	–	–	–	–	–	–	–	–	–	–	–	–
op 4	85.7%	–	–	–	–	1	180	–	–	1	180	–	–	–	–	–	–	–	–	–	–
op 5	38.1%	0.5	40	–	–	–	–	–	–	–	–	0.5	40	–	–	–	–	–	–	–	–
op 6	61.9%	–	–	–	–	–	–	–	–	–	–	–	–	1.5	195	1	130	–	–	–	–
op 7	61.9%	–	–	–	–	–	–	–	–	–	–	–	–	–	–	1	130	–	–	–	–
op 8	100%	0.8	168	–	–	–	–	–	–	–	–	–	–	–	–	0.8	168	0.5	105	1	210
Utilization		266%		19%		71.5%		171.5%		62%		9.5%		46%		102%		25%		50%	
Capacity/day:		7h																			

Figure 7: Capacity harmonization table for process module instantiation

The instantiation process results in a redundant design if one PM instance per PV is not sufficient to satisfy all capacity requirements of the assigned production operations (Equations 4 and 5). The redundancy adds to the system's failure and routing flexibility. As the redundancy only occurs for PM instances, and the one-to-one FR-DP-PV mapping is kept, it does not violate AD's Axiom 1.

$$\begin{Bmatrix} PV1 \\ PV3 \\ PV8 \end{Bmatrix} = \begin{bmatrix} xxx00 \\ 000x0 \\ 0000x \end{bmatrix} \begin{Bmatrix} PM1.1 \\ PM1.2 \\ PM1.3 \\ PM3 \\ PM8 \end{Bmatrix} \tag{4}$$

$$\begin{Bmatrix} CA1 \\ CA2 \\ CA3 \end{Bmatrix} = \begin{bmatrix} x00 \\ 0x0 \\ 00x \end{bmatrix} \begin{bmatrix} x00 \\ 0x0 \\ 00x \end{bmatrix} \begin{bmatrix} xxx00 \\ 000x0 \\ 0000x \end{bmatrix} \begin{Bmatrix} PM1.1 \\ PM1.2 \\ PM1.3 \\ PM3 \\ PM8 \end{Bmatrix} = \begin{bmatrix} xxx00 \\ 000x0 \\ 0000x \end{bmatrix} \begin{Bmatrix} PM1.1 \\ PM1.2 \\ PM1.3 \\ PM3 \\ PM8 \end{Bmatrix} \tag{5}$$

The utilization is an average value. It must be validated regarding its dynamic behavior in the system design evaluation. The harmonization table maintains the information about alternatives to the selected PVs. This enables an iteration to PV selection if an alternative assignment of production operations suggests the likeliness of a better harmonization.

The overall matrix production system does not run on a balanced system cycle time. Nevertheless, it is desirable to achieve a harmonized distribution of PM utilization. In a highly harmonized system, the overall work-in-process inventory (WIP) in the system's buffers is reduced and all PM are able to operate simultaneously. The instantiation process of PMs doubles over-utilized PMs. This almost certainly leads to some under-utilized PMs. The ultimate target is a high utilization of value-add PMs, especially if they are operated manually or if they are built from investment cost-intensive systems. There are two options to deal with under-utilization and disharmonizations:

1. Multi-PM operation: For manually operated PM, it is possible to share operators to prevent idle times of operators. Respective PMs need to be located close to each other, as the operator will have to traverse between them. Multi-PM operation is an excellent flexibility reverse, if space-availability is not a constraint. A scale up of volumes is possible by adding operators and changing from multiple- to single-PM operation.
2. Individual run-times: Due to the flexible linkage of PM in the matrix production system it is possible to schedule individual run-times of certain PMs, if several same-function PM exist, or if certain PM are exclusively dedicated to specific, rarely required production operations. Due to the one-piece-flow concept of a matrix production system, temporary shutdowns of PMs must be considered in the buffer dimensioning of the PMs.

For the design of the layout, the modified triangle method [23], or other heuristic or analytical methods for factory layout design may be applied.

4.2.1. System design evaluation. The evaluation reveals how well the input constraints of the system design are met. A comparison is made possible for several alternative design embodiments. If the evaluation does not have a satisfactory result, an iteration of the system design is necessary. Alternative system designs can be generated by alteration of design choices during the design process.

The maximization of changeability and productivity are seen as general input constraints for a matrix production system design [16]. Actual maximum or minimum limits of productivity parameters, and further, project-specific constraints are defined in the data preparation for the design project.

For the evaluation of system changeability, the changeability system ranges of the overall system are compared in the dimensions of the change barriers and related to the product variants relevant to the design project. The evaluation focuses purely on the excess of system ranges, which are regarded as a positive surplus of changeability, if productivity constraints are nevertheless achieved.

For the evaluation of productivity, personnel productivity, asset productivity, and area productivity are analyzed (Equation 6). Consumed area and investment costs are static parameters and can be obtained from the design embodiment documentation. Output quantities and the number of personnel depend strongly on the production program and the chosen harmonization strategy of the system. A static calculation is hardly possible. A discrete material flow simulation helps to determine output quantities of different production program scenarios. The simulation model must map the respective production system design embodiment by its organizational and process structure and, at the same time, serves to review the material flow and buffer dimensioning.

$$Area\,prod. = \frac{Output\,[\frac{qty}{time}]}{Area\,[m^2]};\ Asset\,prod. = \frac{Output\,[\frac{qty}{time}]}{Fixed\,Cost\,[€]};\ Pers.prod. = \frac{Output\,[\frac{qty}{time}]}{Personnel\,[\frac{€}{time}]} \quad (6)$$

An overall system design evaluation is only possible if the system is designed altogether, including material flow process modules, raw material supply and production control processes.

IOP Conf. Series: Materials Science and Engineering 1174 (2021) 012022 doi:10.1088/1757-899X/1174/1/012022

5. Validation use case: matrix production system for servomotors design

The validation was done at a manufacturer of servomotors. The manufacturer offers motors with a modular product architecture which can be configured according to individual needs (e.g. power, inertia, brake) and which contain customer-individual parts (e.g. length of shaft, type of shaft-hub joint). The company is a rapidly growing small or medium-sized enterprise (SME) that continuously adds new product families to its portfolio. With high uncertainties of volume growth, product variants and customer requirements, a respectively flexible and reconfigurable production system was needed.

The analysis of all manufacturing and assembly operations revealed several technical and organizational change barriers (Table 1). It was considered important to enable automation, as only very few of the operations were designed for automation. Another important change barrier were the competencies of operators, as there were certain production operations that only certain operators were able to execute due to their experience or special process knowledge.

Table 1. Change barriers of the validation project (excerpt)

	Change barrier	Related product or process property
Value-add	automated assembly	design of parts and processes
	high-investment equipment	specific production technologies
	competencies of operators	specific process requirements
	load limit of manipulators	weight of parts
	…	

From the change barriers, 13 clusters of value-add operations were identified as CAs at first. After the FR-DP analysis, two more CA cluster were built by separation of existing clusters, to resolve a coupling between oven and shrinking processes (Table 2).

To build the knowledge base of functions, all machinery, production equipment, tools, and workstations that were in place at the validation partner's production system at the time of the validation project were qualified according to their flexibility and reconfigurability system ranges relative to the change barriers. Furthermore, automation concepts and new oven concepts from system integrators, previously requested by the validation partner, were included as respective solution elements in the knowledge base of functions. For the validation project, the validation partner enquired state-of-the-art workstations, material supply systems for workstations, stripping and crimping machines, and manually operated electrohydraulic and mechanical bench power presses.

Table 2: CA of the validation project (excerpt)

	Operations	FR changeability specification	FR process capability
CA1	Turning operations	$dr_{technology}$ = {turning}; $dr_{competencies}$ = {machinist}	Dimension tolerances
…	…	…	…
CA8	Balancing operations heavy rotor	dr_{weight} = [3kg,7kg]; $dr_{technology}$ = {balancing}; $dr_{competencies}$ = {balacing routine}	Measure precision
CA9	Balancing operations leight rotor	dr_{weight} = [0kg,3kg]; $dr_{technology}$ = {balancing}; $dr_{competencies}$ = {balacing routine}	Measure prescision
…	…	…	…
CA13	Leight manual assembly & press	dr_{weight} = (0kg, 10kg]; dr_{press} = (0,25kN]; $dr_{technology}$ = {other joining*};	---
…			

*other joining = {screwing, inserting, circlip, pressing, wire prep, cleaning, labeling, lubrication, adjust}
dr… design range

For the system design of the validation project, functionally pre-instanced process modules were configured from the knowledge base of functions and saved in a knowledge base of process modules. For many of the configured process modules, the existing machinery and equipment of the manufacturer were incorporated to reduce overall needed investment.

To achieve a capacity harmonization of PMs, a combination of multi-PM operation and individual run times was selected as a harmonization strategy of the validation project. The design of the validation project foresees three different strategies to implement multi-PM operations:

1. Single operator, multiple PMs: One operator is responsible for more than one functionally independent PM. The operator himself directs capacity to the PM, depending on the length of the order-queue in the input buffer of the different PMs.
2. Multiple operators, multiple PMs: Several operators are responsible for several, often sequential, PMs. Operators are self-organized, according to the principles of a production group, or else the operators conduct the production line balancing approach of a lean production rabbit chase.
3. Multiple operators, one shared PM: Several operators are assigned to a specific PM, but additionally share a common supplying PM. Operators have to self-organize the operation of the shared resource.

An ideal layout was created using the modified triangle method. The ideal layout was adapted to the building restrictions of the validation partner. The result was a U-shaped main material flow from the warehouse to the wrapping area, with the warehouse and the preassembly areas forming one flank of the U and the main assembly, testing, painting and wrapping areas forming the second flank (Figure 8).

The matrix-design embodiment of the validation project achieved better KPI values of personnel and area productivity than the validation partner's status-quo production system setup and an acceptable asset productivity. The denominators (number of operators, investment, usage of production area) for the calculating of productivity KPIs were derived from the layout and the design documentation. The dynamic parameter of output quantity was determined with the help of a discrete material flow simulation model, mapping the current status quo of the validation partner's production system and the two design embodiments of the validation project.

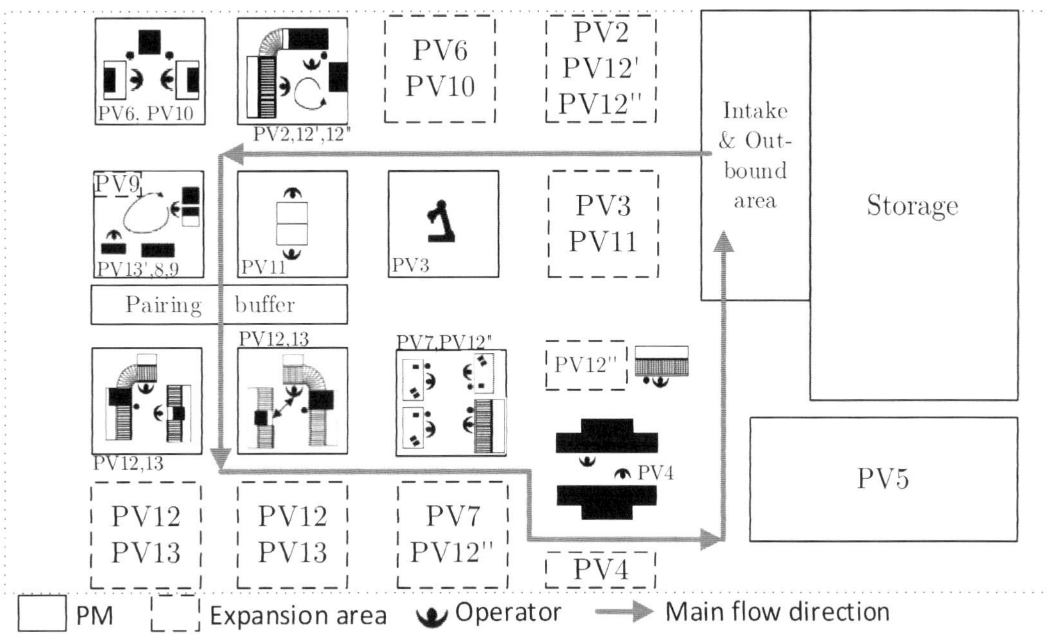

Figure 8: Validation case production system layout.

6. Conclusion and outlook

The paper presented a design method for the design of value add process modules for matrix production systems, to tackle the challenges of personalized production in a turbulent business environment. Matrix production systems are able to integrate all needed processes to produce a wide range of product variants into one production system. The presented method puts the functional process module design prior to the overall system design, to allow for a process-driven design approach and the creation of a system's functional concept independent of specific variants and output quantities.

A matrix production system, designed with the presented method, contains several process modules of different functional scope in a modular layout, linked by a flexible material flow. Value-add process modules can be distinguished by their technology, their technical equipment or by specific organizational or infrastructure requirements. They are specialized if they contain certain technical equipment that cannot be integrated or operated with other equipment, if the union of different material flows makes an automated process module economically feasible or if built-in high-investment equipment or high operational cost of a process module require a high utilization of a specific function. Non-specialized value-add process modules, on the other hand, integrate multiple functions for a variant-universal execution of processes. A process module may be a single instance of a functional process module type, or it may be one of multiple instances of the same functional scope, to cover differing capacity needs of different functions. To ensure a high and balanced utilization of operations, the system design method introduces multiple machine operation and individual run times of process modules to its matrix production concept. On process modules with multi-process ability, an order is processed as long as possible. It is only transported to the next process module if all production operations are executed, that the prevalent process module is functionally capable to execute, depending on the precedence of operations.

The main focus of the paper was the segmentation of system functions and their capacity-considerate instantiation and relation as value-add process modules. A technical design of process module elements, the production order control processes, and the actual design of the material supply systems are not part of the paper. Further research is consequently needed in these fields.

List of used symbols and parameters

CA	Customer Attributes as defined by Axiomatic Design. In the presented method, CA represent a cluster of production operations.
FR	Functional Requirements as defined by Axiomatic Design. In the presented method, FR quantify the functional process needs of the clustered production operations.
DP	Design Paramters as defined by Axiomatic Design. In the presented method, DP represent the actual processes and activities to execute the production operations.
PV	Process Variables as defined by Axiomatic Design. In the presented method, PV is a functionally configured process module.
PM	A process module instance of a PV.
PT [time]	Process time, quantifying the duration of a process or production operation.
operation n; $n \in \{1,...,K\}$	Specific production operation of a production system design project, with n as an integer value.
$I_{overall}$	Overall information content of a requirement-solution pairing.
$I_{process\text{-}capability}$	Information content of a requirement-solution pairing, quantifying the fulfillment of the permissible tolerance requirements.
$I_{changeability}$	Information content of a requirement-solution pairing, quantifying the fulfillment of flexibility and reconfigurability requirements.
dr	design range, quantifying the requirements of a design.

References

[1] OECD, 2020 *Industrial production (indicator): Manufacturing / Construction: 2015=100, Q3 2020 or latest available* (OECD Publishing) URL: https://data.oecd.org/industry/industrial-production.htm Accessed on: 01/21/2021

The World Bank 2021 *Data: Manufacturing, value added (constant 2010 US$): World Bank National Accounts data, and OECD National Accounts data files.* (Washington D.C: The WorldBank)URL: https://data.worldbank.org/indicator/NV.IND.MANF.KD?end=2018&start=1990 Accessed on: 01/21/2021

[2] Koch P, Niemeier S, Faßbender H et al. 2010 *Willkommen in der volatilen Welt: Herausforderungen für die deutsche Wirtschaft durch nachhaltig veränderte Märkte* (Frankfurt am Main: McKinsey & Company) p.21

[3] Kinkel S 2005 *Anforderungen an die Fertigungstechnik von morgen: Wie verändern sich Variantenanzahl, Losgrößen, Materialeinsatz, Genauigkeitsanforderungen und Produktlebenszyklen tatsächlich?* (Karlsruhe: Fraunhofer ISI) Mitteilungen aus der Produktionsinnovationserhebung, 37. URL: https://www.isi.fraunhofer.de/content/dam/isi/dokumente/modernisierung-produktion/erhebu ng2003/pi37.pdf Accessed on: 01/21/2021, p.3

[4] Luckert M, Schiffer M, Wiendahl H-H, Schöllhammer O and Köpler J 2018 *Zulieferer vor der Zerreißprobe: Wie Zulieferer im Automobil- und Maschinenbau den Wandel durch Industrie 4.0 meistern können.* (Stuttgart: Industrie- und Handelskammer) URL: https://www.ipa.fraunhofer.de/de/Publikationen/studien/studie-zulieferer-zerreissprobe.html Accessed on: 01/21/2021, p. 27

[5] ElMaraghy H and Wiendahl H-P 2009 *Changeability - An Introduction* In: ElMaraghy H (ed.): *Changeable and Reconfigurable Manufacturing Systems* (Guildford: Springer London) pp. 3–23, ISBN 978-1-84882-066-1, S. 6

[6] Koren Y 2010 *The global manufacturing revolution: Product-process-business integration and reconfigurable systems* (Hoboken: Wiley) ISBN 978-0-470-58377-7, S. 33-37

[7] Bauernhansl T and Miehe R 2020 *Industrielle Produktion - Historie, Treiber und Ausblick* In: Bauernhansl T (ed.): *Fabrikbetriebslehre 1: Management in der Produktion* (Berlin, Heidelberg: Springer Vieweg) pp. 1–33, ISBN 978-3-662-44537-2, p. 24

[8] Abele E and Wörn A 2004 Chamäleon im Werkzeugmaschinenbau: Rekonfigurierbare Mehrtechnologiemaschinen. *ZWF Zeitschrift für wirtschaftlichen Fabrikbetrieb* **99** (4), pp. 152–156; DOI: 10.3139/104.100756, p. 152

Abele E, Versace A and Wörn A 2007 *Reconfigurable Machining System (RMS) for Machining of Case and Similar Parts in Machine Building* In: Dashchenko A I (ed.): *Reconfigurable Manufacturing Systems and Transformable Factories: 21st Century Technologies* (Berlin, Heidelberg, New York: Springer) pp. 327–339, ISBN 978-3-540-29391-0; p. 327

[9] Pouget P M 2000. *Ganzheitliches Konzept für rekonfigurierbare Produk-tionssysteme auf Basis autonomer Produktionsmodule* (Düsseldorf: VDI) Fortschritt-Berichte VDI Reihe 2, Fertigungstechnik, 537. Zürich, ETH, Diss., 1999; ISBN 3-183537-02-8

Drabow G 2006 *Modulare Gestaltung und ganzheitliche Bewertung wandlungsfähiger Fertigungssysteme* (Garbsen: PZH) Berichte aus dem IFW, 2006-5. Hannover, Univ., Diss., 2006; ISBN 3-939026-13-1

[10] Koren Y, Heisel U, Jovane F, Moriwaki T, Pritschow G, Ulsoy G and van Brussel H 1999 Reconfigurable Manufacturing Systems *Annals of the CIRP* **48** (2), pp. 527–540; DOI: 10.1007/978-3-642-20617-7_6629, p. 528

[11] Maier-Speredelozzi V, Koren Y and Hu S, 2003 Convertibility Measures for Manufacturing Systems *Annals of the CIRP* **52** (1), pp. 367–370; DOI: 10.1016/S0007-8506(07)60603-9, p. 367

[12] Koren Y and Shpitalni M 2010 Design of reconfigurable manufacturing systems *Journal of Manufacturing Systems* **29** 4, pp. 130–141; DOI: 10.1016/j.jmsy.2011.01.001, p. 134-137

[13] Youssef A and ElMaraghy H 2006 Modelling and optimization of multiple-aspect RMS configurations *International Journal of Production Research* **44** (22), pp. 4929–4958; DOI: 10.1080/00207540600620955

Youssef A and ElMaraghy H 2007 Optimal configuration selection for Reconfigurable Manufacturing Systems *International Journal of Flexible Manufacturing Systems* **19** (2), pp. 67–106; DOI: 10.1007/s10696-007-9020-x

Navaei J and ElMaraghy H 2014 Grouping Product Variants based on Alternate Machines for Each Operation *Procedia CIRP* **17**, pp. 61–66; DOI: 10.1016/j.procir.2014.01.124

[14] Manns M, Urbanic R J and ElMaraghy H 2008 A Design Approach for Reconfigurable Assembly Systems. In: 2nd CIRP Conference on Assembly Technologies & Systems (CATS). Toronto, Canada, 09/21-23/2008, pp. 231–243

[15] Fries C, Fechter M, Ranke D, Trierweiler M, AlAssadi A, Foith-Förster P, Wiendahl H-H and Bauernhansl T 2021 *Fluid Manufacturing Systems (FLMS) – A novel approach for versatility in production* In: Weißgräber P, Heieck F, Ackermann C (eds.): *Advances in Automotive Productin Technology – Theory and Application* (Wiesbaden: Springer Vieweg) DOI 10.1007/978-3-662-62962-8

[16] Greschke P I 2016 *Matrix-Produktion als Konzept einer taktunabhängigen Fließfertigung* (Braunschweig: Vulkan) Schriftenreihe des Instituts für Werkzeugmaschinen und Fertigungstechnik der Technischen Universität Braunschweig. Braunschweig, TU, Diss., 2016; ISBN 978-3-8027-8344-9

[17] Kern W, Rusitschka F, Kopytynski W, Keckl S and Bauernhansl T 2015 Alternatives to assembly line production in the automotive industry *23rd International Conference on Production Research* (ICPR). Manila, Philippines, 07/31-08/02/2015, pp. 1–9; URL: https://www.researchgate.net/publication/317279321_Alternatives_to_assembly_line_production_in_the_automotive_industry, Accessed on: 01/14/2021

Kern W, Rusitschka F and Bauernhansl T, 2016 Planning of workstations in a modular automotive assembly system. *Procedia CIRP* **57**, pp. 327–332; DOI: 10.1016/j.procir.2016.11.057

[18] Foith-Förster P and Bauernhansl T 2019 Generic Production System Model of Personalized Production. *MATEC Web of Conferences* **301**, DOI: 10.1051/matecconf/201930100019

[19] März L and von Langsdorff P 2001 *Flexibilität und Marktorientierung in der Montage* In: Westkämper E, Bullinger H-J, Horváth P, Zahn E (eds.): *Montageplanung - effizient und marktgerecht* (Berlin: Springer) pp. 3–10; ISBN 3-540666-47-8, p. 3

[20] Suh N P, 2001 *Axiomatic design: Advances and applications* (New York: Oxford Univ. Press.) The MIT-Pappalardo series in mechanical engineering ISBN 0-195134-66-4

[21] Foith-Förster P, Wiedenmann M, Seichter D and Bauernhansl T 2016 Axiomatic Approach to Flexible and Changeable Production System Design. *Procedia CIRP* **53**, pp. 8–14; DOI: 10.1016/j.procir.2016.05.001

[22] Brown C 2005. Teaching Axiomatic Design to Engineers - Theory, Applications, and Software. *Journal of Manufacturing Systems* **24** (3), pp. 186–195; DOI: 10.1016/S0278-6125(06)80007-5; p. 189

[23] Schmigalla H 1968 *Methoden zur optimalen Maschinenanordnung* (Berlin: Verlag Technik)

Design of Neural Networks

Pam Mantri[1], **John Thomas**[1]

[1]Cognitive Tools Ltd. LLC, P.O. Box 695; 255 North Ave; New Rochelle, NY 10801, USA

E-mail: pammantri@cogtools.com

Abstract. Artificial Neural Networks (ANN) and Deep Learning (DL) are used to solve complex problems including image recognition, speech recognition and have applications in new technologies for autonomous driving, facial recognition, detecting cancers from imaging samples among others. Various design considerations are involved in the design, training, and testing of Artificial Neural Networks (ANN). These include the design of the input/output layer, the structure and number of hidden layers, the data/data-structures of variables, the transformative functions embedded in the network, the optimizers being considered, the learning rate and its systematic adjustment, the prudent usage of dropout, the parallelism-related batch-size, the number of epochs, the adaptive logic for systematically changing the network for better fit, etc. While all these methods and techniques are sensible and relevant, there lacks an overarching framework for the needed design. This paper considers the design of an ANN from an Axiomatic Design (AD) perspective that parallels the biological inspiration for ANN's in the first place, i.e., the brain. The axiomatic design approach is used for explicating and extricating the form, function, and adaptive evolution of the underlying network.

Keywords: Artificial Neural Networks, Biological Neural Networks, Axiomatic Design, Deep Learning, Perceptron

1. Introduction

In less than a hundred years, work on Artificial Neural Networks (ANN) has progressed to the extent where ANNs are used in a multitude of disciplines. ANNs can recognize objects, translate speech to multiple languages, convert voice to text and vice-versa, identify cancer lesions from imaging studies, predict stock prices, beat humans at games like chess and GO, assist in autonomous driving to name just a few applications.

In essence, a Neural Net consists of input variables that through a series of intermediate hidden layers, each containing hidden nodes, can output values for the output variables. Research over the years has focused on techniques to make the ANN model accurate, generalized, efficient, robust in the problems it can solve and the solutions it predicts. The design of the Artificial Neural Net plays an important role in addressing this. Design architecture includes considerations of the input/output layer, the structure and number of hidden layers, the data/data-structures being submitted and its appropriate scaling, the transformative functions embedded in the network, the optimizers being considered, the learning rate and its systematic adjustment, the prudent usage of dropout, the parallelism-related batch-size, the number of epochs, the adaptive logic for systematically changing the network for better fit, use of convolutions and a multitude of other architectural considerations. The design of ANNs happens more at the abstract, architectural level instead of at the fine-grained detailed level. This is because of the self-organizing property of the ANN's at the fine-grained levels. The system is supposed to fill in the fine-grained details in a coordinated orchestration of self-organization.

Whether artificial or biological in origin, neural networks exhibit a hierarchical structuring which agrees well with Axiomatic Design (AD). This paper catalogs some key historic milestones in ANNs and traces the hierarchical "form-follows-function" (i.e., Design Parameter follows Functional Requirement, DP-FR) mapping of the ANN logic.

2. History of Artificial Neural Networks

2.1. Early foundational work

The history of ANNs is less than a hundred years. It can be traced to the work of McCulloch and Pitts [1], in 1943, on threshold logic-based computational model for basic OR/AND/NOT functions. Using this model, binary inputs result in an output of 1, when the input sum exceeds a threshold, otherwise, the output is zero. Another key insight was the Hebbian learning model, which in 1949 postulated synaptic firing to be the mechanism by which the brain learns (i.e., "neurons that fire together wire together"). Adaptively changing firing patterns impacted the weight of the neuronal connection, thus providing synaptic plasticity [2]. In 1957, Rosenblatt conceived the Perceptron model to explain how the neuron in the brain might operate [3]. He then prototyped the Perceptron idea in hardware and applied it for the classification of 20 X 20-pixel data. The perceptron creates a linear-weighted aggregate of all input signals (similar to linear regression) and modulates it with a non-linear activation function. A single perceptron classifies all the inputs it receives pto the binary state of zero or one. For classification into multiple categories, a single layer of perceptrons may be used such that each perceptron receives all the inputs but was responsible only for one part of the classification. This is similar to ANN. A few years later in 1960, an early example of adaptive "ADALINE" neuron [4] using chemical "memistors" was proposed by Widrow. It was an ANN without a thresholding function which made it differentiable. Calculus could then be used to minimize the error. In 1962, Hubel and Wiesel published their theory of the architecture of hierarchical convergence of simple and complex cells in the cortical cells of the cat brain, during visual perception [5]. The work on neural basis of visual perception has inspired many neural networks (NN) and deep learning (DL) architectures.

2.2. AI Winter

Early breakthroughs of perceptrons led to hyped expectations in the media about outlandish futuristic possibilities from Artificial Intelligence (AI). But in 1969, Minksy and Papert [6] published an influential, critical, and skeptical view about perceptrons. Minsky/Papert showed that a single perceptron layer of the Rosenblatt algorithm could not represent the XOR computation. Single layer perceptrons were computing the weights by adjusting them based on the expected correct output. For XOR to work with perceptrons, multiple layers would be required. Many (including Minksy/Papert) knew that multiple layers could solve the XOR problem. But there lacked a self-organizing algorithm to help train the ANN. The backpropagation algorithm that could train the ANN was invented 17 years later. Over the next few years, a period of AI winter ensued.

For ANNs, since the supervised method only specifies the correct final output, the correct weights expected for the layers prior to the output layer needed to be determined. The method for that computation had not yet been identified. The presence of multiple layers, some of which are hidden, is of importance for finding features in the data. Prior to the use of Neural Nets, feature engineering was used to extract relevant features for the model. The great thing about hidden layers is that features can be extracted from the data by the algorithm. So, a leap of computational efficiency was achievable, if the key to determining the weights of the hidden layers could be addressed.

IOP Conf. Series: Materials Science and Engineering 1174 (2021) 012023 doi:10.1088/1757-899X/1174/1/012023

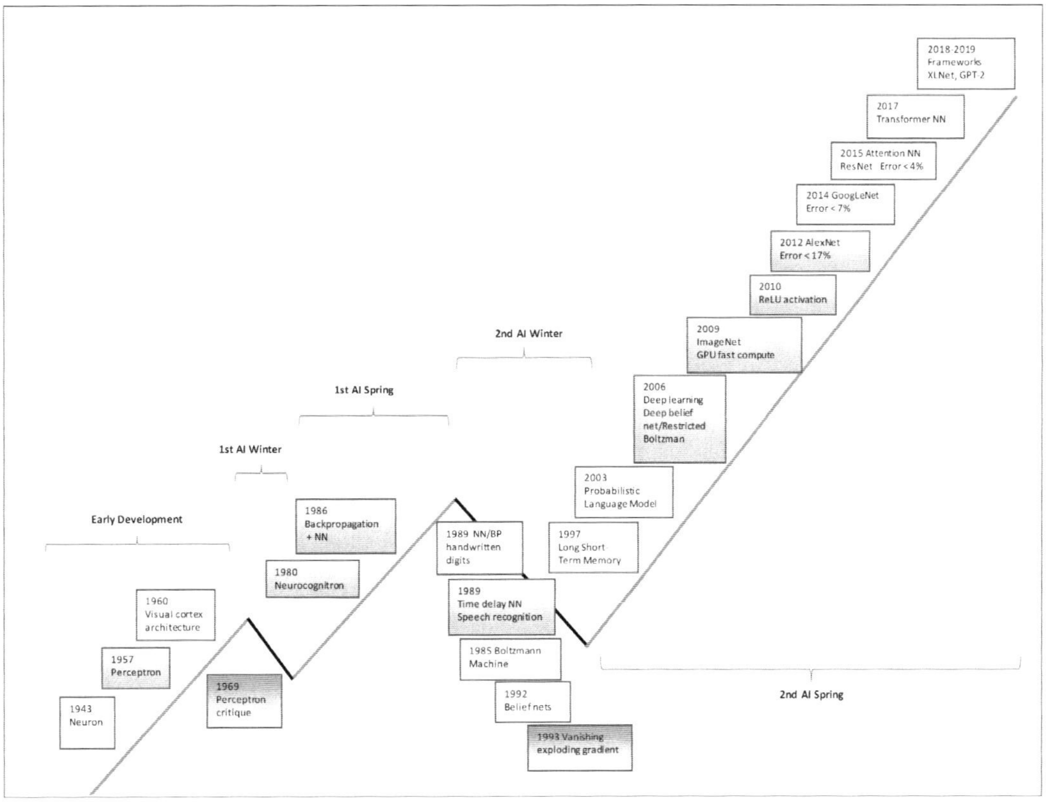

Figure 1. Landmarks in Neural Network history

Backpropagation was identified in the early 1970s by Linnainmaa [7] and postulated to be useful in neural nets by Werbos [8] in 1974. It uses the chain rule from calculus to compute the gradients for the prior layers, as well as weights for the intermediate layer under computation. Not many researchers embraced this finding right away due to the stigma for Neural Nets during the first AI winter. However, a decade after the discovery of backpropagation, in 1986, Rumelhart, Hinton and Williams, used backpropagation [9] and addressed the problems raised by Minsky to decisively give NNs a boost of life.

2.3. AI Spring

From the 1980s to the 2000s research on Neural Nets takes off again. Inspired by the work of Hubel and Wiesel's hierarchy model of the visual cortex, Fukushima, in 1980, proposed a neural network called Neurocognitron [10]. The goal of the network was to use self-organization to auto-detect patterns, and to recognize patterns that were not impacted by a change in position or small distortions. Backpropagation which was already discovered in the 70s was now being applied to neural nets. Once backpropagation was used to train NNs successfully, ambitious targets were set, including the use of Multilayer feedforward networks as universal approximators in 1989.

Fukushima (in 1980) and Mozer (in 1987), found that since features in an image could be at any location, it was important to preserve the approximate position to ensure that higher-order complex features could be detected [10]. This moved the design from backpropagation to deep learning. The first hidden layer of the NN was convolutional. Convolutional Neural Networks

(CNNs) were inspired by the neuronal architecture of the animal visual cortex where cortical neurons respond to their corresponding receptive field which is a part of the visual field. There are many such receptive fields, and they cover the entire visual field by partially overlapping with each other. Layers past the first hidden layer also work in a similar way but they take local hidden features from the previous layer to construct larger subsets of the image to arrive at patterns. The final layers use normal neural nets to discern higher-order larger features generated by convolutional layers to discharge the task of classification. This methodology was used by LeCun et. al. from AT&T Bell Labs, to achieve a major milestone for ANN/CNN, demonstrating that neural nets and backpropagation could be used by a computer to detect and classify handwritten digits [11].

New research ideas were explored in the domain of Energy-based models. Inspired by statistical physics and thermodynamics, Hinton and his team, introduced Boltzmann Machines, wherein a simple algorithm is used to discover features in data that contained binary vectors [12]. These are neuron-like and make probabilistic decisions using symmetrical connections about being on/off. A network of Boltzmann machines forms graphical models and resembles NN with hidden units and hidden variables. However, a network of Boltzmann machines did not have anything corresponding to layers. One problem with these was the speed of training. In 1992 Neal introduced belief nets which had directed forward connections and could be trained faster than Boltzmann's machines [13]. Despite being faster than Boltzmann machines, the computation was still slow and newer approaches were investigated. In 1995, Dayan, Hinton, Neal and Zemel came up with the Helmholtz Machine, wherein separate weights were used for inferring hidden variables from visible variables, as well as the reverse - inferring visible variables from the hidden variables [14].

Separately in the 1980s and 1990s research was also progressing on the use of NNs via reinforcement learning to perform complicated tasks and for playing games like chess and Go. At that time, even in the better cases, hardware limitations prevented NNs from scaling and providing truly forward-looking optimal results.

Speech recognition was another area where active research was ongoing. Speech is challenging because, unlike a batch of images in computer vision, language involves a temporal sequence of spoken words. Yet another problem with speech, is the innumerable pronunciations possible. Work by Waibel et. al (in 1989) looked at using a moving window to process a temporal input subset, and then using the same computation for each sequence of words processed. This work led to the development of time-delay neural networks (TDNN) [15]. Separately, Recurrent Neural Networks (RNNs), where the links form a temporal directed graph, are better suited for this task type, as they can leverage the information in past input weights by using cycles in the graph of the network. Additionally, RNNs can be adaptive to the information being processed.

TDNNs are similar to CNNs, however the window for CNN moves across the entire input image, rather than being time-based. CNNs have become the NNs of choice for images. Even though RNNs outperformed static networks, Bengio (in 1993) found that RNNs worked poorly with long-term dependencies, because they settled in suboptimal solutions with short-term preferences. This limited RNN's use in learning applications that had long-term dependencies such as language [16].

Backpropagation worked well on NNs with few layers, but with deep NNs the problem of vanishing or exploding gradients was encountered. This results from the cumulative error being backpropagated, which unfortunately increased or decreased exponentially with the addition of layers. The exploding/vanishing gradient problem was encountered for ANNs and RNNs, with multiple time layers, that used backpropagation to compute intermediate weights. A breakthrough to solving the problem for RNNs was achieved, in 1997, by Hochreiter and Schmidhuber, with work on Long Short-Term Memory (LSTM). Using this technique, select units of the network, use an identity activation function, and a self-loop with a weight of one, to deal with the vanishing gradient problem [17].

2.4. Second AI Winter

Despite the breakthrough with LTSM, the perception of the limitations of backpropagation and RNNs and the effectiveness of other methods such as Support Vector Machines (SVMs) and Random Forests, coincided with a second AI winter for neural networks resulting in the drying of research funding in the 90s.

2.5. Second AI Spring

In 2003, Bengio published Neural Probabilistic Language Model to model language using NNs [18]. NN research had encountered resistance during the second AI winter, and in 2006, Hinton and others rebranded NNs as DL. In the same year, Hinton and his colleagues also published an algorithm to train deep belief nets faster [19]. They used Restricted Boltzmann machines to initialize weights instead of using random weights. In 2007, Bengio and Hinton showed that deep learning methods were useful for solving difficult problems, that were difficult to solve using shallow networks or SVMs.

In 2006, a few publications by the groups of Hinton, Bengio and LeCun illustrated how DL could learn high-level representation of the data with low error rates. To do this, some methods used multiple layers of computing units using simple gradient descent applied to convolutional networks. Another method used layer-by-layer unsupervised learning, followed by gradient descent. Overfitting was addressed using dropout (i.e., randomly subsampling the data). Fueled by new research ideas, and the confluence of big data and computational power of GPUs there was a new impetus for NN, DL, and the overall field of AI to take off again in the twenty-first century.

ANNs are extremely data hungry. To make full use of their computational power, substantial amounts of data are needed. Whereas MNIST (a database of handwriting digits) was useful for training models for handwriting recognition, newer datasets such as ImageNet by Fei Fei Li's group were published in 2009 [20]. Currently, ImageNet has over 14 million images in over 20,000 categories, which are then further divided into subcategories. But in 2009, it was one of the largest datasets available for training, and therefore a key asset that helped shape the CNN revolution.

Another piece of the puzzle was the required computing power. Since the 1990s computers were getting faster as per the Moore's law. But the CPU parallelism required for computing millions of weights for the nodes in the multitude of layers of the CNNs exceeded what could be effectively done using CPUs. In 2009, Andrew Ng's group reported using GPUs for obtaining multiple orders of magnitude faster computation [21].

In 2010, more insights were gained into the reasons why backpropagation produced vanishing and exploding gradients. Glorot and Bengio identified that the random weights, per se, were not an issue, but rather the logistic sigmoid or hyperbolic tangent activation functions were unsuitable for deep networks [22]. Rectified linear unit (ReLU) activation function was found to be useful for preventing vanishing/exploding gradients. ReLU produces a sparse representation, and only key neurons have a non-zero output for an input, leading to computational efficiency, and parsimonious and robust information.

By 2011, in addition to research in the universities, the power of big data and deep learning was recognized by companies such as Microsoft, Google and IBM. Andrew Ng and Jeff Dean from Google formed Google Brain to build out a huge computational resource of 16,000 CPU cores that could train 1 billion weights. This was three orders of magnitude more than the previous Deep Belief Network (DBN) breakthrough that trained 1 million weights. Separately, thus far, many applications used supervised learning, but one of the applications by Google Brain trained YouTube videos that were not labeled. Using unsupervised learning, at scale, made sense since far more unlabeled data exists today as compared to labeled. In 2012, Andrew Ng and Jeff Dean published research from Google brain to detect cats using high-level features and unsupervised learning [23, 24].

Over the next few years previous ImageNet competition benchmarks were beaten, first in 2012 by AlexNet [25] using a CNN architecture by Krizhevsky, Sutskever and Hinton and having a top-five error rate of 17% - a 9% improvement over the previous year's benchmark. AlexNet was larger and deeper than previous ANNs and used a stack of CNNs on top of each other. Two years later, GoogLeNet [26], by Szegedy et al from Google, won the 2014 ImageNet competition with a top-five error rate of less than 7%. In 2015, ResNet or Residual Network developed by Kaiming He [27] won the ImageNet challenge having a top-five error rate of less than 3.6% (better than human performance estimated at 5.1%). Its architecture included an extremely deep CNN with 152 layers. Some of these highlights of neural network history are captured in Figure 1.

In less than a hundred, starting at the nascent beginning's the field of AI, including ANN and leading to DL, has made huge strides in terms of the performance for games such as chess, GO, identifying objects with better than human performance, RL, etc. Despite many breakthroughs, there are limitations in terms of what exactly it is that the model learns, leading to brittleness as demonstrated by Adversarial examples in Machine Learning. Humans observe and perceive the world, form concepts as they understand it, and integrate their knowledge using the conceptual building blocks and a process of integration. The concepts we form are rich and span many dimensions. In contrast, research efforts continue to discern exactly what a machine learns.

Work is being done in the space of "common sense" learning and Artificial General Intelligence (AGI). Despite the limitations of AI/ANN/DL, the field is embraced in almost all disciplines and applications exist in humanities, genomics, medicine, autonomous driving, finance to name just a few fields. Frameworks such as PyTorch (initial release in 2016), Tensorflow (first released in 2015, with v1.0 in 2017) have been developed making these technologies easily accessible to researchers and practitioners. Mimicking biological neural organization, Hinton, in 2017, proposed Capsule Neural Network to better model hierarchical relationships [28]. Attention models [29] and Transformer NN [30] proposed in 2015 and 2017 provided further breakthroughs in the NLP and language translation. More recently transformer-based models such as XLNet, RoBERTa, GPT-2, and ALBERT have broken previous language benchmarks [31, 32].

Research continues to advance the technology and make the intelligence discerned by the machine comparable to intelligence present in living organisms.

3. Biological Neural Networks (BNN)

Biological Neural Networks (BNNs) started evolving around 600 million years ago. BNNs consist of neurons connected functionally. Neurons are highly specialized cells transmitting information unidirectionally from one part of the body to another. A neuron consists of dendrites, a cell body, and a single axon. The axon terminates at the axon terminus (Figure 2). The end of a neuron can communicate with the dendrites of another neuron through synapses. Neuronal signals are excitatory or inhibitory and are transmitted along the axon using action potentials created by a chemical gradient of sodium and potassium charged ions.

Action potential is an all-or-nothing event that can be considered as a binary signal [33]. The strength of the signal is governed by the number of signals firing, their timing, the location and whether the signal is inhibitory or excitatory. Once the signal reaches the end of the neuron, it communicates with other neurons across the synapse. At the synapse, chemical neurotransmitters are released by the presynaptic neuron, and the signal is passed on to the postsynaptic neuron. Neurons transmit their signals in one direction and can transmit signals over large distances (as for example, in the case of sciatic nerves). There are many types of neurotransmitters that are used in the synapse. They may be excitatory or inhibitory such as epinephrine, norepinephrine, acetylcholine, serotonin, γ-amino butyric acid (GABA), dopamine to name a few. The signal is

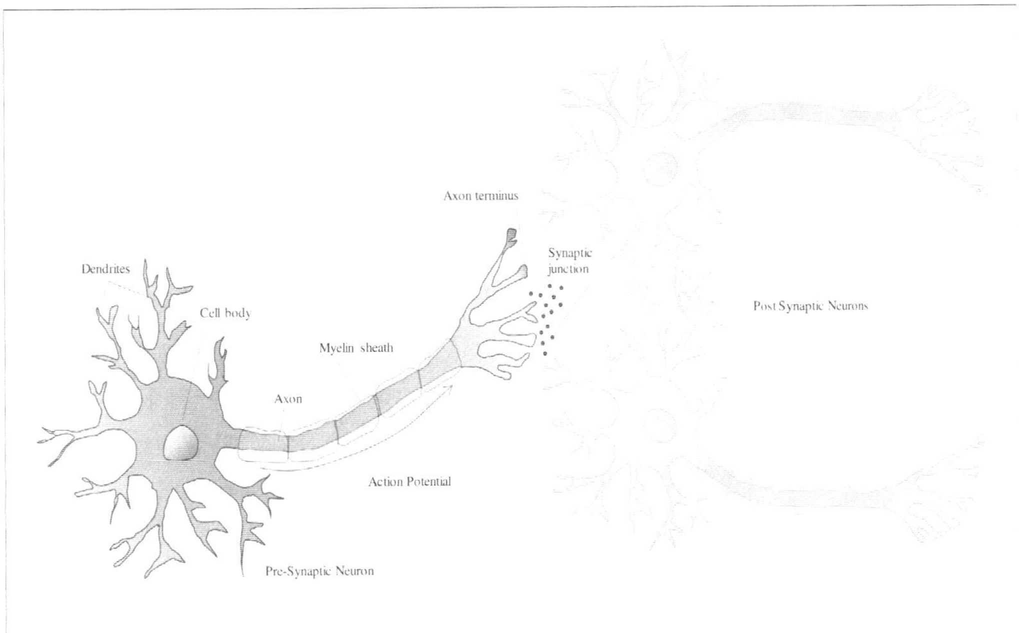

Figure 2. Structural components of Neuron

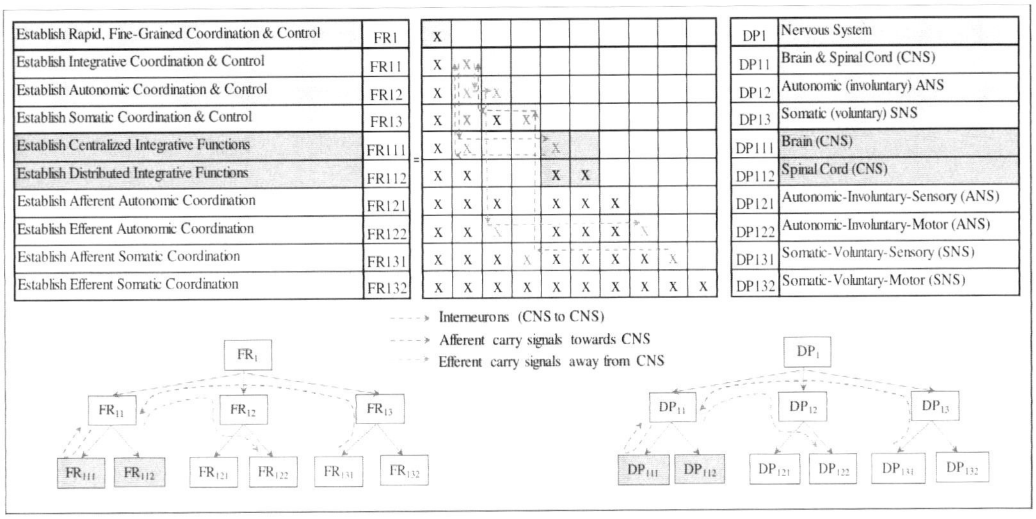

Figure 3. Hierarchic Structure of Biological Neural Systems & Subsystems

terminated by degradation or reuptake of the neurotransmitter.

Functionally, there are three types of neurons - Sensory neurons that carry signals from sensory organs such as eyes, tongue, skin etc to the brain or spinal cord, motor neurons that receive signals from the brains and spinal cord and effect muscle contraction or secretory organs and finally, interneurons connecting neurons to other neurons.

Research estimates there are eighty-six billion neurons in the human brain [34]. Of these, the cerebral cortex alone contains 16 billion neurons linked by about 100 trillion synaptic connections [35].

Primate neural systems are hierarchically organized [36] as shown in (Figure 3). Neural controls are fine-grained and execute rapidly and with high specificity. As discussed earlier, electrical signals (within the neuron) as well as chemical neurotransmitters (between neurons) are used for signaling. In contrast, endocrine controls are coarse-grained (i.e., they have a widespread effect) and use hormones sent via the circulatory system (for example, pancreas that controls the uptake of sugar via insulin secretion).

There are two distinct hierarchies when considering the inherent structure of any of the design-matrices: 1) cross-level, decompositional versus 2) in-level, dominance. For example, when we consider the reverse-engineered, primate neural-system design trace (Figure 3), the decompositional hierarchy is as shown in the tree diagram at the bottom. Each level of this decompositional structure is in effect, the total system. Thus, the totality of the nervous system at level 1 ($FR_1 : DP_1$) is equally captured as the $\{\{FR_{11}, FR_{12}, FR_{13}\}$: $\{DP_{11}, DP_{12}, DP_{13}\}\}$ complex that pertains to the Central Nervous System (CNS), the peripheral Autonomic Nervous System (ANS), as well as the peripheral Somatic Nervous System (SNS). CNS decomposes at the next level into the brain & spinal cord. Automated ANS controls include heart rate modulation, blood pressure modulation, digestion, etc. It occurs below the level of consciousness. Voluntary SNS controls include posture, motion, sense organs (odor, temperature, taste, vision, pain), etc. It occurs within conscious awareness. Within each level, there also exists a dominance hierarchy. Thus, the CNS occupies the dominant (top-left) location within the decoupled green sub-region of the design matrix. This is followed by ANS, and only then by SNS. In-level dominance may be traced by considering the underlying functions that are being performed. Damage to the brain has a widespread effect on the body; in contrast, damage to the peripheral nervous system (i.e., ANS/SNS) generally has a limited, local effect. However, this is not always true; for example, heart-rate modulation failure can directly impact CNS brain function, thereby reversing the in-level dominance. In other words, in-level dominance needs to be considered on a case-by-case basis as the systems are further decomposed. But by and large, the dominance pattern in most cases would be CNS > ANS > SNS.

Now consider how the signaling occurs across the neural architecture. Imagine a hiker who is traversing a wooded mountain landscape as shown in Figure 4b. Suddenly she hears rusting of leaves followed by loud threatening animal noises from behind the trees. These noises trigger her DP_{131} (i.e., her Somatic-Voluntary-Sensory hearing controls) to send afferent alarm signals up the chain of neural hierarchy, and into the emotion-processing CNS component, the amygdala. The amygdala integrates the threat signal with its coded past experience (which also includes evolutionary information) and triggers the adjacent hypothalamus that commands the autonomic nervous system into the "fight-or-flight" mode. Her heart rate increases to supply the blood vessels with increased oxygen. The circulatory system that services the body musculature is dilated while that which supplies the digestive system is constricted. This action diverts the blood to the muscles where it is urgently needed. Eyes dilate to increase light capture and improve visual acuity for better threat perception. The bronchi in her lungs dilate to increase the oxygen capture. In effect, many of her DP_{122} controls (i.e., her Autonomic-Involuntary-Motor controls) are now fully geared up for the "fight-or-flight" action.

Consider now the visual threat processing by the hiker. The visual cortex processes visual input and has inspired many NN, DL and CNN architectures. The entry point for the visual signal is the eye. Here, the five cell types in the retina convert photons into electrical signals. It starts at the photoreceptors and ends in Retinal ganglion cells (RGC). RGCs are involved in edge extraction from an image. The axons of the ganglion cells in each eye use the optic nerve to transmit signals to the primary visual cortex (V1) via the lateral geniculate nuclei (LGN).

V1 is responsible for the first stage of image processing. V1 has six layers that receive inputs

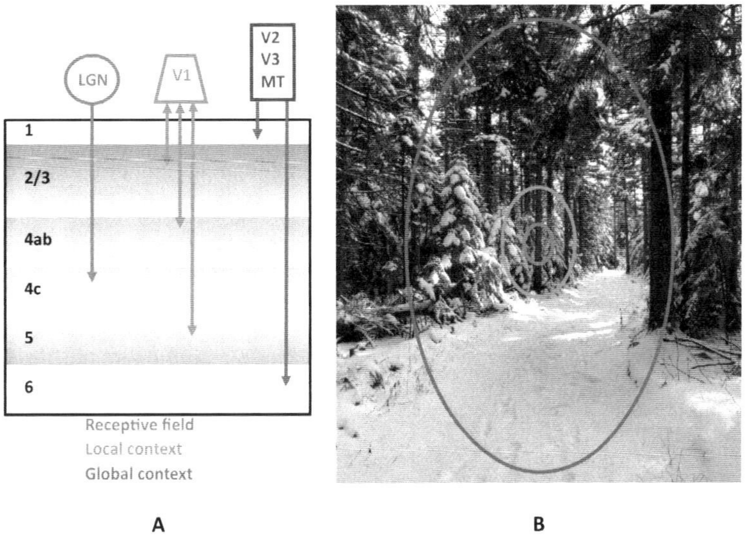

Figure 4. A: Layers of primary visual cortex inputs B: Visual scene with Receptive field, Local and Global surround

from different parts of the brain. These layers are stacked and are hierarchically connected. Receptive fields in layer 4c of V1 are activated in response to the smallest stimuli receiving input from LGN (Figure 4A) [37]. This is depicted using the red circle in Figure 4B. To understand the visual scene being processed, the local context is created by long-range horizontal connections within V1 (green circle in Figure 4B) that add to the information of the receptive field. Higher-level cortical structures such as V2, V3 and MT that receive higher-order visual information directly or indirectly from V1, also provide spatially extensive feedback to layers 1 and 6 of V1, thus enabling it to construct the global information of the scene (colored blue in Figure 4A, B).

4. Evolution of the Multi-Layer Perceptron

As mentioned in the historical review, in 1958 Rosenblatt proposed the perceptron model to emulate the neural computing machinery. He was then the director of the Cognitive Systems Research Program at the Cornell University Ithaca campus. In 1960, the model that Rosenblatt proposed was implemented in hardware as Mark 1 Perceptron. On July 8th, 1958, the U.S Office of Naval Research demonstrated the machine. After a few trials, the machine was able to separate out punched cards marked on the left versus those on the right. Based on this demonstration, The New York Times [38], reported that "[it] learned to differentiate between right and left after fifty attempts. The Navy said the perceptron would be the first non-living mechanism 'capable of receiving, recognizing, and identifying its surroundings without any human training or control'."

A decade later, in 1968, Minsky and Papert [6] evaluated the perceptron model and highlighted certain edge-case deficiencies (such as the replication of the XOR gate, or the issue of recognizing connected figures). The Minsky/Papert critique was based on a simplified model (i.e., a single layer neural network) of Rosenblatt's original perceptron which was not likewise limited. But the mathematical rigor of the critique was sufficient to dry up the funding for Rosenblatt's approach that attempted to emulate the nervous system in a bottom-up fashion of sensation to perception to conceptions. Instead, post-Minsky/Papert critique, research funding moved into a top-down symbolic approach to AI. This later led to the AI winter as the top-down symbolic approach ran into its own difficulties. It has taken 60 years to recognize the pioneering foundations laid by Prof. Rosenblatt. As recounted in [39] by the faculty at Cornell: "Today, many believe Rosenblatt has been vindicated. The principles underlying the perceptron helped spark the modern artificial intelligence revolution. Deep learning and neural networks - which can classify online images, for example, or enable language translation - are transforming society. 'The perceptron was the first neural network,' said Thorsten Joachims, Professor in CIS".

In this section, we review the perceptron logic and link it up with the Axiomatic Design (AD) approach. Fundamental principles of neural network design may be gleaned by reviewing Prof. Rosenblatt's perceptron design using the AD framework. The focus is on the bottom-up formation of the FR-DP hierarchy as it relates to the XOR/XNOR roadblocks that the Minsky/Papert critique highlighted. As reported in [40] by Mengistu et al: "Hierarchical organization-the recursive composition of sub-modules-is ubiquitous in biological networks, including neural, metabolic, ecological, and genetic regulatory networks, and in human-made systems, such as large organizations and the Internet.... However, an open, important question in evolutionary biology is why hierarchical organization evolves in the first place." By including the cost of connection along with performance in solving an AND-XOR-AND via ANN, they were able to show that [40]:

(i) Networks that do not price connection costs do not evolve to be hierarchical.
(ii) But with connection-costs being priced in into the build-up of the ANN, the "networks evolve to be both modular and hierarchical."
(iii) They exhibit higher "overall performance and evolvability (i.e., faster adaptation to new environments)."
(iv) Also, "hierarchy independently improves adaptability after controlling for modularity."
(v) In other words, minimizing "the cost of connections-promotes the evolution of both hierarchy and modularity."

But from an AD perspective, the cost of network connection is a proxy for system-level information content as captured in Axiom II. This seems to suggest that Axiom II can give you both modularity and hierarchy; if so, why then do we need Axiom I? In other words, is Axiom I redundant perhaps? And the answer is no. This is because it is Axiom I that helps bridge the is-ought gap between the FR and DP domains. Without the functional domain, form is meaningless. Also, with hierarchy comes the need for cascade induction. Again, without Axiom I, we cannot do the necessary abductive cascade along the design trace (See Figure 5). This agrees with the design hierarchy as well as the zig-zag rule between the FR and DP domains as outlined in AD.

Axiomatic Design highlights the hierarchical structure of design. As Prof. Suh highlights in two separate instances in [41]:

- "Everything we do in design has a hierarchical nature to it. That is, decisions must be made in order of importance by decomposing the problem into a hierarchy.... When such a hierarchical nature of decision making is not utilized, the process of decision making becomes very complex."
- "The designer must recognize and take advantage of the existence of the functional and physical hierarchies. A good designer can identify the most important FRs at each level of the functional tree by eliminating secondary factors from consideration. Less-able designers often try to consider all the FRs of every level simultaneously, rather than making use of the hierarchical nature of FRs and DP's."

Hierarchies often show up in both biological [42] as well as sociotechnical systems [43]. Capturing the functional roots of these hierarchies may be of significance.

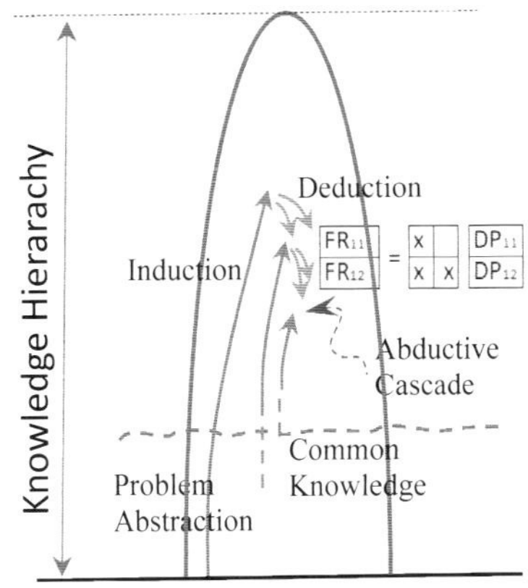

Figure 5. Design Trace along the Knowledge Hierarchy

x1	x2	p1	p2	p3	p4	p5	p6	p7	p8
0	0	0	1	0	1	0	1	0	1
0	1	0	0	1	1	0	0	1	1
1	0	0	0	0	0	1	1	1	1
1	1	0	0	0	0	0	0	0	0
Symbolic		(x1 ∧ ¬x1)	¬(x1 ∨ x2)	¬(x2 → x1)	¬x1	¬(x1 → x2)	¬x2	(x1 ⊕ x2)	¬(x1 ∧ x2)
Equivalence		0	¬x1 ∧ ¬x2	(x2 ∧ ¬x1)	¬¬¬x1	(x1 ∧ ¬x2)	¬¬¬x2	(p3 ∨ p5)	¬x1 ∨ ¬x2

Linear Separability

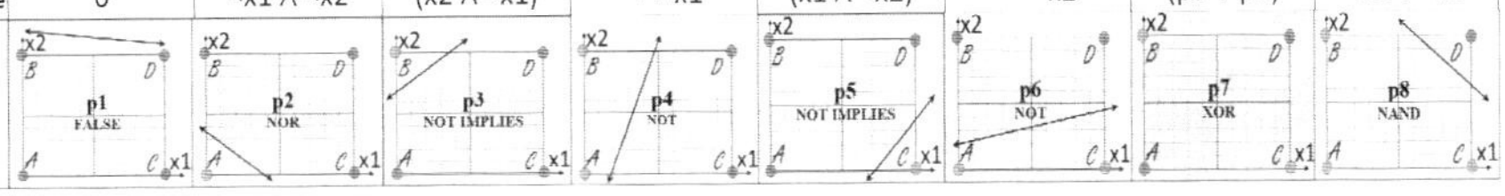

x1	x2	p9	p10	p11	p12	p13	p14	p15	p16
0	0	0	1	0	1	0	1	0	1
0	1	0	0	1	1	0	0	1	1
1	0	0	0	0	0	1	1	1	1
1	1	1	1	1	1	1	1	1	1
Symbolic		(x1 ∧ x2)	(x1 ≡ x2)	x2	(x1 → x2)	x1	(x2 → x1)	(x1 ∨ x2)	(x1 ∨ ¬x1)
Equivalence		(x2 ∧ x1)	(p2 ∨ p9)	x2 ∧ 1	¬(x1 ∧ ¬x2)	x1 ∨ 0	¬(x2 ∧ ¬x1)	(x2 ∨ x1)	1

Linear Separability

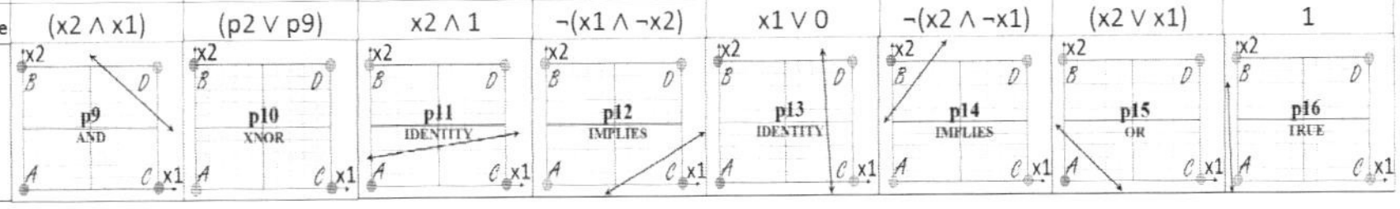

Figure 6. Linear Separability of Single-Layer, Planar Perceptron

In both BNN's and ANN's there are concrete artifacts that correspond to the higher-level design constructs. This is because, unlike the design of a hammer or a screw, these artifacts have adaptability, intelligence and control embedded in them. In other words, the whole design trace is actively being orchestrated and often being modified on an ongoing basis. For example, the central nervous system that sits at a higher level (as compared to the subordinate peripheral nervous system) is not just an abstraction on a conceptual hierarchy; it has a real biological presence. It is in this sense that such design traces may be considered as intelligent, dynamic and alive than being static, final and "once and for all" as is the case with an artifact such as a hammer that lacks embedded intelligence.

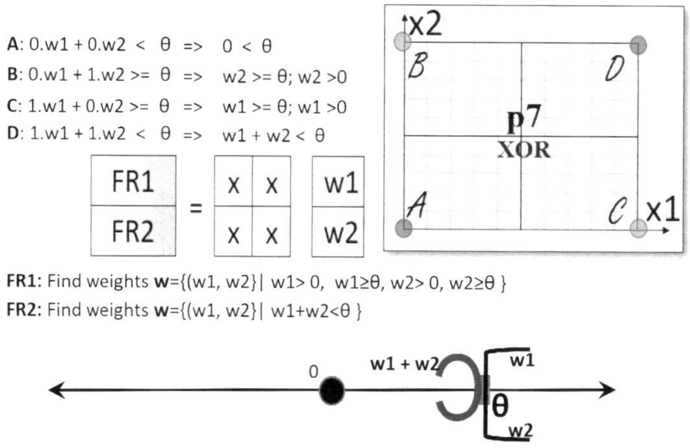

Figure 7. Coupled Single-Layer Perceptron XOR Design

Since nature does not do conceptual induction (and therefore, neither cascade induction), could it not be said that Axiom I is redundant as far as nature is concerned? While it is true that nature does not indulge in conceptual induction, the Darwinian evolutionary principle of "survival of the fittest" captures the essence of Axiom-I's "form follows function" (i.e., DP follows FR). But it is true that nature does not do this in a conceptual sense; instead, it does it painstakingly across billions of years of slowly evolving hit-and-trials, that uses and discards millions of species which have now come and gone. The amount of time it takes biological systems to evolve and adapt to changing natural conditions (even taking into consideration the punctuated equilibrium-style expedited evolution [44] takes around 100,000 years) is orders of magnitude more than what the conceptual approach offers. Using tools such as ANN, this is a definite cost that humans could help short-circuit via conceptual induction (i.e., both direct and cascade) in order to aid Axiom-I. One of the intriguing design questions that the above research as reported in [40] triggers is the following: Does the optimum number of hidden layers correspond to the optimum number of hierarchic levels? For example, with respect to the XOR/XNOR gate problem (which is discussed in detail, as below), it is established that it cannot be solved with a single perceptron layer; it needs at least two. But it may also be solved with more than two layers. Thus, (similar to Occam's Razor) is there a corollary to Axiom II that suggests that we minimize the number of hierarchic levels to the barest minimum required?

With the above context in mind, lets now return to the bottom-up formation of the FR-DP hierarchy as it relates to the XOR/XNOR roadblocks that the Minsky/Papert critique highlighted. A perceptron is a computing element with a set threshold θ and n inputs x: x1, x2, ...xn, each of which is associated weights w: w1, w2, ...wn such that:

- If w . x $\geq \theta$, then the perceptron outputs 1 (i.e., it fires)
- If w . x $< \theta$, then the perceptron outputs 0 (i.e., it is dormant)

Geometrically, the perceptron separates the input space x into two half-spaces; one part that triggers the perceptron firing while the other keeps it dormant. For a two-dimensional input vector, the 16 possible perceptron elements are as shown in Figure 6. To simplify the diagrams, the vectors x1 and x2 are limited to 0,1 (i.e., they compose to form the four corner joints A, B, C, D). Thus, perceptron p2 is a straight line that separates out corner A (which is coloured green

IOP Conf. Series: Materials Science and Engineering 1174 (2021) 012023 doi:10.1088/1757-899X/1174/1/012023

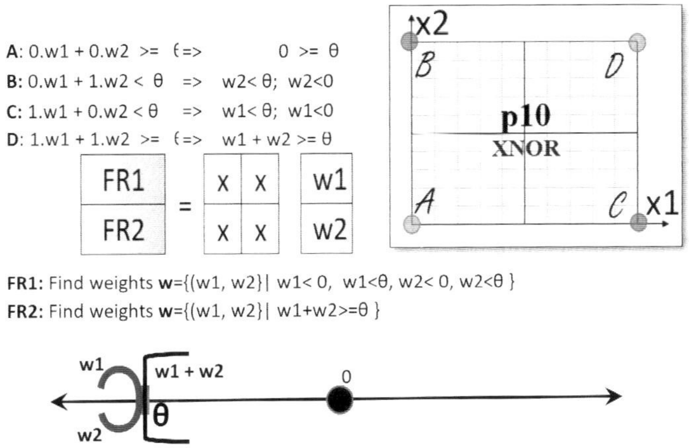

Figure 8. Coupled Single-Layer Perceptron XNOR Design

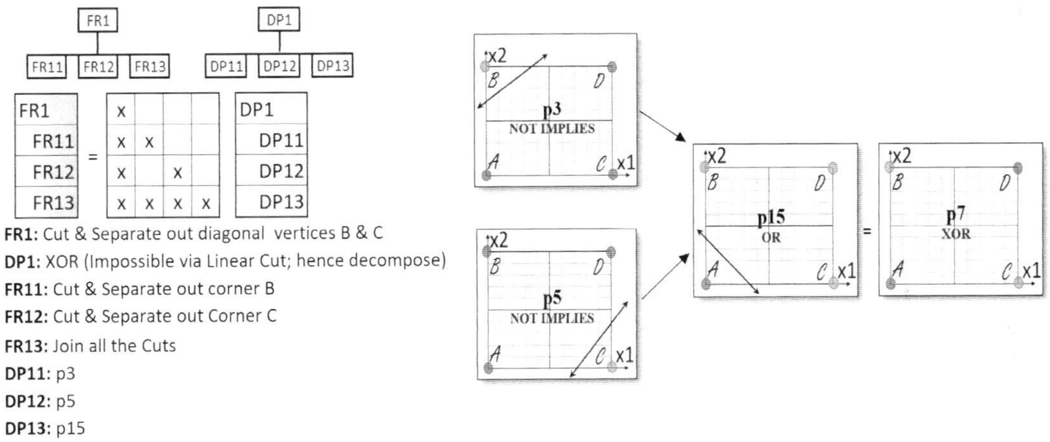

Figure 9. Two Layer Decomposed and Decoupled XOR Perceptron

to indicate that it fires) as against corners B, C & D (which are coloured red to indicate that they will not fire). For a three-dimensional input space, the separator perceptron is a dividing plane (and, so on). As indicated in the perceptron-grid in Figure 6), perceptron's p7 & p10 are different in the sense that there is no linear element that could select the two green cornered diagonals on one side while leaving the rejected red corners to the remaining half. This is one of the problems that the Minsky/Papert critique highlighted.

When analysed using AD (as shown in Figures 7 and 8), both p7 and p10 turn out to be coupled designs with the added frustration that there is no feasible design-space.

But once the constraint of a single layer perceptron is lifted, it is now possible to design a two-layered perceptron by judiciously combining some of the 16 elements (as shown in Figure 9). In step 1, the required corners are appropriately selected. Then they are combined with a simple OR gate. As shown in Figures 9 and 10, the two designs are hierarchical and decoupled. The

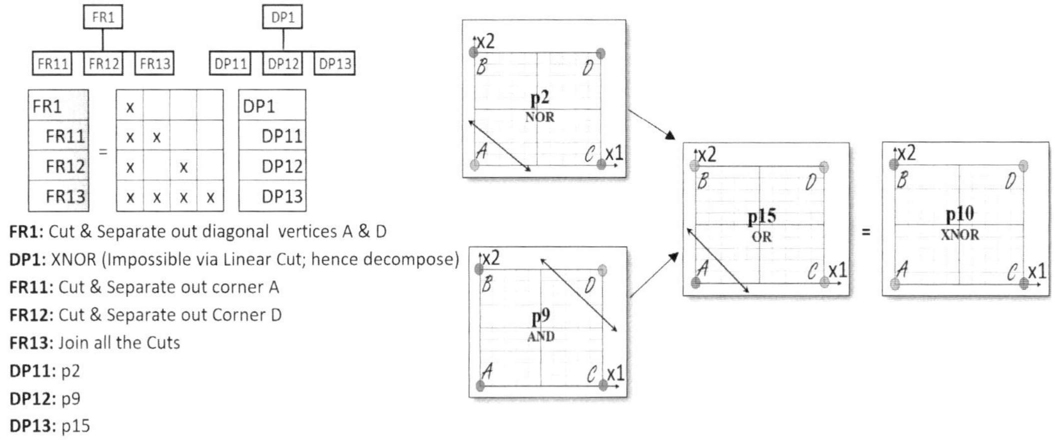

Figure 10. Two Layer Decomposed and Decoupled XNOR Perceptron

fact that the designs are hierarchical and arise out of the need to overcome the blocks that XOR/XNOR gates pose, is salient.

5. Conclusions

In their concluding remarks on Perceptrons, Minsky & Papert state the following [6]: "Theorems are theorems, and the history of science amply demonstrates how discovering limiting principles can lead to deeper understanding."

In this paper, we wanted to show the strategic advantage of viewing the neural-network design from an AD perspective. Some of the useful insights and research directions about the design of neural networks (or for that matter, design in general) that may be gleaned from the above exercise include:

(i) Hierarchies in the design-trace are naturally triggered when existing designs exhibit inherent limits.

(ii) Tracking both form and function (i.e., DP's & FR's) as suggested by AD could aid the evolution of the design trace.

(iii) There may be overarching design laws & corollaries that provide advice regarding the minimum/maximum number of hierarchical levels.

(iv) The number of levels in the hidden layer of the Artificial Neural Network may be linked to the number of hierarchic layers in the design trace.

(v) How would the above correlate with the current understanding about hierarchies in Biological Neural Networks?

Items ii-v above need to be further explored. We hope to cover these in a follow-up report.

References

[1] McCulloch W S and Pitts W 1943 A logical calculus of the ideas immanent in nervous activity *The bulletin of mathematical biophysics* **5** 115–133
[2] Hebb D O 1949 *The organization of behavior: A neuropsychological theory* (Science Editions New York)
[3] Rosenblatt F 1957 *The perceptron, a perceiving and recognizing automaton Project Para* (Cornell Aeronautical Laboratory)

[4] Widrow B 1960 *Adaptive "adaline" Neuron Using Chemical "memistors".* (Stanford Electronics Laboratories Technical Report 1553-2)
[5] Hubel D H and Wiesel T N 1962 Receptive fields, binocular interaction and functional architecture in the cat's visual cortex *The Journal of physiology* **160** 106–154
[6] Minsky M and Papert S A 1969 *Perceptrons: An introduction to computational geometry* (MIT press)
[7] Linnainmaa S 1970 The representation of the cumulative rounding error of an algorithm as a taylor expansion of the local rounding errors *Master's Thesis, Univ. Helsinki*
[8] Werbos P 1974 Beyond regression: new tools for prediction and analysis in the behavioral sciences *Ph. D. dissertation, Harvard University*
[9] Rumelhart D E, Hinton G E and Williams R J 1986 Learning representations by back-propagating errors *Nature* **323** 533–536
[10] Fukushima K 1980 Neocognitron: A self-organizing neural network model for a mechanism of pattern recognition unaffected by shift in position *Biological Cybernetics* **36** 193–202 https://doi.org/10.1007/BF00344251
[11] LeCun Y, Boser B, Denker J S, Henderson D, Howard R E, Hubbard W and Jackel L D 1989 Backpropagation applied to handwritten zip code recognition *Neural computation* **1** 541–551
[12] Ackley D H, Hinton G E and Sejnowski T J 1985 A learning algorithm for boltzmann machines *Cognitive science* **9** 147–169
[13] Neal R M 1992 Connectionist learning of belief networks *Artificial intelligence* **56** 71–113
[14] Dayan P, Hinton G E, Neal R M and Zemel R S 1995 The helmholtz machine *Neural computation* **7** 889–904
[15] Waibel A, Hanazawa T, Hinton G, Shikano K and Lang K J 1989 Phoneme recognition using time-delay neural networks *IEEE transactions on acoustics, speech, and signal processing* **37** 328–339
[16] Bengio Y 1993 *Advances in Pattern Recognition Systems Using Neural Network Technologies* (World Scientific) pp 3–23
[17] Hochreiter S and Schmidhuber J 1997 Long short-term memory *Neural computation* **9** 1735–1780
[18] Bengio Y, Ducharme R, Vincent P and Janvin C 2003 A neural probabilistic language model *The journal of machine learning research* **3** 1137–1155
[19] Hinton G E, Osindero S and Teh Y W 2006 A fast learning algorithm for deep belief nets *Neural computation* **18** 1527–1554
[20] Deng J, Dong W, Socher R, Li L J, Li K and Li F F 2009 Imagenet: A large-scale hierarchical image database *2009 IEEE conference on computer vision and pattern recognition* (Ieee) pp 248–255
[21] Raina R, Madhavan A and Ng A Y 2009 Large-scale deep unsupervised learning using graphics processors *Proceedings of the 26th annual international conference on machine learning* pp 873–880
[22] Glorot X and Bengio Y 2010 Understanding the difficulty of training deep feedforward neural networks *Proceedings of the thirteenth international conference on artificial intelligence and statistics* (JMLR Workshop and Conference Proceedings) pp 249–256
[23] Dean J, Corrado G, Monga R, Chen K, Devin M, Mao M, Ranzato M a, Senior A, Tucker P, Yang K, Le Q and Ng A 2012 Large scale distributed deep networks *Advances in Neural Information Processing Systems* vol 25 ed Pereira F, Burges C J C, Bottou L and Weinberger K Q (Curran Associates, Inc.) https://proceedings.neurips.cc/paper/2012/file/6aca97005c68f1206823815f66102863-Paper.pdf
[24] Le Q V, Ranzato M, Monga R, Devin M, Chen K, Corrado G S, Dean J D and Ng A Y 2012 Building high-level features using large scale unsupervised learning *Proceedings of ICML* pp 81–88 http://research.google.com/archive/unsupervised_icml2012.pdf
[25] Krizhevsky A, Sutskever I and Hinton G E 2012 Imagenet classification with deep convolutional neural networks *Advances in neural information processing systems* **25** 1097–1105
[26] Szegedy C, Liu W, Jia Y, Sermanet P, Reed S, Anguelov D, Erhan D, Vanhoucke V and Rabinovich A 2015 Going deeper with convolutions *Proceedings of the IEEE conference on computer vision and pattern recognition* pp 1–9
[27] Xie S, Girshick R, Dollár P, Tu Z and He K 2017 Aggregated residual transformations for deep neural networks *Proceedings of the IEEE conference on computer vision and pattern recognition* pp 1492–1500
[28] Sabour S, Frosst N and Hinton G E 2017 Dynamic routing between capsules *arXiv preprint arXiv:1710.09829*
[29] Chorowski J, Bahdanau D, Serdyuk D, Cho K and Bengio Y 2015 Attention-based models for speech recognition *arXiv preprint arXiv:1506.07503*
[30] Vaswani A, Shazeer N, Parmar N, Uszkoreit J, Jones L, Gomez A N, Kaiser Ł and Polosukhin I 2017 Attention is all you need *Advances in neural information processing systems* pp 5998–6008
[31] Yang Z, Dai Z, Yang Y, Carbonell J, Salakhutdinov R R and Le Q V 2019 Xlnet: Generalized autoregressive pretraining for language understanding *Advances in neural information processing systems* **32**
[32] Topal M O, Bas A and van Heerden I 2021 Exploring transformers in natural language generation: Gpt, bert, and xlnet *arXiv preprint arXiv:2102.08036*
[33] Luo L 2015 *Principles of neurobiology* (Garland Science)

[34] Herculano-Houzel S 2009 The human brain in numbers: a linearly scaled-up primate brain *Frontiers in human neuroscience* **3** 31
[35] Azevedo F A, Carvalho L R, Grinberg L T, Farfel J M, Ferretti R E, Leite R E, Filho W J, Lent R and Herculano-Houzel S 2009 Equal numbers of neuronal and nonneuronal cells make the human brain an isometrically scaled-up primate brain *Journal of Comparative Neurology* **513** 532–541
[36] Saladin K 2018 Anatomy and physiology: The unity of form and function
[37] Bijanzadeh M, Nurminen L, Merlin S, Clark A M and Angelucci A 2018 Distinct laminar processing of local and global context in primate primary visual cortex *Neuron* **100** 259–274
[38] Times T 1958 New navy device learns by doing *Psychologist Shows Embryo of Computer Designed to Read and Grow Wiser* https://www.nytimes.com/1958/07/08/archives/new-navy-device-learns-by-doing-psychologist-shows-embryo-of.html (Accessed 2021-04-04)
[39] Lefkowitz M 2019 Professor's perceptron paved the way for ai - 60 years too soon *Cornell Chronicle* https://news.cornell.edu/stories/2019/09/professors-perceptron-paved-way-ai-60-years-too-soon (Accessed 2021-04-04)
[40] Mengistu H, Huizinga J, Mouret J B and Clune J 2016 The evolutionary origins of hierarchy *PLoS computational biology* **12** e1004829
[41] Suh N P 1990 *The principles of design* 6 (Oxford University Press)
[42] Mantri P and Thomas J 2019 Nature's design's: The biology of survival *MATEC Web of Conferences* vol 301 (EDP Sciences) p 00023 https://doi.org/10.1051/matecconf/201930100023 (Accessed 2021-04-14)
[43] Thomas J and Mantri P 2019 Axiomatic cloud computing architectural design *MATEC Web of Conferences* vol 301 (EDP Sciences) p 00024 https://doi.org/10.1051/matecconf/201930100024 (Accessed 2021-04-14)
[44] Gould S J 2009 *Punctuated equilibrium* (Harvard University Press)

IOP Conf. Series: Materials Science and Engineering 1174 (2021) 012024 doi:10.1088/1757-899X/1174/1/012024

Increasing the performance of a Salt Washing Machine through Axiomatic Design

António Gabriel-Santos[1, 2, 5], João Fradinho[1, 2], António Mourão[1, 2], Alberto Martinho[1, 2], Miguel Cavique[1, 3], Gonçalo Vieira[4], António Gonçalves-Coelho[1, 2]

[1] R&D Unit for Mechanical and Industrial Engineering - UNIDEMI, NOVA School of Science and Technology, Campus de Caparica, 2829-516 Caparica, PORTUGAL
[2] Department of Mechanical and Industrial Engineering - DEMI, NOVA School of Science and Technology, Campus de Caparica, 2829-516 Caparica, PORTUGAL.
[3] Naval Academy, Base Naval de Lisboa, Alfeite, 2810-001 Almada, PORTUGAL
[4] Mano – Equipamentos de Movimentação e Automação, S. A., Rua D. Luís de Noronha, 2635-380 Rio de Mouro, PORTUGAL
[5] Corresponding author, mailto:agms@fct.unl.pt

Abstract. Systems with fewer design parameters (DPs) than functional requirements (FRs) are coupled or cannot achieve all their FRs simultaneously, as per Axiomatic Design's theorem 1. In that case, production processes have lower productivity and production rate than uncoupled or decoupled designs. This paper reports the upgrading of a salt washing machine concerning its performance, not only on separation of salt from saltwater but also on washed salt production. The machine analysis showed it is a coupled design, with one more functional requirement than the design parameters. A viable enhancement was the addition of one design parameter, which made it possible to control separation and washing independently. This redesign enforced functional independence and allowed increasing the production rate between 20% and 30%. Furthermore, one could perceive increased productivity, quality of the final product, and reduced water and energy consumption.

1. Introduction

The engineering of production systems encompasses two main areas of development: the production process (manufacturing technologies and materials flow and handling) and the production management (flow of information - signals). One must design these two areas to perform according to the stated needs [1]. Furthermore, a good production management requires that the selected conceptual solutions of the production process have a set of independent functional requirements, which is a means to ensure the robustness of solutions.

On the other hand, any manufacturing system aims to maximize productivity [1], as defined by equation (1), according to [2].

$$\text{Productivity} = \frac{\text{Total added value - Production costs}}{\text{Total investment}} \tag{1}$$

Productivity grows by increasing total added value or decreasing either production costs or investment.

Regarding business operations, the added value of a product (or process or service) is brought by its functions, performances, suitability, quality, and price. The added value results from actions in several areas related to marketing and sales, product development, manufacturing, assembly, quality, and logistics.

As for manufacturing processes, two modes allow obtaining the added value: 1) better use of available time, which lets getting more value per time through the increase of produced quantities; 2) quality of manufacturing operations, which enable reaching the specified functions right at first time, thus making the production time more effective [1].

In a production system where the number of design parameters is less than the number of functional requirements, either a coupled design results or the functional requirements may only be partially satisfied, as per Axiomatic Design's theorem 1.

In such a situation, the production process has lower productivity than a decoupled design. The decoupled design allows the adjustment of each design parameter to obtain the best achievement for each functional requirement.

This paper shows the successful use of Axiomatic Design (AD) to find the source of the low level of productivity of a salt washing machine and to design an improved solution.

2. Axiomatic Design at a Glance

In the Axiomatic Design (AD) terminology, the depiction of the world of any design object is made in any one of four design domains: the customer, the functional, the physical, and the process domains [2]. These domains are shown in figure 1, and their contents can be described as follows:

"Customer Domain": contains the Customer Needs (CNs), *i.e.* the attributes that the customer seeks in the product or in the system that must be designed;

"Functional Domain": contains the Functional Requirements (FRs) of the design object. In a good design, they are the minimum set of independent requirements that completely describe the functional needs of the design solution, which must be defined in a solution-neutral manner;

"Physical Domain": contains the Design Parameters (DPs) of the design solution. The DPs are the elements of the design solution that are chosen to satisfy the specified FRs;

"Process Domain": contains the Process Variables (PVs) that characterize the production process of the design solution, *i.e.*, the variables that allow attaining the specified DPs.

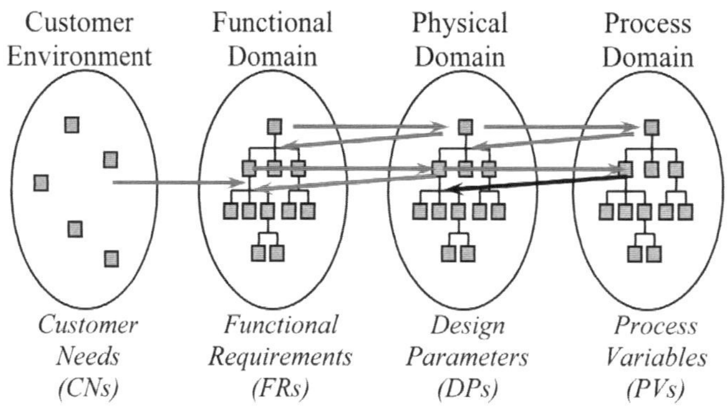

Figure 1. Axiomatic design domains, their contents and relationships (adapted from [3]).

As shown in figure 1, for each pair of adjacent domains, the domain at the left represents "What is required to achieve", or the goals, while the domain at the right represents "How to achieve the goals", or the design solution [2]. That is done by mapping between the goals and the way to achieve them. The extant constraints to this mapping are the bounds for the acceptable solutions and are classified as "input" and "system" constraints [2]. The input constraints are known since the outset of the design process, and the system constraints are found after the outset.

For instance, the mapping between the FRs and DPs is defined by equation (2), the "design equation", where {FR} is the "FR vector", {DP} is the "DP vector", and [A] is the "design matrix"

$$\{FR\} = [A] \cdot \{DP\}, \quad A_{ij} = \frac{\partial FR_i}{\partial DP_j}, \quad i = 1..n, j = 1..m \tag{2}$$

Another unique concept in AD is the hierarchical decomposition through zigzagging between contiguous domains. That advances in a top-bottom way, from the system level and continuing through levels of more detail, as shown in figure 1.

But the fundamental hypothesis of AD is that there are two fundamental principles that govern good design practice [2]:

The Independence Axiom (the first axiom):

Maintain the independence of functional requirements.

This means that each FR should be such that adjusting each DP just disturbs the accomplishment of one of the FRs.

The Minimum Information Axiom (the second axiom):

Minimize the information content of the design.

The purpose of this axiom is to help in finding out the alternative design solution that has minimum information, which is the one that has the highest probability of achieving the FRs.

The independence axiom and the specific constraints should be applied to the design equation during the decomposition, as to ensure that an "uncoupled" or a "decoupled" design solution is obtained at each level of the design process. Uncoupled and decoupled solutions are characterized by diagonal and triangular design matrices, respectively [2]. Any other shape of the squared design matrix corresponds to a "coupled" design solution that should be avoided. Since the design process does not lead to a unique solution, the information axiom should be used to compare the alternative solutions that were previously found and select the alternative design solution with the highest probability of achieving the FRs [4].

3. The Case Study

We aimed at improving the existing salt washing machines in a sea industrial solar saltern. According to AD, one has to carefully define the design's main goals at the onset of the process. The design process can proceed only after stating those goals clearly [5].

3.1. The industrial process of reclaiming and washing salt

In the industrial process of sea salt harvest formed in traditional crystallizer ponds, a mixture of salt crystals and clay is obtained, with the clay particles coming from the bottom of those ponds. Figure 2 shows some overall aspects of the industrial process of sea salt harvesting.

Figure 2. The industrial sea salt harvesting process.

Therefore, there is the need to separate the clay particulate from the salt crystals.

The process of forming crystalline structures warrants there is no clay inside the salt crystals. The typical way to separate the salt from the clay is to dilute the mixture through mechanical agitation in a liquid phase that does not dissolve the particulate. The liquid phase commonly used is an aqueous solution saturated with salt, called brine. Brine cannot dissolve more salt, thus keeping the balance between the salt that will likely dissolve and the salt that recrystallizes. Slow-speed controlled precipitation enables separation.

Figure 3 shortly depicts the salt washing process through shaking. Fast shaking enables diluting the mixture, while slow agitation allows separating salt crystals from clay particulate.

Figure 3. Scheme of the process of washing the salt.

The machines traditionally used to perform salt washing are made up of a tank where the salt with clay is diluted in salt-saturated brine. A screw conveyor shakes the mixture to disperse the clay and the salt crystals in the brine. The screw conveyor rotation motion is also meant to separate the salt from the clay through separate drainage. The salt is removed from the washing tank through the underflow outlet at the higher side of the tank, while the brine with clay leaves the tank through overflow outlet located at the lower side of the tank. Some examples of existing salt washing machines are shown in figure 4.

Figure 4. Examples of salt washing machines with screw conveyors.

3.2. Axiomatic Design analysis

Figure 5 shows the main customer need: to separate the clay from the valuable washed salt.

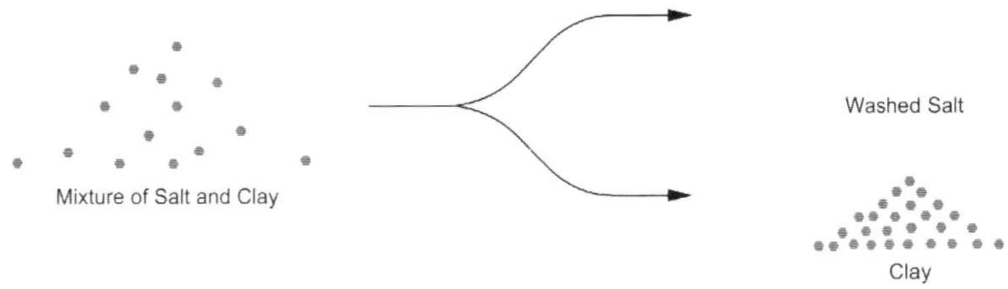

Figure 5. Customer needs to separate the clay from the salt – Customer Domain.

Therefore, the main functional requirements of the salt washing process are the following:

At the first hierarchical level, we have:
FR_1 – Wash salt

At the second hierarchical level, we have:
$FR_{2.1}$ – Accept mixture (salt with clay)
$FR_{2.2}$ – Dilute mixture
$FR_{2.3}$ – Extract clay
$FR_{2.4}$ – Segregate salt

At the third hierarchical level, we have:
$FR_{2.1.1}$ – Collect washed salt
$FR_{2.1.2}$ – Collect clay

As system constraint, we do not want to dissolve the salt.
Schematically, these operations (FRs) can be represented as shown in Figure 6.

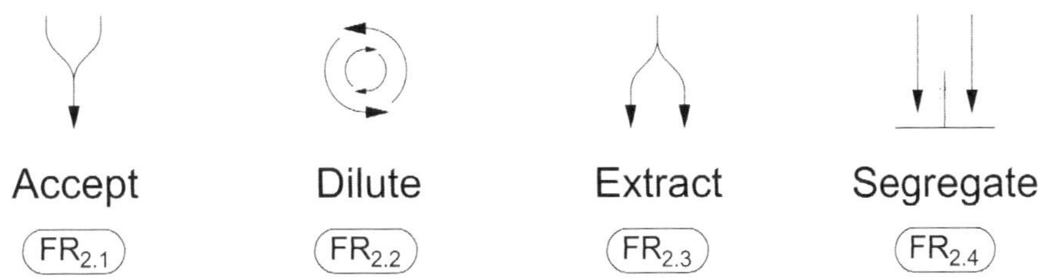

Figure 6. Functional requirements to separate the clay from the salt – Functional Domain.

The perceived main design parameters for the case of traditional salt washing machines are the following:

At the first hierarchical level, we have:
DP_1 – Salt Washing Machine

At the second hierarchical level, we have:
$DP_{2.1}$ – Washing Tank
$DP_{2.2}$ – Brine
$DP_{2.3}$ – Screw conveyor

At the third hierarchical level, we have:
$DP_{2.1.1}$ – Tank's underflow outlet
$DP_{2.1.2}$ – Tank's overflow outlet

IOP Conf. Series: Materials Science and Engineering 1174 (2021) 012024 doi:10.1088/1757-899X/1174/1/012024

Figure 7 represents the solutions (DPs) to perform the above-said actions (FRs).

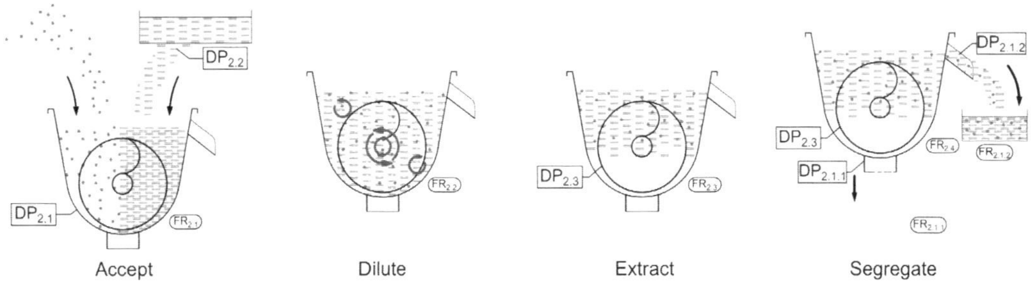

Figure 7. Design parameters and functional requirements of a traditional salt washing machine.

Figure 8 shows the functions as they occur along the salt washing machine.

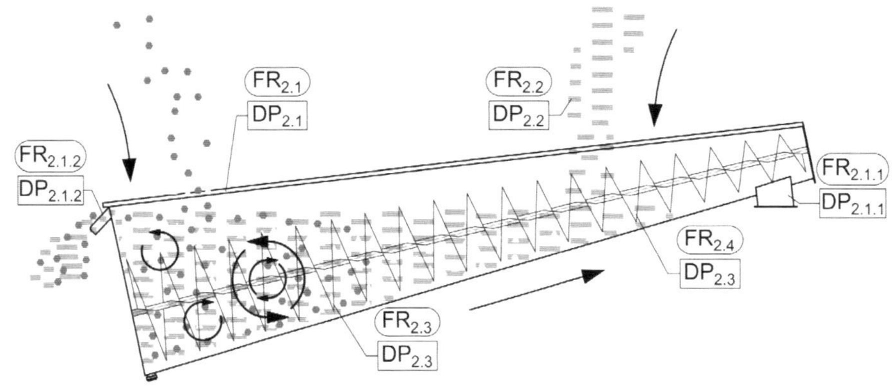

Figure 8. Design parameters and functional requirements along a traditional salt washing machine.

Figure 9 displays a traditional salt washing machine.

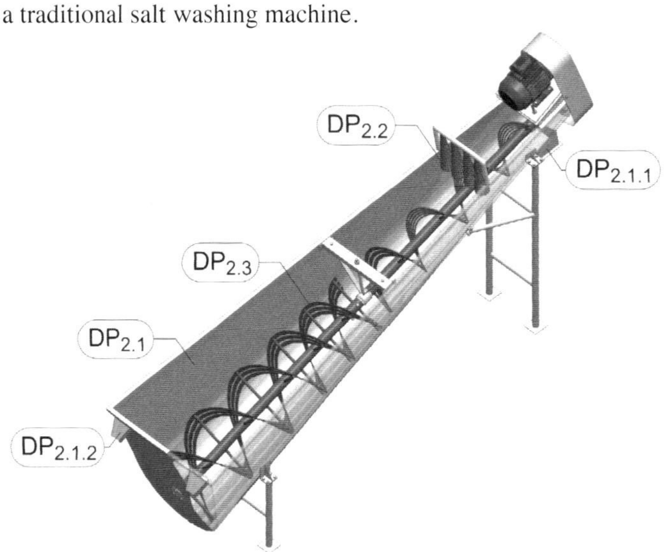

Figure 9. Design parameters of a salt washing machine (Physical Domain).

IOP Conf. Series: Materials Science and Engineering 1174 (2021) 012024 doi:10.1088/1757-899X/1174/1/012024

In traditional washing machines, the same DP, the rotating shaft, implements two FRs. These two functional requirements exhibit a coupling: for extracting clay, shaking the mixture fast is required, while slow shaking allows for segregating salt. That functional coupling forbids physical integration.

As one can see in equation (3), the design matrix for the second hierarchical level, has more FRs than DPs, so that the design matrix is not square

$$\begin{Bmatrix} FR_{2.1} \\ FR_{2.2} \\ FR_{2.3} \\ FR_{2.4} \end{Bmatrix} = \begin{bmatrix} \times & 0 & 0 \\ 0 & \times & 0 \\ 0 & \times & \times \\ 0 & \times & \times \end{bmatrix} \begin{Bmatrix} DP_{2.1} \\ DP_{2.2} \\ DP_{2.3} \end{Bmatrix} \tag{3}$$

The 3rd level FRs are accomplished independently and are not necessary for the following discussion.

AD's theorem 1 (Coupling due to insufficient number of DPs) states, "When the number of DPs is less than the number of FRs, either a coupled design results or the FRs cannot be satisfied." [2].

Therefore, the traditional salt washing machines are coupled designs since the number of DPs is smaller than the number of FRs at the second decomposition level. Their functional requirements are never fully satisfied because one cannot increase the shaft speed sufficiently to warrant effective clay extracting, and the shaft speed cannot be as slow as required to avoid prejudicing salt segregation.

We tried other solutions to improve shaking, for example, by adding transverse fins on the propeller blade to increase the turbulence of the mixture within the clay extraction section of the tank while keeping the conveyor speed low to ensure salt segregation. Another way to keep the conveyor's speed low was to apply compressed air jets inside the extraction section to create a jacuzzi effect, although the improvements were not significant.

3.3. The new design

In the end, the adopted solution was to divide the screw conveyor into two sections with independent rotation movement: a high-speed conveyor in the extraction section of the tank, and a low-speed conveyor in the segregation section. Figure 10 shows the design parameters for a two-screw conveyor salt washing machine.

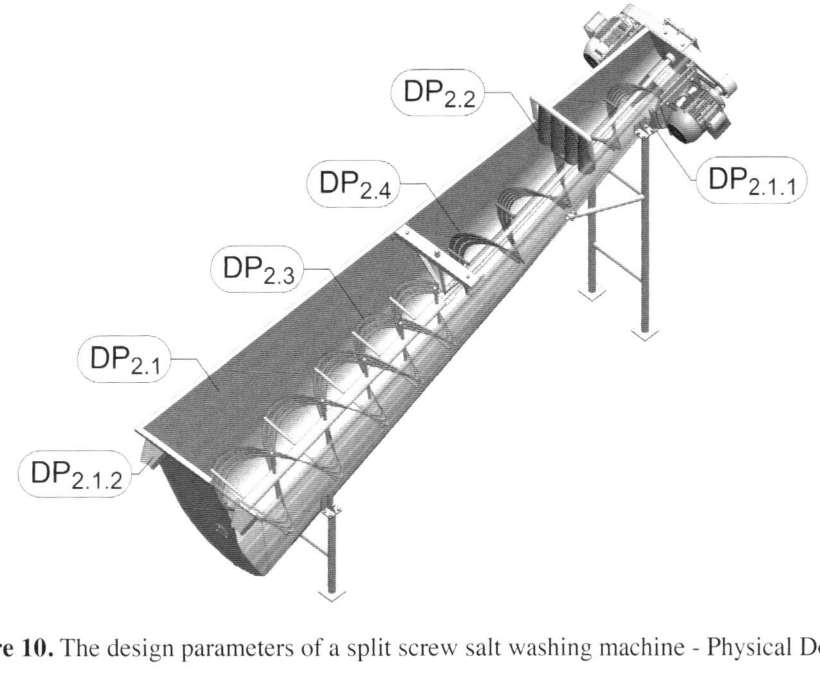

Figure 10. The design parameters of a split screw salt washing machine - Physical Domain.

At the second hierarchical level, the design parameters of the new design are as follows:

$DP_{2.1}$ – Washing Tank
$DP_{2.2}$ – Brine
$DP_{2.3}$ – Screw conveyor 1
$DP_{2.4}$ – Screw conveyor 2

The salt and the clay find their way from conveyor section 1 to conveyor section 2 by themselves. The hierarchical zigzag decomposition is shown in figure 11.

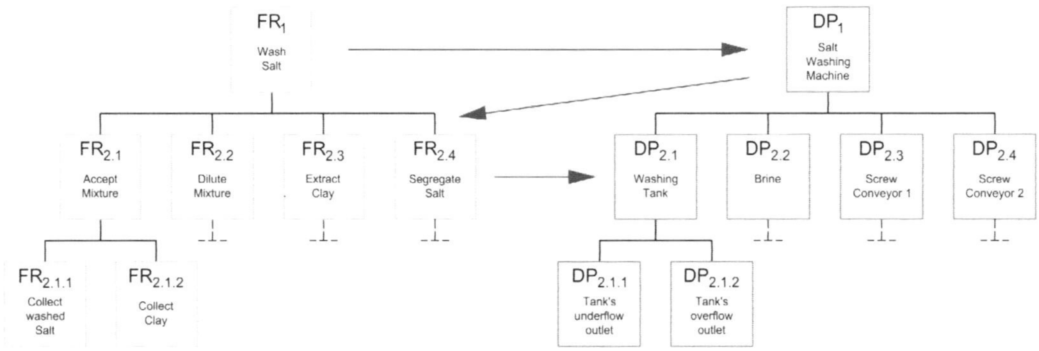

Figure 11. The zigzag path of the hierarchical decomposition of a split screw salt washing machine.

As one can see in equation (4), the matrix became square, and the design became decoupled.

$$\begin{Bmatrix} FR_{2.1} \\ FR_{2.2} \\ FR_{2.3} \\ FR_{2.4} \end{Bmatrix} = \begin{bmatrix} \times & 0 & 0 & 0 \\ 0 & \times & 0 & 0 \\ 0 & \times & \times & 0 \\ 0 & \times & 0 & \times \end{bmatrix} \begin{Bmatrix} DP_{2.1} \\ DP_{2.2} \\ DP_{2.3} \\ DP_{2.4} \end{Bmatrix} \tag{4}$$

4. Results

The proposed solution consists of the addition of one DP to allow controlling the separation and the washing functions in a decoupled way. The final configuration for the new salt washing machine encompasses the division of the screw conveyor into two conveyors with independent rotation speeds, allowing a higher rotation speed in the clay extraction section and a lower velocity in the salt segregation section.

Using the glossary of AD, $FR_{2.3}$ and $FR_{2.4}$ become decoupled and are achieved by sequentially adjusting the speeds of the two screw conveyors. The new solution allowed improving the separation and segregation functions efficiencies while increasing the washing speed by 20% to 30%. Furthermore, the process productivity and the final product's quality improved in a meaningful manner. At last, the same amount of produced washed salt requires less water and energy.

5. Conclusions

In traditional washing machines, the same DP (the screw conveyor) touches two functional requirements so that the number of DPs is smaller than the number of FRs. Following AD's Theorem 1, such peculiarity characterizes coupled designs. The analysis of the design matrix in equation (3) reveals this condition, which usually indicates the poor performance of the corresponding design solution.

The new design resulted from equation (3), to which a new DP was added to accomplish a decoupled design solution. The new DP is a second screw conveyor, which allows adjusting each DP independently to obtain the best fulfillment for each FR. The new design solution reaches:

- improved productivity.
- reduced consumption of brine, water, and energy by about 30%.
- increased production speed by between 20% and 30%.
- increased perceived quality level of the final product.

Axiomatic Design theory enabled analysis and finding the source of the problem, allowing also redesigning a more efficient solution that increases productivity and improves environmental sustainability.

Acknowledgements
The authors acknowledge Fundação para a Ciência e a Tecnologia (FCT, I.P.) for its financial support through the grant UIDB/00667/2020 (UNIDEMI).

References

[1] Gabriel-Santos A, Gonçalves-Coelho AM, Santos P, Fradinho J, Mourão A 2015 Redesigning a manufacturing system based on functional independence: the case of a tree nursery, *9th International Conference on Axiomatic Design, Procedia CIRP* **34** pp 206–211
[2] Suh N P 2005 *Complexity: Theory and Applications* (New York: Oxford University Press) pp 23, 44, 45, 140
[3] Gumus B, Ertas A, Tate D, Cicek I 2008 The Transdisciplinary Product Development Lifecycle Model, *J Eng Des*, **19** 3 pp 185–200
[4] Tate D 2015 Axiomatic Design: Review, Impact, and Future Direction, *Proc. of the ASME 2015 International Mechanical Engineering Congress and Exposition IMECE2015*, Houston, Texas. IMECE2015-51282
[5] Nordlund M, Lee T, Kim S-G 2015 Axiomatic Design: 30 Years After, *Proc. of the ASME 2015 International Mechanical Engineering Congress and Exposition IMECE2015*, Houston, Texas. IMECE2015-52893

IOP Conf. Series: Materials Science and Engineering 1174 (2021) 012025 doi:10.1088/1757-899X/1174/1/012025

Axiomatic Software Engineering

J T Foley[1,2], **M Kyas**[1]

[1]Reykjavik University, Menntavegur 1, Reykjavik 102, Iceland
[2]Massachusetts Institute of Technology, 77 Massachusetts Ave, Cambridge MA 02139, USA

E-mail: `foley AT ru.is`

Abstract. Axiomatic Design and Complexity Theory have primarily focused on refining the process of an abstracted need into a physical solution even under challenging conditions. The concepts have high applicability as guidelines for non-physical design, such as those found in programming and its applied field of software engineering. Software engineering came out of the software crisis of the 1960s: its goal is to deliver safe and secure software that meets its specifications on time and within its budget. The product of software engineering is a program. The program is written in a programming language that looked suitable for the problem at hand. For some time, object-oriented technology (OOT) looked popular, because it promised to address information hiding, composability, and extensibility of programs. In hindsight, OOT failed to deliver on its promises. We analyze how Axiomatic Design leads to solutions that are proposed by current programming technologies. The authors believe that different paradigms such as cohesion also need to be included.

1. Introduction

Software used to be found primarily in highly technical areas: nuclear weapon design, CNC milling machines, and telephone billing systems. Today, software controls every aspect of our daily life: cars, washing machines, refrigerators, coffee machines, light-bulbs, etc. At the same time, the complexity of software has increased. The digital autopilot for the Apollo lunar lander required about 2,000 lines of assembly code [1], while it has been estimated that there are about 1.5 million lines of code in the onboard command and control computers on the International Space Station [2]. Greater needs have arisen for producing software in terms of quantity while maintaining quality. Our reliance on software has not necessarily resulted in the desired reliability evidenced by Microsoft Windows' "Blue Screen of Death" showing up in unexpected areas.

The increased size of software systems created by teams of hundreds of programmers gave rise to specializations in software companies. The teams consist of managers, secretaries, programmers, testers, infrastructure engineers, build engineers, domain experts, and designers. Consequently, the focus of programming has changed from designing and implementing algorithms to constructing software systems bottom-up using an ecosystem of software components. Today's programmers must be skilled in the elements of *software engineering* to fit into such large-scale projects.

The authors focus on the role of Axiomatic Design in modern software engineering. They illustrate the problem and history of software engineering. Then, they argue that object-oriented programming is not the best fit for Axiomatic Design. Instead, they propose modular programming methods which have been rediscovered for modern languages like Go and Rust.

1.1. A Very Brief History of Software Engineering

The term *software engineering* appeared in the 1960s. It was invented to distinguish the process of creating software-intensive systems from computer science. Most textbooks credit the 1968 NATO on Software Engineering conference in Garmisch-Partenkirchen as the origin of the term [3], or at last, it popularized it. However, the name was in use before the conference. Anthony Oettinger uses the term in a letter to the Communications of the ACM in 1966 [4]. Early written uses of the term appear in the June 1965 issue of Computers and Automation [5]. Folklore tells that Margaret Hamilton coined the term in 1963 or 1964 [6]. She was the lead developer for Skylab and Apollo while working at Draper Lab.

The 1968 NATO conference of software engineering addressed the *software crisis:* software projects were delivered late, software was not working to specification, and budgets were frequently overrun. Participants of the conference proposed pioneering solutions. M.D. McIlroy introduced the idea of "mass-produced software components". He likened the production of software to mechanical engineering [7].

Software modules were proposed as one such unit of reuse. In software engineering, a module is a unit of software that encapsulates code and data to implement a particular functionality, provides an interface that lets clients uniformly access its functionality, can work in any context that interacts through the interface, and is usually deployed as a single unit. Examples of such modules are system libraries.

Parnas [8] argues that a module should not represent a part of the program execution, like a block in a flow chart. Instead, a module should encapsulate design decisions. If done correctly, the code that relies on a module is robust to changes in the module's internals. This property, called *information hiding,* enables programmers to work on the functionality of their module without depending on the programmers implementing the surrounding modules. The interface serves as a contract on which each programmer relies when using other modules while ensuring that their implementation conforms to their module's interface.

Barbara Liskov developed the programming language CLU, which supports information hiding as part of the language [9]. CLU defines objects as instances of data types that can only be manipulated through their interface. This language became a key component for teaching MIT's 6.170 Software Engineering course until 1995 when it transitioned to C++.

Early approaches to software development focused on a top-down process. Herbert Benington describes what became known as the Waterfall model in 1956. The process goes through six phases in almost sequential order: Software Requirement Analysis, Preliminary or Abstract Design, Detailed Design, Coding and Unit Testing, Integration, and Testing. This process is then followed by Software Maintenance, in which defects found during use are corrected, and the system is changed to meet evolving requirements. Benington clarified in 1983 that this step-by-step process was defined to describe the specializations of programming tasks and is not necessarily followed in this order [10]. Winston Royce describes this process as a straw-man argument and rejects its as highly risky [11]. He suggests that such a process should always be executed at least twice. It still was standardized by the Department of Defense [12] until it was superseded in 1994 [13], which marks a departure from the strict Waterfall model.

Indeed, the attempt of clarifying all the functional requirements in advance and decomposing the system into modules and tasks resulted in one of software engineering most spectacular failures: Fred Brooks "The Mythical Man-Month", originally published in 1975, attributed the failure of the OS/360 system to this waterfall model [14]. Understanding this failure is important enough that it is required reading as part of MIT's 6.033 Operating Systems course. The manuals of OS/360 describe the user-visible, expected behavior of the system. Brooks describes its volume as a "six-foot shelf of manuals" [14, p. 134]. His lessons include: the functional requirements of modern software systems are complex and never fully understood. The functional requirements of system features may interact in undesirable ways. Such *feature interactions* occur when a feature interferes with the operation of another feature [15]. For example, phone subscriptions may support *call forwarding* and *call waiting*. If both features are enabled at the same time while the user is in a phone call, it is not clear whether a second call may wait or has to be forwarded.

The shortcoming of the Waterfall model for software development leads to the development of iterative processes. The main problem of the Waterfall model is that the later one is in the process, the more expensive it becomes to address mistakes in earlier phases. If a mistake in the requirements is detected during the testing phase, it may result in a complete redesign and a new implementation of the complete system.

Software engineers were commonly not successful in capturing all the requirements of a system correctly. Fred Brooks, for example, suggested developing one software prototype to learn the requirements and how to develop solutions. This prototype is then to be discarded, and a new system shall be developed from scratch. Of course, there is an inherent danger that this replacement may fall victim to the "Second System Effect" and yet another failure [14].

Most software development processes are using an incremental and iterative design (IID) model. According to Larman and Craig, Walter Shewhart laid the foundations for IID by developing "plan-do-study-act" (PDSA) cycles for quality improvement in the 1930s [16]. The first use in software engineering was Project Mercury in the 1960s. The method got popular during the 1980s after the 1970s observed the weaknesses of the strict waterfall model. Since then, IID has been adopted as the dominant design method.

The main idea of IID can be grouped into four phases [17]. *Inception* identifies project scope, requirements (functional and non-functional), and risks at a high level but in enough detail that work can be estimated. *Elaboration* delivers a functional architecture that mitigates the top risks and fulfills the non-functional requirements. *Construction* incrementally fills in the architecture with production-ready code produced from analysis, design, implementation, and testing of the functional requirements. *Transition* delivers the system into the production operating environment. Each phase can be iterated multiple times.

An iteration starts with planning. A feature of the system is selected, and its requirements are formulated. The developers analyze the requirements and design the feature. The feature is implemented and tested. The test results are analyzed, and plans are updated. Already implemented features may undergo a redesign as a consequence of what the team learned from the iteration.

Proponents of object-oriented technology hoped that their programming methods (see below) support encapsulation and modularization as part of the design and implementation process. Objects are supposed to be units of composition. They can be reused and modified by inheritance. Late binding realizes polymorphism by overriding parts of a super-class implementation. Finally, it allows the creation of large systems by composing classes in a bottom-up process. As the authors show below, these promises have not been fulfilled.

This realization is the starting point of Agile Software Development. It is defined by the "Manifesto for Agile Software Development" [18]:

> We are uncovering better ways of developing software by doing it and helping others do it. Through this work we have come to value:
>
> Individuals and interactions over processes and tools
> Working software over comprehensive documentation
> Customer collaboration over contract negotiation
> Responding to change over following a plan
>
> That is, while there is value in the items on the right, we value the items on the left more.

IOP Conf. Series: Materials Science and Engineering 1174 (2021) 012025 doi:10.1088/1757-899X/1174/1/012025

1.2. A History of Object-Oriented Technology

Sutherland pioneered initial ideas of object-oriented programming at MIT. Nygaard and Dahl designed the first object-oriented programming language SIMULA in 1962, which they had been developing since the 1950s [19]. They developed SIMULA to make the programming of discrete event systems easier. The name stands for SIMulation LAnguage.

Influenced by LISP and SIMULA, Alan Kay developed the language Smalltalk and created the name "object-oriented" in 1972 [20]. The concept of a class became a fashionable programming abstraction and led to the creation of C++ [21] and Objective-C [22].

Inspired by the work of Kay and Stroustrup, software engineers placed high hope in object orientation as a tool to address the programming crisis. Software engineering methods based on Object-Oriented programming were developed by Booch [23] and Meyer [24, 25]. This work resulted in the Unified Modeling Language (UML), an amalgamation of Booch [26], Jacobson [27] and Rumbaugh's [28] methods. UML is still a dominant software engineering tool. Agile development methods in current practice have displaced the accompanying Rational Unified Process.

1.3. Axiomatic Design

Nam Pyo Suh developed Axiomatic Design in the 1980s to provide a systematic approach to designing machines, particularly manufacturing equipment [29]. The AD process is relatively straightforward but requires a slight departure from many iterative-focused design processes. Designers develop lists of items in the four domains: Customer, Function, Physical, and Process. The items represent Customer Needs (CNs) mapping to Functional Requirements (FRs) to Design Parameters (DPs) to Process Variables (PVs). These mappings are analyzed in the form of design matrices (a Cartesian product of the elements in the two adjacent domains), or design decomposition trees [30, 31]. The goal of this representation is to show where coupling between elements in a given list are occurring in order to choose designs that fulfill AD's axioms.

Independence Axiom: Maintain the independence of the functional requirements[32, p. 16]

Information Axiom: Minimize the information content I of the design where $I = -\log_2 p_i$ and p_i is the probability of meeting FR_i[32, p. 39]

The Independence axiom indicates that in the matrix representation, there should only be diagonal elements with coupling i.e. a one-to-one mapping. The Information axiom tells the designer to choose implementations that minimize the chance of failure to meet a mapped FR.

The authors would initially adjust these concepts in the context of software to be:

Independence Axiom: Maintain the independence of the functional requirements

Information Axiom: Enforce encapsulation, maintain consistent internal state (representation invariants are often employed here), reduce coupling

Software engineers call an FR the definition of a function that the software system or one of its components has to implement. The function is typically specified by relating valid inputs to outputs.

2. What AD has previously said about software engineering

Suh's primary investigation into the topic of software design appears as chapter 5 in the 2001 book "Axiomatic Design: Advances and Applications"[32]. He details an Axiomatic Design process specifically for software [32, pp. 245–248] based upon [33]:

(i) Define FRs of the Software System

(ii) Mapping between the Domains and Independence of Software Functions

(iii) Information Content for Software Systems to Select the Best Design

(iv) Decomposition of FR, DP, PV, and Software Modules
(v) Object-Oriented Programming and Modules for Leaves
(vi) Design Matrix and Module-Junction Diagrams
(vii) Flow Chart Representation of Software System Architecture
(viii) System Control Command

Much of the details in this chapter focus on the implementation of software using this process that greatly resembles the "waterfall method" previously described where the requirements have been set at the beginning. Of note, AD's approach explicitly acknowledges that a designer will need to revisit previous phases in the form of a zig-zag between domains as part of the decomposition. The applications of this approach in the chapter are from related papers of that era. This includes: library index[33], calculator for the geometry of ribs in an injection molded part[34], and optimization software for CRT geometry optimization[35], Acclaro software for supporting AD development[36],

Acclaro is the only known software product designed using AD still in existence as of 2021. This product has shifted focus from AD to Design for Six Sigma, which is a popular production design methodology in its own right.

Since this publication, there have been additional AD-related publications on AD used in software, primarily focused on database design or development of an application. As an example, Pacifici et al.[37] discussed software deployment for remote patient monitoring software.

3. A critique of the AD approach

To some extent, the approach that Suh takes to the development of software is based on assumptions that software engineering considers to be faulty. He states, "Its framework is based on the idea that software must be designed to satisfy a set of functional requirements and constraints before commencing coding activities."[32, pp. 241] and "It also provides a methodology for design that includes the idea of mapping, decomposition, and the creation of leaves". (The leaves mentioned in this statement are the endpoints of decomposition in contrast to a "branch".) Empirical evidence demonstrates that it is not possible to capture the correct functional requirements and constraints before coding activities start. The risk is that if functional requirements and decomposition are performed before any coding activities, the requirements are not satisfiable, resulting in a redesign of the system.

A second observation is that the definition of software on [32, pp. 241] does not capture what modern software is doing. It is the definition of "algorithm," which is an integral part of software systems, but does not capture software. Instead, we propose to base the definition of software on Harel and Pnueli's definition of reactive systems [38]:

> Software is a set of instructions and data that tell a computer system what to do. The computer system is maintaining ongoing interaction with its environment, without necessarily producing a final result or terminating.

This lack of termination is important even in the simplest case of a small control loop like a PID controller which should *never* terminate in proper usage. Buttazzo, for example, shows that a real-time system can become infeasible if executed on a faster computer than it was designed for![39].

3.1. OOT increases coupling and decreases cohesion

One of the takeaways from Suh's chapter on software engineering [32, pp. 245–248] is that you should use object-oriented technology (OOT) to reduce coupling.

In software engineering, *coupling* is the measure of the degree of interdependence of modules. Tight coupling means that code in one component has intimate knowledge of the internals of other components. Changes in one component ripple through the system and change the behavior of other components. Loose coupling restricts knowledge to published interfaces. Changes in a component do not ripple through the system as long as the interface is maintained.

According to Stevens et al. [40], *cohesion* is a measure of how closely related the responsibilities of data and code in a component are to each other. High cohesion means that unrelated responsibilities are implemented in different components. The responsibilities have to be "glued" together explicitly. The mantra is, "increase cohesion within all components." Changes for one feature should not affect unrelated features. For example, a customer database maintains names and email addresses. Should a customer have a send-email method to send emails to the customer? To achieve high cohesion, we should implement sending emails in a separate component. The customer is glued into it; spammer.send_email(customer.get_email(), text()). AD's Independence Axiom follows the same mindset: changing a single Functional Requirement (send email) should only affect one implementation Design Parameter (customer's email field).

Object-oriented technologies defining mechanisms tend to increase coupling and decrease cohesion. Object-oriented programming languages can be characterized by the following features:

late binding: the system determines at run-time what code it executes.

self-reference: objects in the system can call their code.

open recursion: code in a class can execute code defined in a subclass.

Information hiding is violated and coupling is increased by these features. By looking at a piece of code, we cannot determine if it gets executed or what code it will execute.

Further investigation into such advice reveals a contradiction in the way OOT is often implemented using an inheritance model between classes. The problem is that inheritance introduces a strong dependency on the behavior of the super-classes and the sub-classes. AD and similar methods focus on decoupling the interactions between all the elements of the system. The designer wants to clearly understand the interactions and the interactions be as clear and as little as possible. In OOT inheritance, when you take a class and derive a subclass, whatever happens in the super-class affects the behavior of all the sub-classes. In short, to understand the bottom of the hierarchy of inheritance, you must understand everything above it. This can quickly lead to mental overload and frustration which strongly resembles the elements of Imaginary Complexity.

Another issue that speaks against reuse in OOT is that objects depend on an environment of other objects to do their work. Joe Armstrong describes it as follows [41, p. 213]:

> I think the lack of reusability comes in object-oriented languages, not in functional languages. Because the problem with object-oriented languages is they've got all this implicit environment that they carry around with them. You wanted a banana but what you got was a gorilla holding the banana and the entire jungle.

Older programming languages such a C did not include an inheritance model, causing many programmers to create conceptual equivalents in terms of how functions were named to indicate their relationship. Modern languages wish to enforce a level of modularity, in many ways implementing AD's Independence Axiom, but without creating the coupling problem of inheritance. For example, the robustness-optimized language Rust doesn't have inheritance, instead choosing to only incorporate OOT's method calls. Rust eschews inheritance for "traits". Traits define constraints on types and group methods that can be applied to all types that satisfy the constraints. Static type safety is achieved through a template mechanism with trait bounds.

As we will discuss in the next section, constraints in another sense are an important element in design, especially in the ephemeral world of software.

IOP Conf. Series: Materials Science and Engineering 1174 (2021) 012025 doi:10.1088/1757-899X/1174/1/012025

```
class BoundedBuffer {
    constexpr static size_t N = 64;
    std::mutex lock;
    std::condition_variable not_full,
        not_empty;
    std::array<int, N> buffer;
    size_t capacity, front, rear;
public:
    BoundedBuffer() : lock(),
        not_full(), not_empty(),
        buffer(), capacity(N),
        front(0), rear(0) { }
    void put(const T&);
    int get(void);
};

void BoundedBuffer::put(const int& value)
{
    std::unique_lock<std::mutex>
        lck(this->lock);
    while (this->capacity == 0)
        this->not_full.wait(lck);
    this->buffer[this->front] = value;
    this->capacity--;
    this->front = (this->front + 1) % N;
    this->not_empty.notify_one();
}

int BoundedBuffer::get(void)
{
    std::unique_lock<std::mutex>
        lck(this->lock);
    while (this->capacity > N-1)
        this->not_empty.wait(lck);
    int value = this->buffer[this->rear];
    this->capacity++;
    this->rear = (this->rear + 1) % N;
    this->not_full.notify_one();
    return value;
}
```

Figure 1: A concurrent bounded buffer in C++

3.2. *Multi-threaded Programming Challenges*

An instructive example of the issues that are inherent to object-oriented technology is the fragile base-class problem [42]. The problem arises in multi-threaded programming. The fragile base class problem states that inheriting code from a super-class is often only possible if other, seemingly unrelated methods, need to be redefined. Consider the implementation of a bounded buffer in C++ below.

A programmer needs a bounded buffer with a method getTwo() that waits until at least two elements are available in the buffer and returns two consecutive ones. To implement this, they derive a subclass BoundedBufferTwo that inherits the buffer and the methods put and get from its super-class BoundedBuffer. Such an implementation may appear natural, but this particular implementation exhibits an anomaly. If a process waits in get() and another one waits in getTwo(), the system may block after a call to put(). Input, only one process is notified by the presence of a new element. It may be delivered to the process that waits in getTwo(). This process cannot proceed if only one element is in the buffer. The process that waits in get() never gets notified. To correct this issue, the programmers have to override the put() method to replace the statement **this**->not_empty.notify_one() with **this**->not_empty.notify_all().

To avoid this problem, the programmer needs to conceive all possible reuse of the code. This is at odds with efficiency constraints. In our example, calling **this**->not_empty.notify_one() is much more efficient than calling **this**->not_empty.notify_all().

4. Robustness in Programming Languages

Software Engineering is focused on the premise that it is possible to create reliable software that meets specified requirements on time and with the provided resources. For the sake of our study, it is important to understand that we look at the process of software construction. Software can become unreliable by introducing errors into the system by correcting defects in classes that result in new errors in other classes due to coupling. It can become unreliable by errors in memory management and unsafe memory access (dereferencing null pointers, use after

IOP Conf. Series: Materials Science and Engineering 1174 (2021) 012025 doi:10.1088/1757-899X/1174/1/012025

```
class BoundedBufferTwo : BoundedBuffer {
public:
    void getTwo(int*);
};

void BoundedBufferTwo::getTwo(int* values)
{
    std::unique_lock<std::mutex> lck(this->lock);
    while (this->capacity > N-2)
        this->not_empty.wait(lck);
    values[0] = this->buffer[this->rear+0];
    values[1] = this->buffer[(this->rear+1)%N];
    this->rear = (this->rear + 2) % N;
    this->capacity += 2;
    this->not_full.notify_all();
}
```

Figure 2: Subclass of a bounded buffer

free). The Information Axiom is relevant to this field because higher information content in a program is by definition unable to meet the previous statement.

Memory management in its various forms is a common challenge embedded software engineering. It consists of describing, constructing, transferring, and deleting the memory structures of interest. Doing this consistently and reliably is key to the software itself being reliable. Memory management can be done by a garbage collector that keeps track of when objects (and their associate memory) are no longer in use. This garbage collector incurs processing overhead and creates a non-deterministic execution-time behavior. For time-critical systems, failure to meet the timing requirements reliably would be considered Information Content.

For manually allocated memory, the problem can be even more structural. Careful choice of constructing and deleting can result in a system that runs reliably, but the moment any changes are made to that code in terms of sequence or function, Information Content can arrive in the form of double-allocations, free after use, memory leaks, etc. The program may also be vulnerable to malicious attacks in the form of buffer overflows. It is noted that due to these failures arising from memory issues that map to Information Content that Axiomatic Design would also provide insight into how one might correct it.

The paradigm employed to manage these issues begins with type-checking. Type checking verifies that no statement of the programming language operates on data leads to a run-time error or undefined behavior. Undefined behavior is behavior that is not defined by the programming language but that emerges from how the program is executed. It is not defined what the program does and it may do something undesired.

A type system assumes that all data has a type assigned to it and that every operation has declared input and output types. An operation can be executed safely if the input data conforms to the declared types. Then, the operation is guaranteed to produce a value of the output type.

In a static type system, the compiler verifies that the program never operates on data of the wrong type. If the programmer writes an operation in that the types do not conform, the compiler rejects the program. Dynamic type systems instead perform the checks at execution time, aborting execution if the check fails. Static type systems allow integrated development environments to inform the programmer about what types of data are going to be handed to operations and consequently by choosing an output type and provide information to the next function involved.

Static type systems are considered limiting in terms of expressiveness: programmers often cheat around the type system by casting data to types they want later. Many systems try to perform static type analysis to assist in debugging these type issues, but Rice's Theorem[43] indicates that you can only do a proper analysis of trivial things. Many desirable properties of type systems are not trivial. For example, one can decide the type of an array or if an array index is in its bounds, but never both at the same time. Pierce gives a comprehensive account of types and limitations of type checking [44].

Memory management in the reliability-focused Rust language uses a *linear type system* to track memory. Memory is allocated and initialized at the time of the variable declaration. It is de-allocated when it goes out of scope. This approach, in addition to having immutable variables and objects by default, allows for relatively easy analysis of memory and types. The Rust compiler employs a mechanism called a "borrow checker" to enforce proper memory management; programs that fail to follow the rules fail to compile. In the terms of the Information Axiom, the compiler is reducing the information content of the complete system by ensuring that only compliant code can be executed.

5. Lack of constraints is a problem

Software in its purest form is only constrained by theoretical limitations, unlike designs that are mechanical engineering. In many cases, Computer Science curricula focus on elegant and fast algorithms to solve abstract problems. If processing power or memory becomes something to consider, you can dump it into the cloud and spend more money! This concept is within the purview of the public due to crypto-currency correlating processing energy costs with an income stream. Conversely, edge computing, IoT, and low-power sensing have highlighted that money cannot always solve the processing problem.

Embedded programmers are constrained by the limitations of the target platform. In time-critical systems, the deadlines must be met. In mass-produced, systems, slow and memory-constrained computers are economically beneficial. Also, the domain in which the program executes imposes constraints on what the software does. Sensors are limited by their range. Actuators are limiting by the properties of the domain. These constraints make the design and testing of software systems much easier.

In the even more concrete world of mechatronics, reality quickly creates constraints. A drone software engineer might not know what the physical parameters represent are and is just asked to make it fly. The meaning of those values is from the domain expert: a pilot or avionics engineer. The mechatronics and mechanical engineer might do the AD decomposition, then hand the slots to be filled in by the software guy. This results in occasional mismatches between what the software commands and what is physically possible. One of the authors was tasked with improving the odometry on a larger land-based robot using a new sensor. In the testing process, he accidentally added a 0 to the end of the rotation rate speed, which resulted in the robot spinning at an extremely high rate, scraping his shin, and denting his filing cabinet in an embarrassing episode. In the internal pure model of the software interfacing with the hardware, there is no significant difference commanding a robot to turn at $0.1\,\mathrm{rad\,s^{-1}}$ or $100\,\mathrm{rad\,s^{-1}}$. In the physical world, it becomes painfully clear that this is not true. (In a similar incident that happened on the same day, the author's colleagues input the incorrect forward velocity on a similar robot which caused it to jump out of the test stand and destroy an expensive laptop.) A concrete example of where poorly design software resulted in injury and death is the case of the Therac-25 radiation therapy machine. In this case, a race condition in the user interface and the removal of hardware-based safety interlocks resulted in a very dangerous environment [45].

How can useful constraints be created? Gualtieri et al. [46] have applied AD to the design

```
type Vehicle interface {
  func Drive()
}

type Car struct {}

func (car *Car) Drive() {
   // drive the car
}

type Truck struct {
     car *Car
}

func (truck *Truck) Drive() {
     // delegate the call to the car
     truck.car.Drive()
}
```

Figure 3: Use of delegation to implement code reuse

of collaborative Human-Robot workspaces, particularly focusing on the relevant European workplace standards. This helps create a physically safe environment, but what about a safe environment on the software side? Pedagogically, instructors need to use a computing system that imposes constraints like a Raspberry Pi. This has limited processing power, memory, and IO. Such constraints force the designer to start thinking about resource management earlier and these constraints help reduce the solution space. This is particularly relevant because low-cost microcontrollers such as the Intel 8051, TI MSP430, and ARM7 are still extremely common in modern appliances. It must be noted that making changes later is extremely expensive, an observation that AD often reminds us of.

Another thought is that well-rounded multi-disciplinary education for programmers might provide a better understanding of real-world constraints. Students have trouble with inertia the first time they interface with software motors. They think it is sufficient to turn the motor to full then turn it off to stop. Once they have made this mistake a few times, possibly painfully or with a great deal of smoke, they are highly motivated to improve their software model of the motor system.

6. Modern programming language approaches

Liskov's previously mentioned CLU [9] avoids the complications of object-oriented programming. The language supports information hiding and encapsulation of data. It allows access to the data only through well-defined interfaces (MIT Students in the software engineering course quickly learned to dread the warning "interfaces changed" while compiling.) But the language does not support inheritance. It also does not support open recursion. Operations executed on the current object can be resolved at compile time.

More recent programming languages avoid object orientation for the same reason.

6.1. The Go Programming Language: interfaces and delegation

The Go programming language encapsulates groups of operations into packages. Data types inside packages can be specified through interfaces. An interface declares a set of operations that operate on the data type. Any object that supports all the operations specified in an interface can be used as an instance of this interface. This differs from C++ or Java, where a class needs to specify that it implements an interface.

The Go programming language does not support inheritance. Code reuse can be achieved through delegation as shown in Figure 3. In the code example below we define an interface for Vehicles. We define a data **type** Car that implements the function Drive(). We define a data **type** Truck that implements the function Drive() through delegating the calls to a car object.

6.2. Rust: traits

As previously mentioned, the Rust language deals with object organization differently from common OO languages such as Java and C++. The language incorporates structures and methods. Methods operate on objects of the structure types. Rust uses "traits" to define shared behavior and organize common methods across structures. Klabnik and Nichols [47, Chapter 10.2] describe it thusly:

> Different types share the same behavior if we can call the same methods on all of those types. Trait definitions are a way to group method signatures together to define a set of behaviors necessary to accomplish some purpose.

Traits are not types. Different structured can implement traits, and programmers can thus assume that they can use objects that share a trait the same way.

Let us consider the simple example of a window manager system requiring elements drawing to the screen. The programmer can specify a Draw trait to express the signature of the draw method. The elements implement the Draw trait.

```
pub trait Draw {
  fn draw(&self, scr: &Screen);
}
```

To ensure that the elements implement the Draw trait, they can specify a generic function that only accepts types that implement the Draw trait:

```
fn paint<Elem: Draw>(t: Elem) {
  t.draw(screen)
}
```

The compiler produces a compile error if t does not belong to a trait that implements the Draw trait. The programmers can rely on any such object to have a draw() method.

This mechanism is significantly safer than an inheritance hierarchy because it only needs to consider the structure's signature at the time of compilation to ensure that it meets the desired requirements. In contrast, an inheritance system needs to "walk the tree" to find the appropriate super-class that matches the signature, while implementations are in sub-classes.

7. Summary

Axiomatic Design software practices in 2001 advocated the use of OOT, emphasized inheritance and strict hierarchy following the development of the FR-DP mapping and decomposition. Many researchers have realized this creates complexity in the form of the obscurity of implementation and unreliability. More strikingly, multiprocessors are now standard, creating opportunities and a more challenging environment to program in. Modern languages employ information hiding in objects and methods as prescribed by the Independence Axiom. They also avoid inheritance in favor of a flatter organization modeled on interfaces, delegation, or traits. Software engineering recognizes imaginary complexity as accidental complexity and contrasts it to essential complexity. Elimination of the coupling created by inheritance reduces imaginary complexity from not understanding the class hierarchy involved in the implementation. The Independence Axiom indicates one should Increase coherence, even though it resembles coupling. In a highly coherent system, each component addresses one functional requirement. The authors believe that mapping Axiomatic Design's axioms to modern software engineering paradigms is key to gaining the benefits of an AD mindset's structured creativity while still following coding standards.

References

[1] Clark L 2004 *IEEE Control Systems Magazine* **24** 100–101
[2] Johnson C W 2005 *Formal Methods* pp 9–25

[3] Naur P and Randell B (eds) 1968 *Proc. Nato Software Engineering Conference* (Garmisch)
[4] Oettinger A A 1966 *Communications of the ACM* **9**
[5] 1965 computers and automation: The computer directory and buyer's guide, 1965 URL `https://ia601909.us.archive.org/26/items/bitsavers_computersA_13094490/196506.pdf`
[6] Cameron L 2021 *IEEE Computing Edge* URL `https://www.computer.org/publications/tech-news/events/what-to-know-about-the-scientist-who-invented-the-term-software-engineering`
[7] McIlroy M D 1968 in Naur and Randell [3] pp 138–150
[8] Parnas D L 1972 *Communications of the ACM* **15** 1053–1058
[9] Liskov B and Zilles S 1974 *Proceedings of the ACM SIGPLAN symposium on Very high level languages* (Santa Monica, CA, USA: ACM) pp 50–59
[10] Benington H D 1983 *Annals of the History of Computing* **5** 350–361
[11] Royce W W 1970 *Proceedings IEEE WESCON* (Los Angeles, CA, USA: IEEE) pp 328–338
[12] Department of Defense 1988 *DOD-STD-2167A: Defense Systems Software Development*
[13] Department of Defense 1994 *MIL-STD-498: Software Development and Documentation*
[14] Brooks Jr F P 1995 *The Mythical Man-Month* anniversery ed (Reading, Mass.: Addison-Wesley) ISBN 0-201-83595-9
[15] Bowen T, Dworack F, Chow C, Griffeth N, Herman G and Lin Y J 1989 *Seventh International Conference on Software Engineering for Telecommunication Switching Systems* (Bournemouth, UK: IET) pp 59–62
[16] Larman C and Basili V R 2003 *Computer* **37** 47–56
[17] Jacobson I, Booch G and Rumbaugh J 1999 *The Unified Software Development Process* Addison-Wesley Object Technology Series (Addison-Wesley) ISBN 978-0321822000
[18] Beck K, Beedle M, van Bennekum A, Cockburn A, Cunningham W, Fowler M, Grenning J, Highsmith J, Hunt A, Jeffries R, Kern J, Marick B, Martin R C, Mellor S, Schwaber K, Sutherland J and Thomas D 2001 Manifesto for agile software development Web page `https://agilemanifesto.org/`
[19] Nygaard K and Dahl O J 1978 *History of programming languages* (Los Angeles, CA, USA: ACM) pp 439–480
[20] Kay A C 1993 *ACM SIGPLAN Notices* **28** 69–95
[21] Stroustrup B 1980 Classes: An abstract data type facility for the C language Computer Science Technical Report CSTR-84 Bell Laboratories
[22] Cox B L 1983 *ACM SIGPLAN Notices* **18** 15–22
[23] Booch G 1982 *ACM SIGAda Ada Letters* **I** 64–76
[24] Meyer B 1985 *IEEE Software* **2** 6–20
[25] Meyer B 1988 *Object-Oriented Software Construction* (Prentice Hall)
[26] Booch G 1993 *Object-Oriented Analysis and Design with Applications* 2nd ed (Benjamin Cummings)
[27] Jacobson I, Christerson M and Jonsson P 1992 *Object-Oriented Software Engineering — A Use Case Driven Approach* (Reading, MA: Addison-Wesley)
[28] Rumbaugh J, Blaha M, Premerlani W, Eddy F and Lorensen W 1990 *Object-Oriented Modeling and Design* (Prentice Hall)
[29] Suh N P 1990 *The Principles of Design* (Oxford University Press)
[30] Cochran D S, Foley J T and Bi Z 2016 *International Journal of Production Research* **55**(3) 870–890
[31] Benavides E M 2012 *Advanced engineering design - An integrated approach* (Woodhead Publishing)
[32] Suh N P 2001 *Axiomatic Design - Advances and Applications* (Oxford University Press)
[33] Kim S J, Suh N P and Kim S G 1991 *Annals of the CIRP* **40** 165–170
[34] Kim S G and Suh N P 1987 *International Journal of Robotics and CIM* **3**
[35] Do S H and Park G J 1996 *CIRP Workshop on Design and Intelligent Manufacturing Systems* (Tokyo)
[36] Do S H and Suh N P 2000 *CIRP Annals* **49**
[37] Pacifici B, Parretti C, Girgenti A and Citti P 2017 *11th International Conference on Axiomatic Design (ICAD)* ed Dodoun O (Iasi, Romania: MATEC Web of Conferences) p 01010 sep. 15–18
[38] Harel D and Pnueli A 1985 *Logics and Models of Concurrent Systems* (*NATO ASI Series* vol 13) ed Apt K R (Springer Verlag) pp 477–498
[39] Buttazzo G C 2006 *IEEE Computer* **39** 54–59
[40] Stevens W P, Myers G J and Constantine L L 1974 *IBM Systems Journal* **13** 115–139
[41] Seibel P 2009 *Coders at Work: Reflections on the Craft of Programming* 1st ed (Apress) ISBN 978-1430219484
[42] Matsuoka S and Yonezawa A 1993 *Research Directions in Concurrent Object-Oriented Programming* ed Agha G, Wegner P and Yonezawa A (MIT Press) pp 107–150
[43] Rice H G 1953 *Transactions of the American Mathematical Society* **74** 358–366
[44] Pierce B C 2002 *Types and Programming Languages* (MIT Press)
[45] Leveson N G and Turner C S 1993 *IEEE Computer Magazine* 18–41
[46] Gualtieri L, Rauch E, Rojas R, Vidoni R and Matt D 2018 *12th International Conference on Axiomatic Design (ICAD)* ed Puik E, Foley J T, Cochran D and Betasolo M (Reykjavík, Iceland: MATEC Web of

Conferences) p 01003 october. 9–11
[47] Klabnik S and Nichols C 2019 *The Rust Programming Language* (No Starch Press) URL https://doc.rust-lang.org/book

IOP Conf. Series: Materials Science and Engineering 1174 (2021) 012026 doi:10.1088/1757-899X/1174/1/012026

Mobile High Voltage Power Line Thermometer

F S Sudermann[1,2], K M Kjartansson[2], S Á Jakobsson[2], J T Foley[2,3]

[1]Karlsruhe Institute of Technology, Kaiserstraße 12, 76131 Karlsruhe, Germany
[2]Reykjavik University, Menntavegur 1, Reykjavik 102, Iceland
[3]Massachusetts Institute of Technology, 77 Massachusetts Ave, Cambridge MA 02139, USA

E-mail: fynn20@ru.is kristjank18@ru.is, sigurdurj09@ru.is, foley@ru.is

Abstract. Modern power grids perform systematically important tasks with immense capital investment behind them. Infrastructure monitoring is an important field in this industry to guarantee safety, prevent damages and optimize transmission. Accurate readings of environmental variables also help power operators design new infrastructure. Operators address this by visual monitoring, and through the use of sensors placed on high voltage lines. Traditionally, these sensors are permanently installed and require lines to be powered down for deployment and removal. The required disruption of service is associated with high costs and therefore undesirable. Axiomatic Design Theory highlighted unnecessary coupling between disruption and measurement as an opportunity. This paper details a design for a thermometer unit that can be deployed to and removed from a live power line. The device can provide temperature measurements to be used in conjunction with visual LiDAR scans to model the behavior of high voltage lines at different load levels to gauge line sag. Mechanisms were designed to safely fasten the thermometer to a live high voltage line via a VTOL drone and for it to detach given a preset time. In case of system failure, the device removes itself through a variety of redundant fail-safe mechanisms inspired by the Information Axiom. The success of key design concepts was confirmed through 3D printed prototype testing on a section of a surrogate conductor line in a controlled environment.

1. Introduction

The power grids of most modern countries perform systematically important tasks with an immense amount of capital investment behind them. According to the International Energy Agency, there was a yearly global investment in the range of 273–306 billion USD into electricity networks between 2015 and 2019 [1]. As such, power infrastructure monitoring is an important field in this industry to guarantee safety, prevent costly damage and optimize power transmission. It is important to monitor the health of power equipment to prevent costly damage and optimize maintenance. In addition, getting accurate readings about environmental variables helps power grid operators analyze and manage their grids and design new lines based on evaluations of the usage and utilization of existing power lines.

Modern power transmission grids are the backbone of industry and production. With increased power demand the commission of new power lines is required. These new lines are designed based on data available about lines already in operation. A variety of data are collected during monitoring and surveillance of active power lines.

Power grid monitoring and automation begin forming as a field in the 1920s. The main goal was to connect the operations of power plants and nearby generating stations for operational cost benefits [2]: load and frequency had to be monitored to accomplish this goal.

IOP Conf. Series: Materials Science and Engineering 1174 (2021) 012026 doi:10.1088/1757-899X/1174/1/012026

Requirements have grown since the 1920s, and an important aspect of modern grid maintenance is temperature and sag monitoring. Operators of power grids monitor the wire temperature as increased wire temperatures also mean increased line sag. The now lower hanging power lines can also be an electrocution hazard to individuals [3]. For this reason, most modern countries have legal frameworks in place that define minimum ground clearances for power lines [4]. These regulations aim to make sure that power lines maintain necessary clearances from environmental objects. Line sag is strongly correlated with changes in the wire temperature. The temperature is influenced by outside temperature, load, weather especially precipitation, and the wire's properties. Transmission system operators can use temperature measurements in conjunction with LiDAR scans of the power line to develop an analytical model with the conductor temperature as the input and the line sag as the output. The model is then used during the design and dimension of new power lines.

The current state of the art focuses on monitoring stations performing long-term measurement and securely attaching to the power line; more details will be discussed in Section 2. This concept requires durable devices hardened to survive extreme weather conditions for long periods. In addition, the line must be turned off for the maintenance personnel to attach the device, remove the device, or service it. Due to the long operation period and lost revenue associated with such actions, line operators must budget carefully in terms of time and money. The long operation time also increases the cost and power-budget requirements of such devices. The authors wondered if there might be a way to choose a different paradigm for line monitoring that might make these challenges easier to address. As will be explained shortly, Axiomatic Design provided the key to designing a new concept better able to meet the needs of this industry.

1.1. Methodology

The general process used to design the sensor mainly follows the Generic Product Design Process outlined by Ulrich et al. [5]. However, it also integrates with Axiomatic Design and Collective System Design theory to formally define design elements and Functional Requirements (FRs), and analyze them concerning coupling.

Axiomatic Design (AD) is a methodology created by Nam Pyo Suh [6] to provide a generic method to design mechanisms with a focus on the manufacturing industry. Axiomatic Design uses Design Parameters (DPs) to facilitate the design of modular and robust systems through the application of two axioms. The Independence Axiom states that the designer must eliminate undesired coupling to keep FRs independent. The Information Axiom mandates that the design should minimize information content i.e. minimize the probability that the FRs are not met. This method utilizes a design matrix to detect coupling between FRs and DPs.

Collective System Design (CSD) was developed by David S. Cochran [7] to make Axiomatic Design more approachable to the automobile manufacturing industry.

CSD instead partitions DPs into Physical Solutions (PSs) and quantifiable metrics. The design matrix is replaced with a box and arrow diagram called a Collective System Design Decomposition (CSDD) to detect coupling in the system. Arrows indicate coupling and should never cross; non-vertical arrows indicate undesirable coupling that needs to be eliminated when possible (Independence Axiom). CSD is Cochran's answer to simplify some aspects of Axiomatic design that many find complicated to apply.

After the problem was identified, the design team moved on to identifying CNs to the product and mapping these to FRs. From there, metrics were defined that the product would have to fulfill. The metrics were used to drive concept generation and selection. This part of the process was somewhat iterative as interviews were conducted at different stages of the process and the information gleaned in them had to be used to the fullest.

1.2. Customer Needs

As with any design process, we need to gain insight into the stakeholders to ensure the solution makes sense. The target users of our product are Icelandic Transmission System Operators (TOS) and Service Providers. Based upon the feedback from these stakeholders i.e. "customers", the product is likely to be appropriate worldwide.

The Generic Product Design Process[5] method was used to develop the customer needs (CNs) for such a product. In general terms, this method revolves around identifying explicit, unfulfilled, and latent CNs through interviews and customer interaction. Results are analyzed critically to prioritize, consolidate, and build a hierarchy. The following explicit and/or unfulfilled CNs were compiled from discussions with the local service providers and operators, Svarmi [8] and Landsnet [9]. The top-level need identified was "CN_0 Measure the surface temperature of high voltage conductor lines" which can be decomposed into:

CN_1 Accurate data about conductor temperature.

CN_2 Gather data over up to two weeks.

CN_3 Deploy it quickly and without too much hassle.

CN_4 Have no worries about the equipment

CN_5 Remove any equipment from a line reliably and easily

CN_6 Analyze the data reliably

CN_7 Have no worries about my safety or damage to my surroundings.

From the development of these needs, it is clear that the customer is interested in increasing convenience. We continue with a survey of existing products for power line sag evaluation. This survey gives the context of what the customer's current offerings are.

2. Prior Art

Currently, there are several standardized methods to monitor the sag of power lines:

- Sending out in-person teams that pass along the power lines to monitor the sag.
- Drone surveillance with a camera, LiDAR and/or temperature sensors[10]
- SAR (Synthetic-aperture radar) satellite data[11]
- Sensors on the line that measure position and/or temperature

Customers considered sending out teams to be outdated due to inefficiency and high personal cost. Drones surveillance is an exciting prospect, but are not able to measure for sustained periods in a single location; they are more appropriate to check problem areas on a routine basis. Direct drone measurement also requires an operator for the drone during the measurement. Radar measurement is an exciting new technology, but it is not universally available. The customers contacted currently consider sensors on the line to be the best mechanism to gather the needed data so the authors focused on investigating existing products in that category.

The ASTROSE® sensor (Figure 1a) is a power line sensor developed by the German Fraunhofer Institute for Reliability and Microintegration IZM [15][16]. It can measure the inclination, torsion, and current of the power line it is permanently attached to. Sensors are up to 500 m apart and the signal is sent through the chain of sensors to a base station. The sensors harvest the power from the power lines and do not need any battery.

LAKI is an Icelandic startup that focuses on power line monitoring and surveillance [17]. Their main product is a sensor station (Figure 1b) that is attached to a power line. The station can be equipped with various sensors like cameras, thermal sensors, etc. They claim their station can harvest significantly more power from the attached line than their competitors. The camera feed can be used to detect sag or ice on the line.

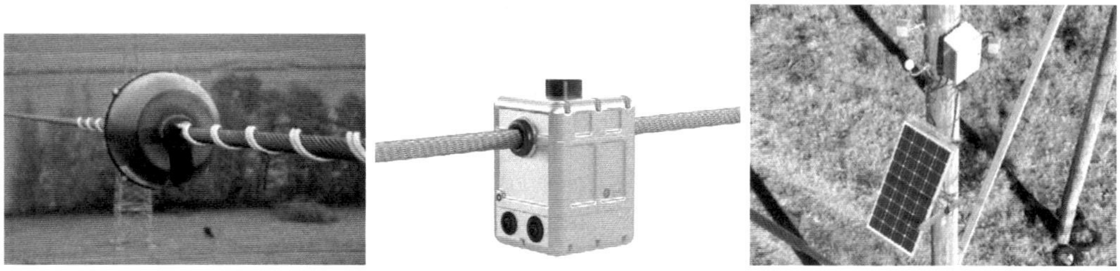

(a) ASTROSE® sensor with smooth casing [12] (b) Laki sensor station harvesting power from the line [13] (c) Kadtronex sensors with node [14]

Figure 1: A selection of power line monitoring systems

Kadtronex is a British company that builds basic temperature sensors (Figure 1c) that can be attached to a power line [18]. The sensors themselves are connected to a node at the base of the pole and communicate with other nodes or a cloud-based server. The data can then be monitored in real-time.

All the investigated products can measure and monitor the surface temperature of power lines. Their general concept is very similar: you attach a sensor permanently to a power line and monitor the temperature. You can either read out the data in real-time or read out their logs afterward. None can address CN_3 and CN_5 regarding easy deployment and retrieval. The authors were not able to find an existing commercial solution that can meet those requirements in particular. This indicates that there is an opportunity to innovate with a new design that can meet all of these needs.

2.1. Functional Requirements

With the background of the product and the CNs now defined, it was possible to define Functional Requirements (FRs) for our product. The authors investigated a variety of concepts and considered them in the context of AD. It was quickly realized that there was an assumed coupling in the Customer Domain between a secure installation (CN_4) and having an operator interact with the live power line (CN_3,CN_7). The proposed concept that was able to eliminate this coupling was to make the sensor deployed by a drone and retrievable without additional intervention This "temporary sensor" would achieve the same results without the need to install the sensor permanently. The goal is to create a self-detaching thermometer, inspiring the top-level "FR_0: Measure power line temperature for up to two weeks without service interruption" which is decomposed into:

FR_1 Logs data with 0.5 °C accuracy

FR_2 Discharges battery over two weeks of operation

FR_3 Deploys with the help of a drone within 10 minutes

FR_4 Survives a fall of 20 meters

FR_5 Removes itself from the line

FR_6 Transmits its data after usage

FR_7 Does not damage people or the power line

In addition to desired functionality, constraints were identified, though the exact tolerances were not known. Constraints come in two types[19, p. 21]: input constraints from the design specifications and system constraints that relate to the operating environment.

CON_1 Cost must be below X ISK. (Input)

CON_2 Survive extreme weather environments including thermal cycling, sub-zero temperatures, high-velocity wind, and precipitation in all formats. (System)

CON_3 Materials and components commercially available in Iceland. (Input)

CON_4 Measurement must not be affected by the strong electromagnetic field emitted by the alternating current high voltage transmission line. (System)

3. Concept Phase

These FRs led the team to the concept of a mobile temperature sensor that can be temporarily deployed and removed from live high voltage lines with the help of an external mechanism. Temperature readings from the conductors can be used in conjunction with LiDAR scans [20] of power infrastructure to optimize the allowable load on lines and predict the behavior of new power lines during the design phase. The device would deploy to the line with the assistance of a VTOL drone or other mechanism safe to use near high-voltage lines. Of note, LiDAR scans are already performed via a VTOL drone. The sensor would then log line temperature data over a predetermined period and fall off the line due to gravity at the end of this period. Multiple design elements need to be taken into consideration around high voltage lines suspended many meters up in the air.

The Information Axiom was considered for FR_3 and it was quickly realized that it might be very challenging for a drone pilot to line up an attachment area with the power line and trigger a clamp. A guiding rod and ramp geometry was added to orient the device with the line so that a drone can lower it onto a power line. A clamping assembly using an infrared reflection sensor triggers the device to start closing and an end stop indicates when the power line is secured.

There was also a concern about the robustness of temperature measurement. Data collection was handled with three digital thermometers for redundancy in case of error. These separate data streams can be evaluated together to check for corruption.

The Information Axiom guided the team to realize that there might be high information content on the release mechanism, so an additional effort was put into it. Customers indicated that even in the case of complete malfunction, it was critical for the device to end up on the ground to avoid service interruption while retrieving a faulty device. This resulted in a variety of fail-safe mechanisms that trigger on software, hardware, or battery failure: it is optimized to "Fail-open". A real-time clock ensures proper timekeeping and triggers release upon matching the specified release time even if there is a period of low power.

3.1. Implementation

The chosen approach named "Line Dancer" was now ready for implementation. The PSs that were derived to fulfill the FRs are summarized by the top-level "PS_0: Self-attaching and self-detaching thermometer"

The subsequent PSs describe how the overall physical design should be implemented:

PS_1 Contact thermometer

PS_2 Compact battery with capacity at least 10000 mAh. Low power components drawing less than 100 mA

PS_3 Claw detects wire for attachment, guide rod

PS_4 Shock absorbent casing, Acrylonitrile styrene acrylate (ASA) filament for 3D printing

PS_5 Two independent mechanical/electrical fail-open emergency systems that detach unit in case of error

PS_6 Internals shielded for electromagnetism with metal lining. Redundant sensors for error correction

PS_7 Indicative safety through strong colors and warning noises.

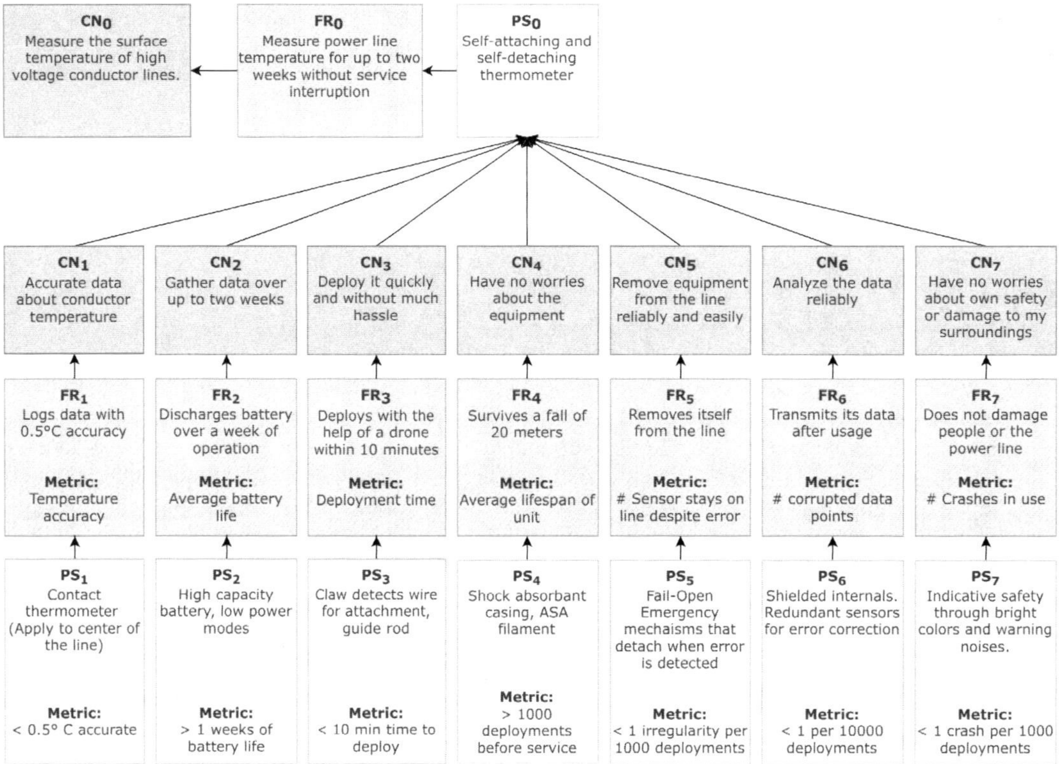

Figure 2: Collective System Design Decomposition

The list of CNs, FRs, and PSs compiled into a CSDD diagram (Figure 2). Analysis of the CSDD indicates that this is an uncoupled system due to the lack of diagonal arrows. Metrics for FRs and PSs are used to check whether the prototype fulfills that element.

3.2. Concept Testing and Detailed Design

The next step in the process was concept testing where different deployment designs were shown to stakeholders. The clear preference from the stakeholders was deployment with a drone. The team decided to move forward with a design depicted in Figure 3a.

Next, a non-functional sketch model (Figure 3b) was created to gain a better understanding of how the unit would behave physically and the sizing of components. The prototype showed the need for an alignment rod and defined the boundary for the modules contained within.

3.3. Prototype

Based upon the CSDD, the team considered the appropriate modularization to build an integrated prototype. This prototype was intended to be a functioning proof of concept that would show the mechanisms without including the environmental hardening required for deployment in the field. The modules selected were:

Chassis: 3D printed housing, with shock-absorbent material particularly on the "head" which will be oriented down during release.

Real Time Clock: To ensure that the controller knows when it is time to release

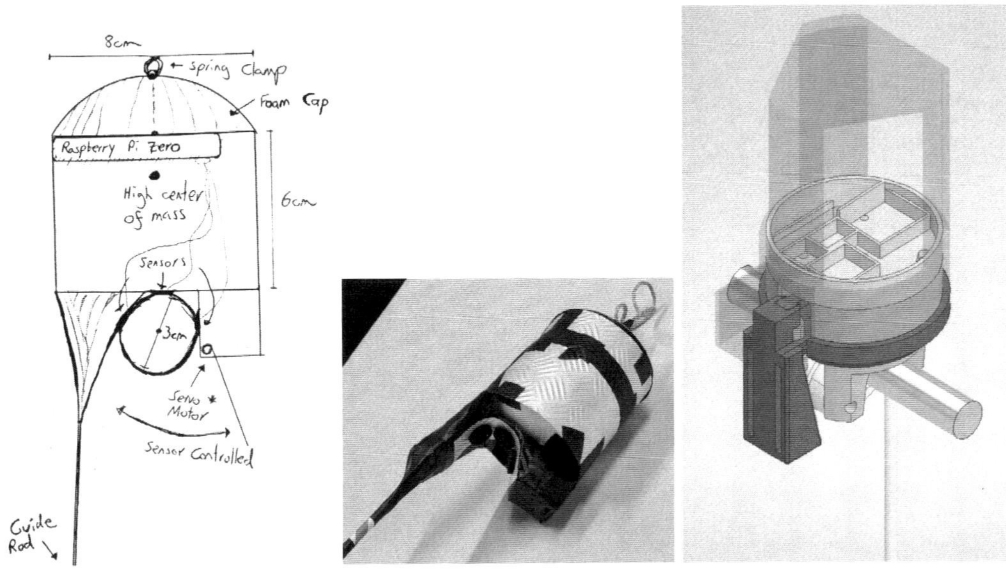

(a) Sketch of selected concept: drone deployed

(b) Sketch model

(c) CAD model: Chassis overview and locking mechanism

Figure 3: Progression from sketch to model

Micro Controller: Adafruit Feather M0 Express Microcontrollers (Atmega SAMD ARM-based)

Battery: 10000 mAh LiPoly battery pack

Sensor-Wire Clamp: 3D printed transmission with DC Motor connected to worm drive

Temperature Sensors and Attachment Module: Multiple digital temperature sensors allow for detection of anomalies.

Watchdog Module and Electrical Release: To check if the main controller is functioning normally.

Physical Release: To force release via drone in case both sets of electronic fail.

The form of the final device was modeled in Autodesk Inventor 2022 (Figure 3c). The circuit diagram shown in Figure 4a shows how the components are wired together. To implement controller redundancy, two Adafruit feather microcontrollers were used. One acts as the main controller while the other one acts as a watchdog that triggers an emergency release routine when the main controller does not send a heartbeat signal for too long. A simple state machine and pseudo-code were used to guide what functionality the code needed to implement. The final version follows the control flow in Figure 4b.

The sensor-wire clamp and emergency release mechanism are pictured in Figure 5. The worm drive (threaded shaft) allows the clamp position to be maintained even when the motor is not powered. The spur gear is pressed against the worm drive by a monofilament fishing line holding the transmission together. The device releases from the power line via three independent ways:

(i) The main controller can reverse the clamp motor to open the clamp.

(ii) The watchdog controller can turn on a relay to a thermal wire to melt a polymer string that holds the clamp's transmission together. Once the transmission separates into two pieces, the clamp can move freely.

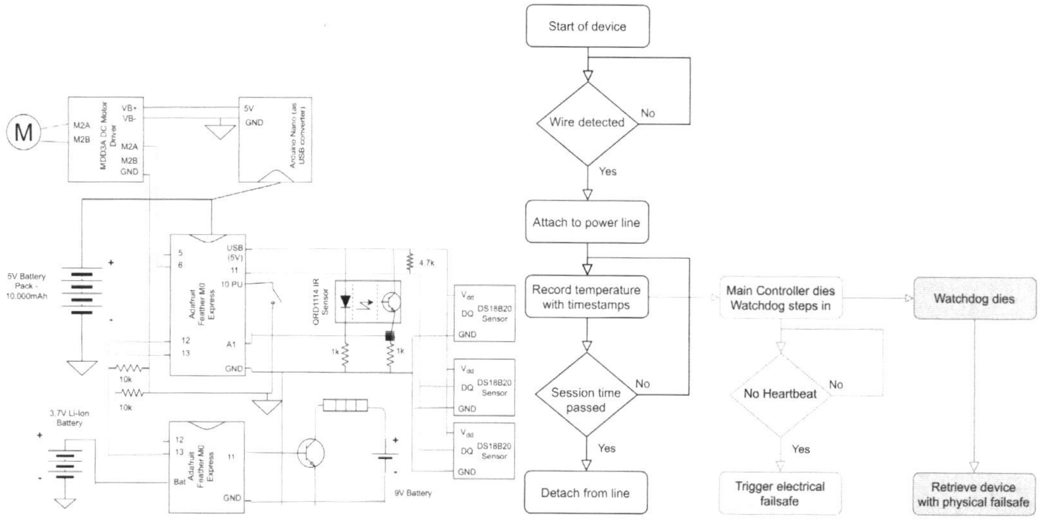

(a) Circuit Diagram. Main controller on top and watchdog on the bottom.

(b) Software Control Flow

Figure 4: Line Dancer's electronics and software design

Figure 5: Side view of clamp assembly. Monofilament fishing line keeps worm drive transmission engaged until it is melted or cut.

(iii) A drone can grab onto a hook on the alignment rod, which pulls a cutting blade against the polymer string to release the clamp (See bottom of Figure 6b).

The components were 3D printed using the model. The final prototype was then assembled from the components printed and modules build previously. The 3D printed components went through several iterations to optimize functionality. Figure 6 shows the final prototype in the two operational orientations.

4. Results and Discussion

Once the prototype was ready the team moved on to testing. An analytical prototype was developed in parallel with the physical one which both focused on the durability of the clamping mechanism to the power line. The prototype and accompanying tests had to confirm that the geometry was suitable for attaching the whole unit to the line. The device also needs to reorient

(a) Deployment orientation highlighting electronics placement and impact absorption cap. Hooks on the end of the guide rod trigger the physical emergency release.

(b) Measurement orientation with closed attachment mechanism

Figure 6: The Line Dancer integrated prototype

when detaching to "land on its head" i.e. the shock absorber. Fail-safe detachment in case of software, hardware, or battery error also needed testing. Finally, the clamping force needed testing to prove that the device would stay securely on the line.

An important input into the analytical models was the mass of the system and its center of mass. The prototype weighs 954 grams. The center of mass is correctly placed to enable the Line Dancer to invert itself upon release of the clamp by pivoting around the wire: this is accomplished by positioning the center of mass high enough to create sufficient torque. The next prototype's mass is likely to be heavier once metal shielding is incorporated into the casing for durability and magnetic field protection. This new mass will be used to calibrate the shock-absorber geometry for the next generation.

With the original cardboard sketch model, it was already possible to test out the physical geometries and specifications needed. The 3D printed physical prototype allowed us further testing whether the geometry and component selections were valid choices to meet the FRs. A basic FEA was performed with the analytical prototype which was derived from the CAD model of the suggested solution. The focus again was on the sensor to power line clamp. From the stress analysis (Figure 7), the design of the claw can withstand 80 N.

Subsequently, the team performed stress tests with the 3D printed physical model by attaching a force gauge to the device, while it was deployed to a metal rod and pulling until the device fell off. The unit detached under a load of approximately 25 N at which point the spur gear would slip against the worm gear. This force is significantly below the target value of 80 N and will require stiffer interfaces holding the worm drive against the spur gear; it is also possible this is a limitation of the 3D printed plastic.

The power consumption had to be analyzed to ensure that the device can operate for the desired interval of two weeks. The current usage could be found from data sheets of the various components and current measurements on the device. Battery capacity was generally listed in milliampere hours (mAh). Given the capacity of a battery and the current consumption, it was possible to calculate the battery discharge time through the relation $t = C/I$ where t is time, C is capacity, and I is the current draw. Table 1 shows the calculations for the system at an idle state and maximum load. For the main apparatus, a 10000 mAh battery was chosen due to availability. The capacity in the deployment environment is likely to be lower due to lowered

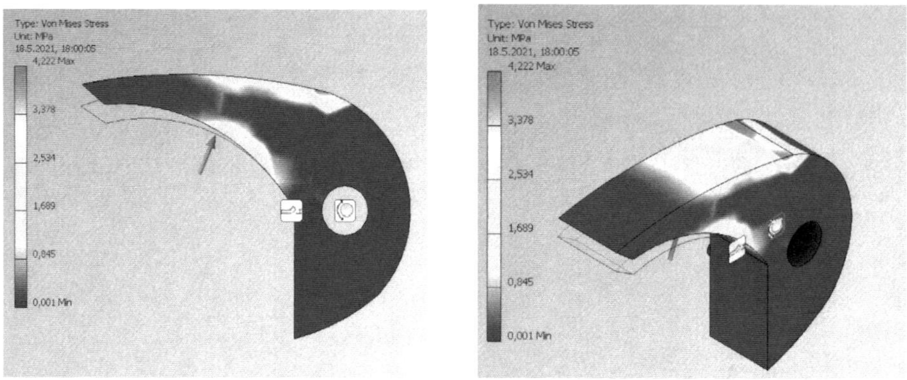

Figure 7: Stress analysis conducted with Autodesk Inventor. Point load of 80 N applied at the arrow.

Table 1: Prototype Power Consumption

		Idle		Max	
	V	A	W	A	W
2×Adafruit Feather M0 [22]	5	0.0146	0.073	0.0146	0.073
Micro Metal Geared motor w/Encoder	5	0.03	0.15	0.1	0.5
3×DS18B20 Digital Thermometer [23]	5	0.0045	0.0225	0.0045	0.0225
Infrared LED	5	0.004	0.02	0.004	0.02
Total		0.053	0.265	0.123	0.615

temperatures [21]. If the device spends most of its time running at the idle current consumption of 53 mA, this would result in 188 hours ≈ 7.8 days. This is lower than the session time of two weeks specified in FR_2 but a reasonable measurement period because that the device can be retrieved, recharged, then redeployed again.

A future version will employ multiple strategies to meet FR_2. The main controller will control a solid-state relay to cut power to the motor controllers and sensors during idle periods. The current battery can maintain the microcontrollers in idle mode for 666 hours ≈ 27.8 days, which would meet FR_2's duration. Idle periods will be increased to take measurements at 1-minute intervals or longer as the measurement session requires. Finally, the code will take advantage of the SAMD-based microcontrollers' low-power modes to further reduce current draws.

An integration test was performed to evaluate the functionality of the system as a whole. A length of pipe the same diameter as a high voltage line was placed into the clamping area. The IR distance sensor was able to detect the "wire" and begin closing. If the wire was moved away before it finished closing, the clamp began opening. When the clamp completely closed, the end stop was triggered and the motor turned off. The device then measured the temperature on the line and was able to detach after a predetermined amount of time. Due to limited time, the prototype was not able to be tested with a drone deployment, but the prototype should be able to perform in the same way in such a situation.

At the stage of testing that was reached, temperature accuracy was secondary to the application of the device. The temperature sensors rated for 0.5 °C accuracy which is believed to be sufficient. The three separate measurements are processed by the microcontroller to generate a single temperature value even in the case of multiple sensors failing. Further testing is needed with these sensors in the presence of strong electromagnetic fields and to calibrate them with an official calibration thermometer.

4.1. Conclusion

The system dubbed "The LineDancer" can now successfully clamp to wire surrogates of various diameters in a controlled environment through manual placement. It can log temperatures even in the case of multiple sensor failures. Additional shielding is required to operate these sensors on a 220 kV powerline which will be addressed in future work. There are two fail-safe releases implemented in the Line Dancer, one physical and one electric. The physical release is trigger by an outside force and the electric one is triggered by the watchdog micro-controller. AD-driven design enabled the independent development of modules due to the FRs not being coupled.

The general design was able to measure temperature on a wire-like surrogate (FR_0). The three digital thermometers with error correction ensure reliable measurement (FR_1). The chosen LiPo battery does not currently meet the measurement duration but can be modified with less frequent measurements (FR_2). Based upon automatic closing of the clamp in the integration experiments, drone deployment should be quick (FR_3). The 3D-printed casing was not durable enough to survive a fall in its current iteration. Additional shock-absorbing material and a sturdier casing are in the works (FR_4) FR_5 was realized with a standard clamp release procedure and dual failsafe. The Adafruit Feather microcontrollers provide easy data extraction via USB or microSD (FR_6) The warning signals, reflectors, and bright colors were unable to be realized in this phase of the project (FR_7)

Overall, AD identified opportunities that became central to this product's innovative aspects. We find that the current state of the project is a success. The prototype is not ready for production but highlights the benefits of a temporarily attached thermometer for line sag models. Overall the team believes that it has designed a unique product that has proved many key concepts it wanted to see working in action. Even with a very limited development cycle of 15 weeks, these results provide a springboard to further work in the area of high voltage engineering and drone-assisted measurement.

References

[1] Waldron M, Arboleya L, Bennet S, Gonzalez P, Gould T, Nobuoka Y, Olejarnik P, Pospiech R and Schulz R 2020 World energy investment 2020 Tech. rep. IEA
[2] Smith H L 2010 *Electric Energy T&D Magazine* **14**
[3] Gupta S D, Kundu S and Mallik A 2012 *International Journal of Recent Technology and Engineering* **1** 43–45 ISSN 2277-3878
[4] Landsnet 2021 Legal requirements power lines URL `https://www.landsnet.is/flutningskerfid/um-flutningskerfid/raforkumannvirki/`
[5] Ulrich K T, Eppinger S D and Yang M C 2020 *Product Design and Development* 7th ed (McGraw Hill)
[6] Suh N P 1990 *The Principles of Design* (Oxford University Press)
[7] Cochran D S 2009 *The Journal of RMS (Reliability, Maintainability, Supportability) in Systems Engineering* **Winter** 6–10
[8] Svarmi Corporate website URL `https://svarmi.is/`
[9] Landsnet Corporate website URL `http://www.landsnet.is/`
[10] Delair SAS 2020 Delair: Aerial intelligence corporate webpage URL `https://delair.aero/`
[11] Iceeye Corporate website URL `https://www.iceye.com/`
[12] Fraunhofer IZM Astrose sensor image URL `http://astrose.de/de/astrose-system/jcr:content/contentPar/teaser_overview/teaserPar/teaser/image.img.2col.large.png/1474303757173/astrose-knoten-02.png`
[13] Laki Power Laki power sensor station URL `https://www.lakipower.com/products`
[14] Kadtronex Kadtronex sensors with node URL `http://www.kadtronex.co.uk/wp-content/uploads/2018/10/TORINO01.png`
[15] Fraunhofer IZM ASTROSE: Perfect monitoring for grid operators URL `https://www.astrose.de/en.html`
[16] Fraunhofer IZM Corporate website URL `https://www.izm.fraunhofer.de/en.html`
[17] Laki Power Corporate website URL `https://www.lakipower.com/`

[18] Kadtronex Corporate website URL http://www.kadtronex.co.uk/
[19] Suh N P 2001 *Axiomatic Design - Advances and Applications* (Oxford University Press)
[20] Svarmi Drone LiDAR survey URL https://svarmi.is/services-with-drones/#lidar
[21] Zhang S, Xu K and Jow T 2003 *Journal of Power Sources* **115** 137–140 ISSN 0378-7753 URL https://www.sciencedirect.com/science/article/pii/S0378775302006183
[22] Atmel 2015 *SMART ARM-Based Microcontroller*
[23] Maxim Integrated 2008 *Programmable Resolution 1-Wire Digital Thermometer* rev. 042208

IOP Conf. Series: Materials Science and Engineering 1174 (2021) 012027 doi:10.1088/1757-899X/1174/1/012027

Collaborative Engineering definition: Distinguishing it from Concurrent Engineering through the complexity and semiotics lenses

Goran D. Putnik[1,2], **Zlata Putnik**[2], **Vaibhav Shah**[3,2], **Leonilde Varela**[1,2], **Luis Ferreira**[4], **Helio Castro**[5,2,6], **Catia, Alves**[1,2], **Pedro Pinheiro**[2]

[1]Universidade do Minho, Department of Production and Systems Engineering, 4800-058 Guimarães, Portugal
[2]ALGORITMI Research Centre, Universidade do Minho, 4800-058 Guimarães, Portugal
[3]Universidade do Minho, Department of Information Systems, 4800-058 Guimarães, Portugal
[4]2Ai—School of Technology, IPCA, 4750 - 810, Barcelos, Portugal
[5]School of Engineering (ISEP) - Polytechnic of Porto, 4200-465, Porto, Portugal
[6]INESC TEC, 4200-465, Porto, Portugal

E-mail: putnikgd@dps.uminho.pt

Abstract. A contribution to the definition of Collaborative Engineering (ColE), together with the definition itself, are presented. The need for a more rigorous definition is due to the fact that the Collaborative Engineering (ColE) definitions in literature within the so-called "classical" discourse in fact could be reduced to the Concurrent Engineering (CE), making synonymy between the two terms. However, the need for distinguishing the semantic contents of the two terms and concepts, Collaborative Engineering and Concurrent Engineering, could be understood to be driven mainly by new, emergent, organizational and management concepts that refer to new features required for the engineering design organizational and management approaches. The two new, emerging theories and paradigms, that inform design and practice of new, emerging engineering design theories, approaches and practices, and based on which the definition of the Collaborative Engineering is proposed are the complexity theory and semiotics, in particular the complexity management in organizations and organizational semiotics.

1. Introduction

The paper's objective is to contribute to the definition of Collaborative Engineering (ColE). The need for this definition is due to the fact that the definitions (of ColE) that could be found in the literature are many times confusing referring to the same features as the Concurrent Engineering (CE) features, meaning that for the same concept two different terms are used. While this situation is acceptable in some contexts, other contexts require clear distinction of meanings by two terms.

The need for distinguishing the semantic contents of the two terms, ColE and CE, could be understood to be driven mainly by new, emergent, organizational and management concepts that refer to new features required for the engineering design organizational and management approaches.

These new, emergent, organizational and management concepts, that open new space for defining new approaches to engineering design, are mainly based on two main sources.

The first source of the need for definition of a new engineering design organizational and management approach is emerging complexity thinking paradigm, that is, in its model oriented to "chaos and complexity in organizations" called Chaordic System Thinking (CST). The second source of the need for definition of a new engineering design organizational and management approach is semiotics, that is, organizational semiotics. Organizational semiotics is closely related to the complexity thinking paradigm, i.e. both share a number of similar and/or equal concepts.

One of the questions is how to name this new engineering design organizational and management approach. The authors propose, that this new engineering design organizational and management approach to be named Collaborative Engineering.

In this paper a definition of ColE through the lenses of Complexity thinking and Semiotics will be presented, which will make a clear distinction from well-known CE. In this due, the paper will present the related concepts, as well as the features of the ColE, as a new engineering design approach.

2. Collaborative Engineering (ColE) and Concurrent Engineering (CE) are synonyms within the "classical" discourse

Considering the differences between the semantic contents of the terms ColE and CE, i.e. between their definitions, the first step is to recall the definitions of the term *definition* itself. Historically, the perception of the meaning of the term *definition* has evolved within the areas of logic and linguistics.

The "classical" approach considers that the role, or function, of a definition of some concept, phenomenon, object - denominated by a term or syntagm (such as the terms and/or syntagms for denominations of the concepts we are addressing in this paper, i.e. of the concepts of "concurrent" and "collaborative", "concurrent engineering", "collaborative engineering") - is to describe precisely that concept, phenomenon, object. This type of definition is called "normative" or "prescriptive" definition (the first discussion on the topic by the authors of this paper is published in [1]). The normative, or prescriptive, definitions are those in which it is prescribed how something should be in reality. Normativity denotes theory, ordering and truth.

As described in [1], when analyzing the descriptions, their semantics adopted in common language, of the terms "concurrent" and "collaborative" taken from the commonly used dictionaries, namely from the on-line Merriam-Webster's dictionary [2], it is easy to see that the terms "concurrent" and "collaborative" are usually considered as synonymous.

Consequently, the syntagms "Simultaneous Engineering", "Concurrent Engineering" and "Collaborative Engineering" should be considered to denominate the equivalent concepts, i.e. the respective terms/syntagms are synonymous, see [1].

The "common features" referred to both CE and ColE are e.g. 1) approaches oriented to, we would call it, "conventional" engineering/design environment, which is a company. The "company", as the organizational, management and phenomenological environment implies 2) the common goal, meaning practically without exception the common product, to the (members of) engineering/design team. In other words, the goal product is given[1]. 3) Another implicit engineering/design feature implicit to "company" is that the main management goal is some kind of "efficiency", e.g. coordination, or conflict resolution in engineering/design activities [1]. Phenomenologically, whenever "multifunctional teams", or "stakeholders along the product life-cycle" within the "company" are referred, "conflicting interests", the ColE could be, virtually, reduced to, or interpreted as synonymous to CE [1] (we should not forget that we are talking about the definitions within the "classical" approach/discourse).

One of the typical examples of using two different terms for the same concept is in [3]. Citing from [1]: "in [3], the ColE is defined as 'A Formal Definition of Collaborative Engineering facilitates the

[1] When discussed the feature of the "common goal" and/or "given goal" one of the international experts, participant in discussion, said that he never participated in any engineering/design meeting without "common" or "given" goal.

communal establishment of technical agreements among a team of interdisciplinary stakeholders, who work jointly toward a **common goal** with limited resources or **conflicting interests**'[2], and (specifying the decision making methodology) "*Engineering Collaboration via Negotiation (ECN)*". However, the suggested 'new' decision making methodology 'negotiation', is just a tool for conflict resolution in CE.

3. Non- "classical" discourse for distinction between the Collaborative Engineering and Concurrent Engineering definitions

However, today, it is known that the semantics of the terms, or syntagms, of the concepts by itself is not sufficient, but it (the semantics) should be considered within the context in which it has been defined and in which (context) it is used, considering that each phenomenon is context sensitive, and changing the context the semantics could (but not necessarily) change as well. This is especially notable when considering new scientific discoveries, research results and technological developments and innovations, especially in engineering sciences. These new discoveries, in fact, generate motivations, and needs, to change the semantic contents of the previously "established" definitions, and, consequently, their change or defining new definitions.

This reality led to "abandoning" of "normative" or "prescriptive" approach to definition of *definitions*, and to the formulation of, so called, "descriptive" definitions. The descriptive definitions are those in which the reality is described and analyzed in different "spheres" (domains, fields, use, applications and circumstances) and contexts [1].

Actually, a number of problems arise. These could be resumed as follows [1]:

1) keeping the same denominating term, or syntagm, for different semantic contents, implying different definitions, and

2) using different denominating term, or syntagm, for the same semantic contents, implying same definitions.

Both situations happens in relation with the terms/syntagms: "concurrent" and "collaborative", i.e. with the "concurrent engineering" and "collaborative engineering", see [1].

Is this situation acceptable?

From the point of view of the co-called "classical" (normative, prescriptive) approach this is not acceptable. From the other side, from the non-"classical" (descriptive) approach this is acceptable, and even desirable. In fact, we need both approaches.

The theory "behind" this position, that we need both approaches, could be the view from semiotics on development of "signs" [4]. Paraphrasing Guiraud's (1975) scheme [5], Figure 1 presents an explanation of the sign, i.e. the definitions, changes and creation and changes of 'fields of observations', or 'information fields', and influence of the 'individual/community' and 'society' on the "collaborative (engineering)" paradigm definition, through the enrichment or 'impoverishment' of the potential for innovation and novelty of the "collaborative (engineering)" definition and of the potential for the "collaborative (engineering)" effectiveness or efficiency.

The scheme from the Figure 1 in fact means that the needs for the knowledge and "codification", i.e. it could be said, the needs for the "normative" definitions, grows with the need of society, i.e. for the communication within the society, while when there are needs for change, originality, innovation, novelty, the need for the "normative" definitions diminish and the need for learning and "de-codification" grows.

In other words, with the development of new approaches to the engineering design, new organizational and management forms emerges and the need for changing the semantic contents of the terms and syntagms of the "collaborative" and "collaborative engineering" usually considered within the "classical" approach, or within the "classical" discourse, especially in relation to the terms and syntagms "concurrent" and "concurrent engineering", emerges as well.

So, to define "Collaborative Engineering" as distinct from "Concurrent Engineering" (because '*nomen est omen*'), the question is which are the new, emerging engineering/design theories, approaches

[2] Bold letters by the authors of this paper

or practices where there is no "competing/concurrent/conflicting interests" over the common goal product to the group of engineers that "collaborate"? [1]

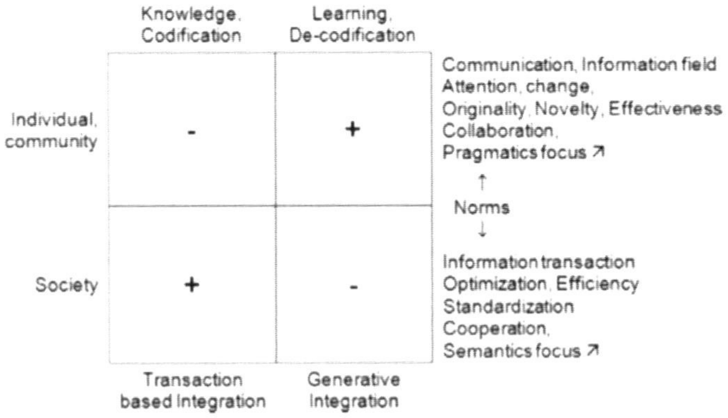

Figure 1. Influence of the 'individual/community' and 'society' on the paradigm characterisation and definition [4].

4. Chaordic System Thinking (CST) and Semiotic frameworks for defining a new engineering design organizational and management approach

The new, emerging theories and paradigms, that could inform design and practice of the new, emerging engineering design theories, approaches or practices, and that could represent an instrumental framework for their development, are without doubts those that are changing, and transforming the "traditional" view in many other areas. These are, at the first place, the complexity theory and semiotics, in particular the complexity management in organizations and organizational semiotics. From the other side, both complexity and semiotics paradigms, or theories, serve as well as the explanatory frameworks for a number of emerging practices in the area of engineering design not yet "codified" as the theoretical models.

4.1. Chaordic System Thinking (CST) and Complexity Management framework for new engineering design organization and management systems

Chaordic System Thinking (CST) is a model of complexity thinking and of complexity science applied to the complexity management in organizations. Complex systems are distinguished from the "classical" Newtonian systems by a number of features. At the very high level of abstraction, the complex systems are characterized by [6][7]:

> "a great many independent agents are interacting with each other in a great many ways, (...) [*in which*] systemic interactions can lead (...) to spontaneous self-organization, [*and*] learning which takes place through feedback." [7]

These features, together with more detailed features of CST, are critical for design of the new engineering design organizational and management approach. All these features, of the systems' complexity and CST, are well presented in a number of references, especially in [6], [8], [9].

A CST based system, has the capacity for creation of true novelty, whether novel products or novel technologies. The capacity for creation of the "novel" product is one of the main requirements for the new engineering design organization and management systems. From the other side, the capacity for creation of the "novel" technologies refers to the completely new forms of work and management of the engineering design organizational system. The term "novelty" is defined as something completely new. The basic idea is that both the *what* (Goal) (the subject matter to be created) and *how* (Path) (the process by which the subject matter is to be created) are unknown, in advance [10]. This is just another description of the above referred feature of the capacity for "spontaneous self-organization".

To provide this capacity for novelty creation (for novel ideas), the new engineering design organization and management system should implement the CST five chaordic properties: *Consciousness*, *Connectivity*, *Indeterminacy*, *Dissipation*, and *Emergence*. These are presented in more

IOP Conf. Series: Materials Science and Engineering 1174 (2021) 012027 doi:10.1088/1757-899X/1174/1/012027

details, including the implicit design rules for the new engineering design organization and management system, in [11], [9] (cited in [9]).

Selected features of the complexity / CST based approach to inform the design of the new engineering design organization and management system are given in the first part of the Table 1.

4.2. Semiotic frameworks for new engineering design organization and management systems

The second fundamental and qualitatively new feature of the new engineering design organization and management system is employment of semiotics as an instrument.

A view on use of the semiotic-based approach for manufacturing systems, in particular for the manufacturing systems integration / interoperability, together with an overview of the semiotics basic properties, is presented in [4]. The proposed semiotics-based framework for Manufacturing Systems Integration and interoperability (S-MSI) in [4] considers a number of S-MSI determinants with their values on two main levels: syntactic-semantic level and pragmatics level. Although the referred framework was proposed primarily for MSI, referring in fact to interoperability, its two aspects – 1) generality and paradigmatic nature of semiotics, making it applicable virtually to any system, and 2) the interpretation of the system "integration" as the system design – or features, make various determinants of the framework characterizing and explanatory, instrumental and applicable, for the new engineering design organization and management system.

It is worth referring to [5], [12], [13] (cited also in [4]) which give excellent descriptions of the consequences based on the semiotics-based view on system, i.e. on engineering systems, and in particular, on the engineering design organization and management systems.

Again, paraphrasing Guiraud (1975) [5] (see [4]), the new, semiotics-based, engineering design organization and management system, could be seen as — the generator of the creative power in the design systems, i.e. as the 'maker' and 'inventor' of the new, semiotics-based, engineering design organization and management system, providing higher levels of coherence with the environment, and social reality, and promoting space for the individual and community-based generation and interpretations of signs. Figuratively, semiotics is the *poiéte* ('maker') of the new engineering design organization and management system. In other words, 'making', and the 'inventing' of the new engineering design organization and management system, is grounded in pragmatics.

Implications of the semiotics-based view on the system design and management, are that actually it is hardly possible to exist an 'absolute', common and universal, interpretation of reality, but, rather, there are multiple interpretations by multiple communities and in different times. It means that semiotics inform us that the new engineering design organization and management system is conditioned by, and employing at the same time, multiple interpretations of the same reality in the same and in different times, differently by different communities. In other words, rooted in the social process of communication, as excellently described by Saludadez and Taylor in [12].

A similar position is expressed by Moor and Weigand in [13]: information systems should be approached "much more as communication systems than computation systems (…). The information systems built on the information field paradigm do not produce sterile data, but aim to generate and communicate information that can lead to knowledge that helps people to perceive, understand, value, and act in the world". This position is well summarized in two tables presented in [13], and cited in [4], which we will use for our presentation of the new engineering design organizational and management approach, i.e. for the new definition of Collaborative Engineering.

Selected determinants, adopted from the S-MSI framework [4], and from [13], of the semiotics-based framework for 1) guiding the development of the new engineering design organization and management system, and 2) for evaluating a concrete engineering design organization and management system solution's compliance with the semiotics-based paradigm, are given in the second part of the Table 1.

IOP Conf. Series: Materials Science and Engineering 1174 (2021) 012027 doi:10.1088/1757-899X/1174/1/012027

5. Characterization of the new engineering design organization and management approach

5.1. The new engineering design organization and management approach characterization

The new engineering design organization and management approach characterization, as indicated in the Section 3, follows the characterization of any systems that could be considered as a complex one, and which employs instruments of complexity management and semiotics based framework for system development, design, management and evaluation.

Table 1 presents a list of the features that characterize ColE as the new engineering design organization and management approach and comparison with the "classical" design engineering approach represented by CE. The list of features in Table 1 is a compilation of the CST and semiotics frameworks features for systems development, design, management and evaluation presented in the Section 4, including as well features referred in [1].

Table 1. Characterization of ColE as the new engineering design organization and management approach and comparison with the "classical" design engineering approach of CE

Perspective	**Feature**	**"Classical" engineering design organization and management approach –** ***(CONCURRENT ENGINEERING)***	**The new engineering design organization and management approach –** ***(COLLABORATIVE ENGINEERING)***
Complexity	*A great many independent agents are interacting with each other in a great many ways*	No	Yes
	Systemic interactions	No	Yes
	Spontaneous self-organization	No	Yes
	Learning which takes place through feedback	Individual learning	Generative learning, Collaborative learning
	Collective aspirations are set free	No	Yes
	Planning	device for controlling the future	a mean for priority setting in the present
	Dissipation - active dissolving of winning design teams, and agile unlearning	No	Yes
	Self- organization, self- reference, and self-transcendence	No	Yes
	Complexity	To be eliminated	To be nurtured
	Organization structure	Fixed, stable structure	"Flow" - as a dynamical process that is traveling an attractor- basins' landscape
	Capacity for true novelty (path and goal unknown in advance)	No	Yes
Semiotics	*Focus*	Information	Communication
	Support	Transaction process	Communication process
	Design objects	Clear specification	'Fuzzy' process definition
	Development process	Single project	Continuous project
	Developers	Elite development team	Many stakeholders
	Change	Static	Dynamic

	Responsability	Anonymous	Individual responsabilities
	Design process	Representation	Interpretation
	Objective	Control	Perceive, understand, value, act
	Control Logic	Rules	Norms
	Environment	Static, linear	Dynamic, non-linear, ´chaotic´
	Process	Computation	Semiosis
	Cognition	Knowledge	Learning
	Performance measure	Efficiency	Efficacy, creativity, novelty
	Design paradigm / process	Prescriptive	Descriptive, co-design, co-creation
	Semiotic field focus	Semantics	Pragmatics
Implementation and operation	***Common goal – PRODUCT***	Yes	No
	Common goal – LEARNING, NOVELTY CREATION	No	Yes
	Simultaneous processes	Yes	Yes
	Development effort focused on the initial phase of the product development life-cycle	Yes	Yes/No
	Relationship graph or network type	Complete / regular network	Complex network
	Group type	Team (common goal product)	Community (no common goal product)
	Conversation type	Discussion	Dialogue
	Principle engineering/design process Management objective	Conflict Resolution	Learning

5.2. Naming the new engineering design organization and management approach as "Collaborative Engineering"

If the new engineering design organization and management approach is clearly characterized, or defined in the generic form, the following question is: how to name it?

That is, what is the name of the concept we are defining?

This is a very important question (reminding the proverb: '*nomen est omen*'), which is already addressed in the Section 3. At the moment we have several terms/syntagms in use that refers the engineering design approaches. These are: Sequential Engineering (SeqE), Simultaneous Engineering (SE), Concurrent Engineering (CE) and Collaborative Engineering (ColE) (see [1])[3]. As the definitions of SeqE and SE are not critical (although many authors use the term/syntagm 'simultaneous engineering' as the synonymous for the term/syntagm 'concurrent engineering', see [1]), we have to decide on terms/syntagms of 'Concurrent Engineering' (CE) and 'Collaborative Engineering' (ColE). As already discussed in Section 2, in common use the 'Collaborative Engineering' could be, virtually, reduced to, or interpreted as, synonymous of 'Concurrent Engineering'. However, we cannot ignore that the two terms/syntagms of 'Concurrent Engineering' (CE) and 'Collaborative Engineering' (ColE) are different,

[3] There are some other terms/syntagms related to the engineering design approaches, such as *Set-based Engineering*, *Design Thinking*, etc., but these will be not discussed in this paper as it could be shown that these follow the principles of one of the approaches referred in this paper, usually follows the principles of concurrent engineering, which could be the subject of another paper.

and therefore, it is expected to denominate different concepts. This observation is of interest for naming of the new engineering design organization and management approach.

So, we have to decide if the new engineering design organization and management approach will be named Collaborative Engineering (ColE) (removing the usual synonymy with the Concurrent Engineering) or to define a new term.

The hypothesis in this paper is that the new engineering design organization and management approach should be named Collaborative Engineering (ColE).

Proving this hypothesis is based on literature review on collaboration theories, i.e. on selected definitions of what does it mean "collaboration".

It is important to say, that a unified theory on "collaboration" does not exist (see a review in e.g. [14]). This fact could justify the practice identified in the Section 2, the practice that created synonymy between CE and ColE. However, there are other models of "collaboration", we would say, more in accordance with emergent system thinking and with other social phenomena, especially related to learning.

For example, in [15] (although not considered as a scientific text) the author clearly stated what "collaboration" is not, addressing the message to the "classical" use of "collaboration", and presented a description of "collaboration", which is almost totally in accordance with the "new engineering design organization and management approach " identified in Sections 4 and 5.1 above):

> "But what is collaboration? Many people prefer the idea that collaboration is "working together to do something." While this is a reasonable starting point, the problem is that there are many different forms of working together. (…) *Collaborating* involves substantial organizational commitment, a very high level of trust, and extensive sharing of turf. The qualitative difference between Cooperating and Collaborating is that collaborating partners demonstrate a public enthusiasm for—and commitment to the value of—learning from each other to become better at what they do collectively. (…) Collaborators are clear that the importance of their partners' success is as great as their own and that their own success depends on their partners' success. Collaborating partners willingly share the risks, responsibilities, resources, and rewards of the work" [15].

The same definitions, in support to our hypothesis, are also found in the scientific literature. For example, in [16] "collaboration" is defined as:

> "Collaboration is the process of shared creation: two or more individuals with complementary skills interacting to create a shared understanding that none had previously possessed or could have come to on their own. Collaboration creates a shared meaning about a process, a product, or an event. In this sense, there is nothing routine about it. Something is there that wasn't there before. Collaboration can occur by mail, over the phone lines, and in person. But the true medium of collaboration is other people. Real innovation comes from the social matrix... [and] is a relationship with a dynamic fundamentally different from ordinary communication (40–41)" [16].

Additionally, important sources to support our hypothesis are presented in a vast literature on "collaborative learning", especially as in CST learning is one of the most important processes and instruments. For example:

> "A learning community is a small group or cohort of students who share common academic goals and work collaboratively in the classroom with one or more professors. At (…), we offer First Year Learning Communities for entering students and Collaborative Learning Communities for all students. When you enroll in a learning community, you will be part of a cohort of 22-24 students. Together, you will take courses that are either "linked" thematically or enhanced by peer learning. By taking classes together and/or engaging in peer-to-peer learning, you get to know each other better, learn from each other, and support each other. And because classes are limited to 22-24 students, you and your instructors will get to know each other better, too!" [17].

In conclusion, "Collaborative Engineering" is the new engineering design organization and management approach, informed by the emerging system thinkings, namely by the complexity management and semiotics.

IOP Conf. Series: Materials Science and Engineering 1174 (2021) 012027 doi:10.1088/1757-899X/1174/1/012027

6. Collaborative Engineering definition

The Collaborative Engineering definition follows from the "new engineering design organization and management approach" description by the features from requirements for it, and their implementation forms.

The definition, in its form presented in this paper, is inspired by the definition of the "learning community" in [17], especially taking into account that the "learning community" for design represents one of the ColE organizational forms (see [18]).

> Definition:
> "**Collaborative Engineering** is the engineering design process (practice) within an intentionally developed community as a complex network, created to promote and maximize the individual and shared learning of its members, through ongoing interaction, interplay, dialogue and collaboration among the community's members to achieve common goals of collaborative learning on design and creation of novel design" [19].

The community, taking form of a complex network of independent designers, usually uses advanced web based environments and platforms for communication, sharing, dialog, learning.

Concerning the common goal, it is important to notice that the common goal is not a particular product (or the product portfolio) "given" by a company, typical for the CE, but the common goal is learning on design and creation of novel design.

7. Conclusions

The definition of the ColE presented is based on the complexity theory and semiotics, in particular the complexity management in organizations and organizational semiotics. This definition is needed as the ColE definitions in literature within the so-called "classical" discourse in fact could be reduced to the CE, making synonymy between the two terms. This situation is not satisfactory because we are witnessing the new, emergent, organizational and management concepts that refer to new features required for the engineering design organizational and management approaches. For these new design organizational and management approaches to design practices, the definitions of the ColE within the "classical" discourse couldn't be used. Therefore, it was necessary to invent a new term/syntagm for these new practices or to redefine the definition of the ColE. The paper's hypothesis was that the term/syntagm Collaborative Engineering is fully adequate. This position was justified by the definitions from the collaboration theory.

The future work will address, in the first phase, a consolidation of the proposed definition of the Collaborative Engineering through investigation of the emergent organizational and management forms and practices in accordance with the definition proposed, investigation of their properties through the lenses of the relevant theories, and in the second phase, development of new organizational and management forms and instruments, using design theories, for example, Axiomatic Design (AD) and other design theories, for designing these new forms and operational instruments, including advanced information/digital technologies. Of course, addressing Industry 4.0 in both phases is understood.

Acknowledgements
This work has been supported by FCT – Fundação para a Ciência e Tecnologia within the R&D Units Project Scope: UIDB/00319/2020

References

[1] Putnik, G. D., & Putnik, Z. (2019). Defining sequential engineering (SeqE), simultaneous engineering (SE), concurrent engineering (CE) and collaborative engineering (ColE): on similarities and differences. *Procedia CIRP*, 84, 68-75.
[2] www.merriam-webster.com/dictionary
[3] Lu, S. Y., ElMaraghy, W., Schuh, G., & Wilhelm, R. (2007). A scientific foundation of

collaborative engineering. *CIRP Annals*, 56(2), 605-634

[4] Putnik, Goran D. and Putnik, Zlata (2010). A semiotic framework for manufacturing systems integration -Part I: Generative integration model. *International Journal of Computer Integrated Manufacturing*, 23: 8, 691–709.

[5] Guiraud, Pierre (1975). *Semiology*. London: Routledge and Kegan Paul Ltd. (first published as: *La Se´miologie*, Presses Universitaires de France, 1971).

[6] van Eijnatten, F. V., Putnik, G. D. (2004). Chaos, complexity, learning, and the learning organization. Towards a chaordic enterprise. *The Learning Organization*, 11(6), 418-429.

[7] Waldrop, M.M. (1992), *Complexity: The Emerging Science at the Edge of Chaos and Order*, Simon and Schuster, New York, NY.

[8] van Eijnatten, F.M. (2004), Chaordic systems thinking: Some suggestions for a complexity framework to inform a learning organization, *The Learning Organization*, 11(6),, pp. 430-449

[9] van Eijnatten, F. V., Putnik, G. D., & Sluga, A. (2007). Chaordic systems thinking for novelty in contemporary manufacturing. *CIRP Annals*, 56(1), 447-450.

[10] Eijnatten, F.M. van, Putnik, G.D. (2005) A Different View of Learning and Knowledge Creation in Collaborative Networks, In: Camarinha-Matos, L.M., Afsarmanesh, H., Ortiz, A. (Eds.), *Collaborative Networks and their Breeding Environments* (pp. 531-539), Springer / IFIP, Berlin.

[11] Fitzgerald, L.A., 2002, Chaos, the Lens that Transcends, *Journal of Organizational Change Management*, 15/4:339-358

[12] Saludadez, J.A. and Taylor, J.R., 2006. The structuring of collaborative research networks in the stories researchers tell. In: F. Cooren, J.R. Taylor, and E.J. van Every, eds. *Communication as Organizing*. Mahwah, NJ: Lawrence Erlbaum Associates, 37–54.

[13] Moor, A. and Weigand, H., 2002. Towards a semiotic communications quality model. In: K. Liu, R.J. Clarke, P.B. Andersen, R.K. Stamper, and E.-S. Abou-Zeid, eds. *Organizational semiotics: Evolving a science of information systems*. Dordrecht: Kluwer, 275–285.

[14] Williams, A. P. (2015) The Development of Collaboration Theory: Typologies and Systems Approaches. In: Morris, J. C., & Miller-Stevens, K. (Eds.). (2015). Advancing collaboration theory: Models, typologies, and evidence. Routledge.

[15] Mashek, D. (2016) *Collaboration: It's Not What You Think – Collaboration is just one of the many ways we can work together*. https://www.psychologytoday.com/us/blog/relationships-intimate-and-more/201602/collaboration-its-not-what-you-think

[16] Montiel-Overall, P. (2005). Toward a Theory of Collaboration for Teachers and Librarians. *School library media research*, v8.

[17] LIUBrooklyn (n.d.) *What Is a Learning Community?*, https://liu.edu/Brooklyn/Academics/Centers/Learning-Communities/What-is-a-Learning-Community

[18] Putnik, G. *et al*. (2021) Collaborative Engineering: A Review of Organizational Forms for Implementation and Operation. submitted for publishing in the *Proceedings of the The International Conference on Axiomatic Design (ICAD) 2021*, Caparica - Lisbon.

[19] Putnik G. (2021) *Concurrent and Collaborative Engineering*. Lectures on Post-graduate course on Industrial Engneering and Management, Department of Production and Systems Engineering, University of Minho. Guimaraes.

IOP Conf. Series: Materials Science and Engineering 1174 (2021) 012028 doi:10.1088/1757-899X/1174/1/012028

Collaborative Engineering: A Review of Organisational Forms for Implementation and Operation

Goran D. Putnik[1,2], Zlata Putnik[2], Vaibhav Shah[3,2], Leonilde Varela[1,2], Luis Ferreira[4], Helio Castro[5,2,6], Catia, Alves[1,2], Pedro Pinheiro[2]

[1]Universidade do Minho, Department of Production and Systems Engineering, 4800-058 Guimarães, Portugal
[2]ALGORITMI Research Centre, Universidade do Minho, 4800-058 Guimarães, Portugal
[3]Universidade do Minho, Department of Information Systems, 4800-058 Guimarães, Portugal
[4]2Ai—School of Technology, IPCA, 4750 - 810, Barcelos, Portugal
[5]School of Engineering (ISEP) - Polytechnic of Porto, 4200-465, Porto, Portugal
[6]INESC TEC, 4200-465, Porto, Portugal

E-mail: putnikgd@dps.uminho.pt

Abstract. A review of organisational forms for implementation and operation of Collaborative Engineering following the Collaborative Engineering organizational and management models based on complexity and semiotics frameworks, is presented. These forms are mainly the forms of communities, which are intentionally created for the purpose of Collaborative engineering. The paper presents basic definitions of three types of communities: learning community, community of practice and open design as the organizational environments for Collaborative Engineering. These communities usually take forms of complex networks of independent designers, and usually uses advanced internet based environments and platforms for communication, sharing, dialog, learning and operation in general. In the second part, some examples of the internet based platforms for supporting the Collaborative engineering communities are presented.

1. Introduction

One of the "keywords" of the Collaborative engineering definition is "community". In fact, by our definition, any organisational form concerning the Collaborative Engineering addresses organizational form of a community. Is a "community" itself an organisational form ? Yes and no. It is in the sense that any group of persons that have some relationship, and interaction, makes an "organization". Even "crowd" could be considered as an organization if some patterns of behavior could be identified. From the other hand, if we consider that an "organization" should have some predefined form or structure, the Collaborative Engineering communities are considered as organisations as these, by the definition, are "intentionally developed". Even in the case of their most "free", or "open", not constrained access and behavior, at least a participant has to register and, in fact, all behavior is controlled by some mediator or similar.

In that sense, communities that are oriented to engineering design activities and resources share, having an implicit or explicit, i.e. prescribed, organizational and management rules, would be considered as Collaborative Engineering organizational forms.

These forms could be classified by diverse criteria, including the purpose, tolls, geography, etc. In the further text, we will refer to three organizational forms by the purpose, namely to the 1) learning communities", 2) "communities of practice" and 3) "open design" communities. All of them share very similar organizational and management properties, in accordance with the Collaborative Engineering organizational and management features, or functional requirements for the design (of the Collaborative Engineering organizational and management forms), but their embodiment in the concrete (organizational forms) implementation (concrete design solutions) could differ in a number of details as well as in scope.

This paper presents in a way a continuation of the paper by the same authors [1]

The further text, is organized as follows: In the section 2, a definition of Collaborative Engineering is presented, following [1], including a list of the Collaborative Engineering organizational and management features, or requirements for design (of the Collaborative Engineering organizational and management forms), is presented. The section 3, will further present the three organisational forms for Collaborative Engineering: 1) learning communities", 2) "communities of practice" and 3) "open design" communities. Finally, in Section 4, some conclusions, and suggestions for future work are given, following by the list of References.

2. Collaborative Engineering definition

The Collaborative Engineering definition follows from the "new engineering design organisation and management approach", based on the complexity and semiotics framework, given in [1].

The definition, in its form presented in this paper, is adapted definition of the "learning community" in [2], especially taking into account that the "learning community" for design is considered as one of the Collaborative Engineering organizational forms. The Collaborative Engineering organizational forms are determined by the features, or by the organizational forms functional requirements, presented in [1].

> Definition:
> "**Collaborative Engineering** is the engineering design process (practice) within an intentionally developed community as a complex network, created to promote and maximize the individual and shared learning of its members, through ongoing interaction, interplay, dialogue and collaboration among the community's members to achieve common goals of collaborative learning on design and creation of novel design" [3]

The community, taking form of a complex network of independent designers, usually uses advanced internet (web) based environments and platforms for communication, sharing, dialog, learning.

Concerning the common goal, it is important to notice that the common goal is not a particular product (or the product portfolio) "given" by a company, typical for the CE, but the common goal is learning on design and creation of novel design

3. Three organisational forms for Collaborative Engineering

In this section three organizational forms for implementation and operation of Collaborative Engineering are presented. These are 1) learning communities, 2) communities of practice and 3) open design communities.

The presentation of each form will include the presentation of the definitions, selected from the literature, followed by some examples of implementations.

It could be said that all definitions (for all three forms) are not accepted universally, but depend on authors and other contexts, i.e. the definitions could be classified by the criteria such as purpose, geography, technology, type of community, etc. For example, the situation with the definitions is well described in [4]: "The term learning communities is used variously within the literature, often without explicit definition." Concerning the criteria by which the definition is determined, or by the focus of each definition in particular, these (criteria) are multiple, such as: human element of communities, purpose, curricular structures, geographical-boundaries, institutions (educational, government bodies, industry partners, public, private and non-profit organizations), community groups.

Although the cited finding is related to the definitions of "learning communities" it could be said that the same is valid for other forms (or community types) as well.

It is worth to remind that, in accordance with the discussion on the concept of *definition* in general [1], all definitions presented could be considered as "descriptive", rather than "prescriptive".

3.1. Learning Communities (LC) for Design

Definitions of Learning Communities (LC) are multiple and differ in details, that is, in considering or not some specific features.

A very good generic description of the main features of the learning community is given is [5], which summarized well the definitions from different sources, i.e. from [6], [7], [8] and others. The learning communities are characterized by the capacities for

> "self-regulation as a critical human capability to deliberate knowledge sharing and collaboration with others through the interactions among individuals and groups. (...) monitoring the cognition, motivation, and behavior in pursuing their personal mastery and prosocial goals, members can effectively learn and share knowledge in the knowledge communities. This may in turn foster a high level of professional ability to collaborate with others and efficacy to emotionally and instrumentally help others." [5].

What is indicative in the above description is what makes the qualitative difference from Concurrent Engineering. This is the feature of "to emotionally and instrumentally help others". The keyword is "help", with all implications on specifications and distinguishing between the Concurrent Engineering and Collaborative Engineering.

Also, an indicative characterization is given by [9], following [10]:

> "... at least two challenges exist when we try to create our own learning communities. First, they must stand for something, meaning they must have a boundary that defines who is and is not a member. People will join a learning community because of its unique identity, mission, goals, or opportunities. But we also know that students often join learning communities to escape other aspects of the university. The challenge, therefore, is to create communities that have strong identities without being elitist, that are defined by their goals and missions rather than by who they exclude.
>
> Second, a learning community must be large enough both to accomplish its goals and to include all members who wish to join. It cannot be so large, however, that an individual is lost within it. When we define a community as a group of individuals committed to shared values and goals, who purposefully come together and work together to reach these goals, then diversity is encouraged not just for ideological reasons but also because different perspectives, experiences, and backgrounds contribute to the collaborative efforts to achieve the community's goals. Unlike formal organizations, which are designed and constructed around well-defined hierarchies and roles, community members and roles evolve over time through a sense of shared commitment, obligations, and resources. Finding the right mix of roles, tasks, and number of members is an ongoing process of trial and error." [9]

Also:

> "any group of people, whether linked by geography or some other shared interest, which addresses the learning needs of its members through proactive partnerships. It explicitly uses learning as a way of promoting social cohesion, regeneration and economic development." [11] (cited in [4])

or similarly (based on [11] but using the word "collaborate"):

"Learning communities are developed where groups of people, linked geographically or by shared interest, collaborate and work in partnership to address their members' learning needs. Learning communities facilitated through adult and community education are a powerful tool for social cohesion, community capacity building and social, cultural and economic development." [12] (cited in [4])

More explicit definition of the Learning Communities found in literature are mostly related to the education domain (the above referred description (definition), by [9] and [10], refers to educational settings as well), referring both to the students' and teachers' learning communities. For example:

"A learning community is a small group or cohort of students who share common academic goals and work collaboratively in the classroom with one or more professors. At (...), we offer First Year Learning Communities for entering students and Collaborative Learning Communities for all students. When you enroll in a learning community, you will be part of a cohort of 22-24 students. Together, you will take courses that are either "linked" thematically or enhanced by peer learning. By taking classes together and/or engaging in peer-to-peer learning, you get to know each other better, learn from each other, and support each other. And because classes are limited to 22-24 students, you and your instructors will get to know each other better, too!" [2].

"Learning communities, according to Lieberman, are 'places in which teachers pursue clear, shared purposes for student learning, engage in collaborative activities to achieve their purposes and take collective responsibility for student learning' (Sparks, 1999, para. 1)" [13]

The question that could be raised is: what is the relationship between the Learning Communities in education and the engineering design and Collaborative Engineering ? First it could be related to the education of (engineering) Design, e.g. [14]. From the other side, students working within a Learning Factory environment (for the concept of Learning Factory see e.g. [15] and [16]) are supposed to interchange directly with companies and in that way to participate in real-life engineering design problems solution.

When referring to the professional domains other than education, learning communities could be referred as "professional learning communities" (PLC) [17] (although many times the professional learning communities are related to the communities of teachers, i.e. to the education domain as well e.g. [18]). The professional learning communjities are "groups of people who share a concern, a set of problems, or passion about a topic, who deepen their knowledge and expertise in this area by interacting on an ongoing basis" [19] (cited also in [17])

Many times, the learning communities, and professional learning communities, are also, referring to the community of practice [17].

3.2. Design communities of practice (CoP)

Communities of practice (CoP) are another form of communities, distinct from the learning communities mainly by the purpose. As the name says, the purpose are improving practices. However, interconnections within the communities of practice generates learning too. In this sense both forms, shares the most of the features, which is in accordance with the assumptions that all these types of the communities are different organisational and management forms of Collaborative Engineering. In many cases, and discourses, the communities of practice are related, or even connected to the communities of learning.

Definitions are multiple as well, basically descriptive, as there is no one definition accepted universally. In one of the most cited works on community of practice these are described as follows [19]:

"Communities of practice are groups of people who share a concern, a set of problems, or a passion about a topic, and who deepen their knowledge and expertise in this area by interacting on an ongoing basis. Engineers who design a certain kind of electronic circuit called phase-lock loops find it useful to

IOP Conf. Series: Materials Science and Engineering 1174 (2021) 012028 doi:10.1088/1757-899X/1174/1/012028

> compare designs regularly and to discuss the intricacies of their esoteric specialty. (…) Frontline managers running manufacturing operations get a chance to commiserate, to learn about upcoming technologies, and to foresee shifts in the winds of power.
> These people don't necessarily work together every day, but they meet because they find value in their interactions. As they spend time together, they typically share information, insight, and advice. They help each other solve problems. They discuss their situations, their aspirations, and their needs. They ponder common issues, explore ideas, and act as sounding boards. They may create tools, standards, generic designs, manuals, and other documents—or they may simply develop a tacit understanding that they share. However they accumulate knowledge, they become informally bound by the value that they find in learning together. This value is not merely instrumental for their work. It also accrues in the personal satisfaction of knowing colleagues who understand each other's perspectives and of belonging to an interesting group of people. Over time, they develop a unique perspective on their topic as well as a body of common knowledge, practices, and approaches. They also develop personal relationships and established ways of interacting. They may even develop a common sense of identity. They become a community of practice." [19]

As it is easy to see, a number of properties are in accordance with our definition of the Collaborative Engineering, e.g. the learning is one of the most important effects, or outcomes, the participants in the community of practice usually do not belong to the same company, they do not have a common goal product. But, again, one word there makes the most qualitative distinguishing form the Concurrent Engineering: "help". Other word for "help", in this context, could be (to) "give".

Some other descriptions, including the definitions, are also interesting to consider. For example:

> "While CoPs were initially conceptualised as a spontaneous phenomenon, marked by informal nature and lack of regulation, shortly afterwards it has become clear that they should be cultivated and managed <…>[1]. Nowadays, they are considered as key components of systematic and deliberate KM strategies <…>. In particular, CoPs can be regarded as a manifestation of a human-oriented approach to KM <…> that sees knowledge as constructed through the joint experience in social networks and, thus, as characterised by a prevalently tacit dimension. In this view, the potential of ICT applications is not neglected (indeed, without ICT, it would not be possible to connect professionals working globally, nor to store and share large amounts of information), but their role is somewhat seen as ''ancillary'': what is deemed vital is the implementation of appropriate organisational structures, processes, and mechanisms that facilitate the sharing of experience, ideas, and suggestions among individuals." [20]

> "Communities of practice are systems of collective critical inquiry and reflection focused on building a shared identity and collective intelligence garnered over time" [21]

Design is also of concerns by the Communities of Practice.

For example, in [22] the authors investigated "whether the traditional team is in fact the most effective methodology for distributed design projects". The conclusions are that "It is clear from these empirical studies that the concept of distributed CoP's can offer insight into the way distributed design teams function" and that "However within present practice the authors observe that there are often barriers to the development of a productive social learning environment".

In [23] the authors analyzed "the biases and barriers of two different types of design communities: *communities of practice* and *communities of interest*." The conclusions are that

> "The complexity of design problems requires communities to address them. (…) a conceptual framework based on different design approaches (…) have identified different design communities. Communities of interest (CoIs) bring different communities of practice (CoPs) together to cope with the complexities of real-world design problems. CoIs provide unique opportunities to bring social creativity alive by transcending individual perspectives. To create a shared understanding and common ground between different CoPs requires boundary objects. In the past, most computational environments

[1] "<…>" substitutes the references used by [20], which may be consulted through [20].

focused on the needs of individual users. Our research has evolved from (1) empowering individuals to (2) supporting CoPs with domain oriented design environments to (3) creating shared understanding among CoIs with the Envisionment and Discovery Collaboratory and the Evolving Artifact approach."

3.3. Open Design (OD)

Open design is the third Collaborative Engineering organisational and management form considered in this paper. Differently from the Learning Community and Communities of Practices, that could be applied to virtually any application domain, the Open Design approach as the name itself says, is directly and exclusively oriented to Design.

Following we are rewriting some considerations from the authors' earlier publication [24]. As a continuum process [25] of design, adjustment and redesign, that could lead to transformation, Open Design (OD) have a direct relation with the involvement of the community. Collaboration, sharing and learning processes [26] among members of a specific community is a must in order to leverage OD, that is supported by a digital platform. Using these premises and based on technical development and dialogue practices, OD is a generative process with a potential high growth rate. Some open initiatives and movement, such as Open Source Initiative [27] and Open Source Hardware [28], can be considered as high-level sharing initiatives, without boundaries (time, place), applying a commitment with a community, in particular, and with the society, in general.

The concept of Open Design (OD) is based on open-source design and in redefining the intellectual and industrial property and rights, through internet, to facilitate collaborative development of hardware and systems [29]. Considering the existing number of projects/platforms, it is notorious that the development in open-source of software design is more active and proliferated than hardware design.

One of the few examples of an OD hardware is the additive manufacturing printing machines. The emergency of affordable 3D printing is leading to the spread of open source physical products, to the involvement of communities related to this type of equipment, and to the democratization of design, manufacture and innovation [30]. Moreover, open-source design is playing an important role in research (the development of research papers in open access is leading for an easier, faster, and more cost-effective research output [31]), the cost for making scientific equipment is being reduced drastically, with an estimation of 90-99% reduction of the traditional costs [32].

As for the previously presented concepts, there are multiple definitions of the Open Design. For example:

"Open Design as the adoption of tools, processes and principles from Open Source software development in the Design discipline" [33].

"Open source hardware is further defined as "hardware whose design is made publicly available so that anyone can study, modify, distribute, make, and sell the design or hardware based on that design" [34].

4. Some platforms for creation and operation of the communities

Communities today are created and operate mostly over internet based platforms. Simple search over internet would inform of hundreds, if not thousands, of platforms for supporting communities, including the communities for Collaborative Engineering.

These platforms are owned by the commercial companies or by the interest groups or by institutions (public or private) or by other type of owners, using the variety of business models that provide the users practically free access and use of services. Depending of the complexity of the platforms supporting software, the users could, or not, create their own services. There is a number of forms the communication among the communities' members is implemented. These forms comprises blogs, forums, hubs, events, repositories (based on cloud or not), etc. Table 1 presents just a small number of internet based platforms, which are well known within the community(ies), without any further classification and/or analysis (that would be the task of a future work). The majority of platforms presented in Table 1 are oriented to the communities in general but with the capacity to, and examples of, supporting the Collaborative Engineering communities.

IOP Conf. Series: Materials Science and Engineering 1174 (2021) 012028 doi:10.1088/1757-899X/1174/1/012028

Table 1. Selected platforms for operating Communities of Practice and Open Design

	Title + ***www***	**Description**
Communities of Practice	General question answer-forum: Quora *quora.com*	"question-and-answer website where questions are asked, answered, followed, and edited by Internet users, either factually or in the form of opinions" (https://en.wikipedia.org/wiki/Quora); "It allows users to create social networks and follow topics that interest them" (https://learn.g2.com/what-is-quora); "Quora is a top-ranked question and answering community that can form the basis of a successful marketing campaign" (https://www.searchenginejournal.com/what-is-quora-why-should-you-care/379341/)
	Subjectwise question answer and knowledge exchange platform *stackexchange.com*	"Launched in 2010, the Stack Exchange network comprises 173 Q&A communities including Stack Overflow, the largest, most trusted online community for developers to learn, share their knowledge, and build their careers"
	Community for coder and developers *stackoverflow.com*	"Stack Exchange network has grown into a top 50 online destination with Stack Overflow and our technical sites serving more than 100 million developers and technologists every month"
	Engineering & Technology community platform by IET *communities.theiet.org*	"... inspire, inform and influence the global engineering community to engineer a better world. As a diverse home across engineering and technology, (...) share knowledge that helps make better sense of the world in order to solve the challenges that matter."
	National Instruments' forum for LabVIEW developers' community *forums.ni.com*	"The NI Community is part of the (...) NI Ecosystem. This is (...) place to network, ask questions, and collaborate on code with users all over the world"
Open Design	Open Design initiatives – Farmbot *https://farm.bot/*	FarmBot is an open source precision agriculture CNC farming project consisting of a Cartesian coordinate robot farming machine, software and documentation including a farming data repository. The project aims to "Create an open and accessible technology aiding everyone to grow food and to grow food for everyone." FarmBot is an open source project allowing hardware, software and documentation modifications and additions from users (https://en.wikipedia.org/wiki/FarmBot)
	Open Systems Lab	"Open Systems Lab is a non-profit R&D lab working on open digital innovation to tackle systemic social, economic and environmental

https://www.opensystemslab.io/ Open Systems Lab	challenges. (…) working to redesign the systems that shape the places where we live. (…) work with private, public and third sector organisations, and with citizens, to design and deploy 21st century systems that work for everyone."
Open Source Ecology *https://www.opensourceecology.org/*	"The vision of Open Source Ecology is a world of collaborative design – for a transparent and inclusive economy of abundance. An open source, libre economy is an efficient economy which increases innovation by open collaboration. To get there, OSE is currently developing a set of open source blueprints for the Global Village Construction Set (GVCS) – a set of the 50 most important machines that it takes for modern life to exist (…) intends to develop a modular, scalable platform for documenting and developing open source, libre hardware – including blueprints for both physical artifacts and for related open enterprises"
RepRap project *https://reprap.org/wiki/RepRap*	"community portal "RepRap is about making self-replicating machines, and making them freely available for the benefit of everyone. (…) using 3D printing to do this (…) Reprap.org is a community project, which means (…) welcome to edit most pages on this site, or better yet, create new pages of your own."
GitHub *https://github.com/*	"GitHub, Inc. is a provider of Internet hosting for software development and version control using Git. (…) GitHub offers its basic services free of charge. Its more advanced professional and enterprise services are commercial. Free GitHub accounts are commonly used to host open-source projects. (…) the largest host of source code in the world"

5. Conclusions

The definition of the Collaborative Engineering presented in [1], based on the complexity theory and semiotics frameworks, provides the capacity of completely new approaches to engineering design organizational and management models. It is interesting that it is not only the proposal for the future but it is already the reality. This new reality many times is not recognized explicitly as the new engineering design organizational and management models. In these terms, this paper contributes to this recognition and the potential, and needed, future renewals of the companies and their design functions.

Looking through the lenses of complexity management in organisations, the Collaborative Engineering, based on the complexity theory and semiotics frameworks, could be considered as an emergent approach with potential for leading to a system New-Thinking, New-Doing, i.e., to a real novelty.

An interesting position concerning renewals (reinventing) of companies is presented in [35]: "Not so long, companies were reinvented by teams. Communities of practice may reinvent them yet again – if managers learn to cultivate these fertile organizational forms without destroying them". Paraphrasing the referred statement, we could say that the Collaborative Engineering will renew, or reinvent, the engineering design organization and management approaches and models.

Concerning the future work, besides the usual directions towards more detailed studies, one of the important issues is the dynamics of the communities. In all definitions of diverse types of communities, although these are descriptive, including our definition of the Collaborative Engineering, the communities are referred as "static", i.e. as the collaboration will always remain in communities. However, the nature of conversation, or communication, which is basically a dialog in the Collaborative Engineering, permits, of course, that the participants in the Collaborative Engineering community could establish business or project partnerships with the defined goal product. In this case, these participants are creating the design project team which starts to behave as the "classical" Concurrent Engineering team. And contrary, it doesn't mean that the participants of a Concurrent Engineering team could not go "back" to join again the Collaborative Engineering community, or to act as the participants of both the Collaborative Engineering community and the Concurrent Engineering team in parallel. Supposedly, this dynamics, which was not found to be referred in the literature, is a virtual source of confusing the concepts / terms / approaches of Concurrent Engineering and the Collaborative Engineering. Consequently, research of this dynamics is identified as one of the important issues of the future work.

Finally, we would like to underline once more the keywords that characterize and distinguish the concept of Collaborative Engineering, and that we have already referred in the above text:

(to) "help", and (to) "give",

and to add one more (for which our colleague Prof. C. Brown always call for attention). This third word is

(to) "serve".

Acknowledgements
This work has been supported by FCT – Fundação para a Ciência e Tecnologia within the R&D Units Project Scope: UIDB/00319/2020

References

[1] Putnik, G. D., Putnik, Z., Shah, V., Varela, L., Ferreira, L., Castro, H., Alves, C., Pinheiro, P. (2021) Collaborative Engineering definition: Distinguishing it from Concurrent Engineering through the complexity and semiotics lenses. submitted for publishing in the *Proceedings of the The International Conference on Axiomatic Design (ICAD) 2021*, Caparica - Lisbon

[2] LIUBrooklyn (n.d.) *What Is a Learning Community?*, https://liu.edu/Brooklyn/Academics/Centers/Learning-Communities/What-is-a-Learning-Community

[3] Putnik G. (2021) *Concurrent and Collaborative Engineering*. Lectures on Post-graduate course on Industrial Engneering and Management, Department of Production and Systems Engineering, University of Minho. Guimaraes

[4] Kilpatrick, S., Jones, T., & Barrett, M. (2003). *Defining learning communities*. Launceston, Tasmania: Centre for Research and Learning in Regional Australia.

[5] Tseng, F. C., & Kuo, F. Y. (2010). The way we share and learn: An exploratory study of the self-regulatory mechanisms in the professional online learning community. *Computers in Human Behavior*, 26(5), 1043-1053.

[6] Margaryan, Anoush; Milligan, Colin; Littlejohn, Allison; Hendrix, Donna and Graeb-Koenneker, Sebastian (2009). Self-regulated learning and knowledge sharing in the workplace. In: *Organizational Learning, Knowledge and Capabilities Conference*, 28 Apr 2009, Amsterdam.

[7] Pintrich, P. R. (2000). The role of goal orientation in self-regulated learning. In *Handbook of self-regulation* (pp. 451-502). Academic Press.

[8] Imants J. (2003) Two basic mechanisms for organisational learning in schools. *European Journal of Teacher Education*, 26 (3), pp. 293-311

[9] Brower, A. M., & Dettinger, K. M. (1998). What is a learning community?: Toward a comprehensive model. *About campus*, 3(5), 15-21.

[10] Gardner, J. (1996) Building Community. *Community Education Journal*, 23(3), 6–9.

[11] Kearns, P., McDonald, R., Candy, P., Knights, S. & Papadopoulos, G. (1999). *VET in the learning age: The challenge of lifelong learning for all* (Vol. 2: Overview of international trends, and case studies). Canberra: National Centre for Vocational Education Research Ltd.

[12] Department of Education. (2002). *Essential Learnings: Framework 1*. Hobart, Tasmania: Department of Education.

[13] Wells, C., & Feun, L. (2007). Implementation of learning community principles: A study of six high schools. *Nassp Bulletin*, 91(2), 141-160.

[14] Khan, R., & Slotta, J. D. (2020). Fostering Learning Communities in Engineering Design

Education-A Critical Discourse Analysis of Engineering Syllabi. In *2020 IEEE Frontiers in Education Conference (FIE)* (pp. 1-5). IEEE.

[15] Abele, E., et al. (2017). Learning factories for future oriented research and education in manufacturing. *CIRP Annals*, 66(2), 803-826.

[16] Putnik G. (2016) *Three short contributions to learning factories theory and practice*, presentation at the CIRP Winter meeting, Paris.

[17] Jacobs, J., & Yendol-Hoppey, D. (2010). Supervisor transformation within a professional learning community. *Teacher education quarterly*, 37(2), 97-114.

[18] Lunenburg, F. C. (2010). Creating a professional learning community. *National Forum of Educational Administration and Supervision Journal*, Vol. 27, No. 4, pp. 1-7.

[19] Wenger, E., McDermott, R., & Snyder, W. N. (2002). *Cultivating communities of practice: A guide to managing knowledge*. Cambridge, MA: Harvard Business School Publishing.

[20] Scarso, E., Bolisani, E., & Salvador, L. (2009). A systematic framework for analysing the critical success factors of communities of practice. *Journal of knowledge management*. Vol. 13 No. 6, pp. 431-447

[21] Creating Communities of Practice (n.d.) What is a community of practice?. http://www.communityofpractice.ca/background/what-is-a-community-of-practice/

[22] Pemberton-Billing, J., Cooper, R., Wootton, A. B., & North, A. N. (2003). Distributed design teams as communities of practice. In *5th European Academy of Design Conference*, Barcelona.

[23] Fischer, G., & Ostwald, J. (2005). Knowledge communication in design communities. In *Barriers and biases in computer-mediated knowledge communication* (pp. 213-242). Springer, Boston, MA.

[24] Castro, H., Putnik, G., Castro, A., & Fontana, R. D. B. (2019). Could Open Design learn from Wikipedia?. *Procedia CIRP*, 84, 1112-1115.

[25] Avital, M. (2011). The generative bedrock of open design. In B. van Abel, R. Klaassen, L. Evers, & P. Troxler (Eds.), *Open design now: why design cannot remain exclusive* (pp. 48-58). Amsterdam: BIS Publishers.

[26] James Tooze, Sharon Baurley, Robert Phillips, Paul Smith, Edwin Foote & Sarah Silve (2014) Open Design: Contributions, Solutions, Processes and Projects, *The Design Journal*, 17:4, 538-559

[27] https://opensource.org/osd

[28] https://www.oshwa.org/definition/

[29] Michael Thomson and Tapio Koskinen (2012). *Design for Growth & Prosperity* - Report and Recommendations of the European Design Leadership Board. DG Enterprise and Industry of the European Commission.

[30] Hannah Forbes, Dirk Schaefer (2017). Social Product Development: The Democratization of Design, Manufacture and Innovation. *Procedia CIRP*, Volume 60, Pages 404-409.

[31] Erin C McKiernan, Philip E Bourne, C Titus Brown, Stuart Buck, Amye Kenall, Jennifer Lin, Damon McDougall, Brian A Nosek, Karthik Ram, Courtney K Soderberg, Jeffrey R Spies, Kaitlin Thaney, Andrew Updegrove, Kara H Woo, Tal Yarkoni (2016). Point of View: How open science helps researchers succeed. *eLife* 2016;5:e16800

[32] Pearce, Joshua. (2017). Impacts of open source hardware in science and engineering. *Bridge*. 47. 24-31.

[33] Massimo Menichinelli (2017) A data-driven approach for understanding Open Design. Mapping social interactions in collaborative processes on GitHub, *The Design Journal*, 20:sup1, S3643-S3658.

[34] The Open Source Hardware Definition [WWW Document]. URL http://www.oshwa.org/definition/

[35] Wenger, E. C., & Snyder, W. M. (2000). Communities of practice: The organizational frontier. *Harvard Business Review*, 78(1), 139-146.

AUTHOR INDEX